Simulation Foundations, Methods and Applications

More information about this series at http://www.springer.com/series/10128

Levent Yilmaz
Editor

Concepts and Methodologies for Modeling and Simulation

A Tribute to Tuncer Ören

Editor
Levent Yilmaz
Department of Computer Science and
Software Engineering
Department of Industrial and Systems Engineering
Auburn University
Auburn, AL, USA

ISSN 2195-2817 ISSN 2195-2825 (electronic)
Simulation Foundations, Methods and Applications
ISBN 978-3-319-38260-9 ISBN 978-3-319-15096-3 (eBook)
DOI 10.1007/978-3-319-15096-3

Springer Cham Heidelberg New York Dordrecht London

Softcover reprint of the hardcover 1st edition 2015

Printed on acid-free paper

Springer International Publishing AG Switzerland is part of Springer Science+Business Media (www.springer.com)

Foreword

The work of Tuncer Ören, the archetypal Renaissance man of the modeling and simulation community, embodies the vision of M&S as the essential enabler of future science and engineering. Ören's vision stretches widely over the whole M&S domain encompassing its fundamental body of knowledge, its methodology, its practice, and its ethics. It also includes the quality of M&S products and specific application domains such as cognitive and emotive social simulation. The authors of this tribute book pursue a few—essential—threads of the many emanating from Ören's core vision. As a consequence of the underlying unity of conception, the book is more than a collection of disparate state-of-the art articles. This integration is enhanced by the fact that every chapter is explicitly connected to pertinent features of Ören's thought. Within this perspective, the book covers cutting-edge topics in simulation methodologies; modeling methodologies; quality assurance and reliability of simulation studies; and cognitive, emotive, and social simulation. Notably, it touches on Ören's intense interest in the body of knowledge of modeling and simulation, with a review of existing M&S literature through the newly emerging techniques of journal profiling and co-citation analysis. This book beckons you, whether theorist or practitioner, generalist or domain application professional, to partake in and contribute to Ören's powerful vision.

Potomac, MD, USA
December 29, 2014

Bernard P. Zeigler

Preface

Concepts and Methodologies for Modeling and Simulation aims to present recent advances in the theory and methodology of Modeling and Simulation (M&S). By connecting these developments to the conceptual, theoretical, and methodological foundations developed by Professor Tuncer Ören, this volume serves as a testimonial that honors Dr. Ören's long-lasting and fundamental contributions to the M&S discipline for over 50 years.

Since 2003, I have had the privilege to collaborate with Dr. Ören, whom I see as my mentor and a titan in our field. The articles in this book are a testament to the diversity and innovativeness of his thoughts. As evidenced by this volume, his influences in the philosophy, theory, methodology, ethics, and the body of knowledge of M&S have numerous connections to recent advancements in our field and continue to provide directions for its further development. This book is largely due to the efforts and contributions of the authors, who shared their recent research in the context of Dr. Ören's seminal contributions to the M&S discipline. I am indebted to them for their contributions to this tribute volume. They are not only authorities in their field but also colleagues of Dr. Ören. Hence, they are qualified and entitled to trace recent advancements in their fields to the most influential concepts and methods introduced by him.

In the area of simulation methodologies, Dr. Ören's earlier work on model-based M&S and detailed categorizations and taxonomies of M&S has been highly influential. In particular, normative views for the advancement of M&S, including synergies of artificial intelligence and systems theories, and his comprehensive and integrative views have provided a sound and thorough framework for the development of advanced simulation methodologies. To explain these contributions, the book starts with my reflections on the recent developments in agent-directed simulation (ADS), providing a framework that explores synergies between simulation and agent technologies. The readers can trace the provenance of various ideas explored under ADS to concepts introduced by Dr. Ören when he first examined and demonstrated how artificial intelligence methods can assist simulation. The second chapter in this section presents how model engineering and service technologies can be leveraged to contribute to System-of-Systems (SoS) Engineering. The third chapter overviews emerging trends and drivers in high-performance simulation systems, while the fourth chapter examines the role of data in the context of dynamic data-driven simulations that connect real-time sensor data to online simulations.

The second section of the book focuses on advanced modeling methodologies. The first chapter in this area focuses on the philosophical fields of ontology and epistemology to delineate the role and use of simulations in relation to the taxonomies and categories of M&S developed by Dr. Ören as part of his contributions to the M&S body of knowledge. The second chapter demonstrates how innovations in modeling formalisms can help manage the challenges in hybrid model composition, especially in the context of agent-based, human, social, and environment models. Specifically, the authors describe the use of a polyformalism model composition approach and highlight its relation to multimodeling strategies that Dr. Ören and I have developed during the early 2000s. The third chapter of this section underlines the importance of a model-based approach to M&S and underlines model building, model-based management, and model processing activities advocated by Dr. Ören. The authors then present a formal, declarative, and visual transformation (model processing) methodology to translate a domain conceptual model to a distributed simulation architecture model.

The third section of the book is devoted to the reliability and quality assurance of models. The section starts with an overview of quality indicators that can be used to support a structured and quality-centered approach to simulation development throughout the entire M&S life cycle. The second paper reviews, summarizes, and describes the influence of important M&S quality assurance papers developed by Dr. Ören. The paper also promotes strategies for the replicability and reproducibility of simulation studies to instill confidence in simulation experiments. The third paper in this section refers to challenges involved in qualitative and quantitative comparisons of agent-based models to calibrated statistical models for the purpose of validation and reproducibility. The last chapter in this section introduces the Generalized Discrete Event System Specification to build more accurate discrete-event models of dynamic systems. This work highlights the need for engineering quality into models to improve their accuracy.

The fourth section of the book focuses on the specification and simulation of human and social behavior, acknowledging Dr. Ören's contributions to model specification language development as well as his recent research in cognitive and emotive simulation modeling including the specification of models of personality, emotions, conflict management, perception, and anticipation. In this section, the first chapter presents work on social science models that benefits from the principles based on Dr. Ören's influences of model specification languages, goal-directed agents, anticipatory simulation, agent perceptions, and multifacetted models. Similarly, the second chapter in this area refers to the multisimulation methodology as a basis to examine bridging human decision processes and computer simulation while also referring to multisimulation as an enabling technology for backtracking and replaying situated simulation histories with altered conditions as well as futures generated before exploring alternative realities in social sciences.

The last section of the book is devoted to M&S body of knowledge work. The chapter presented in this section was inspired by Dr. Ören's work and shares common ground by profiling and classifying M&S publications in terms of techniques, application areas, and their context in a relevant way with the second and third parts of the body of knowledge, which defines the M&S core areas and supporting domains.

Auburn, AL, USA
December 19, 2014

Levent Yilmaz

Contents

Contributors

S.M. Niaz Arifin Department of Computer Science and Engineering, University of Notre Dame, Notre Dame, IN, USA

Jang Won Bae Department of Industrial and Systems Engineering, KAIST, Daejeon, South Korea

Osman Balci Mobile/Cloud Software Engineering Lab, Department of Computer Science, Virginia Polytechnic Institute and State University (Virginia Tech), Blacksburg, VA, USA

C. Michael Barton Center for Social Dynamics and Complexity, Arizona State University, Tempe, AZ, USA

Xudong Chai Beijing Simulation Center, Beijing, China

Paul K. Davis Department of Engineering and Applied Sciences, RAND Corporation, Santa Monica, CA, USA

Paul Fishwick Department of Computer Science, The University of Texas at Dallas, Richardson, TX, USA

Norbert Giambiasi Aix Marseille Université, CNRS, ENSAM, Université de Toulon, LSIS UMR 7296, Marseille, 13397, France

Maamar El Amine Hamri Aix Marseille Université, CNRS, ENSAM, Université de Toulon, LSIS UMR 7296, Marseille, 13397, France

Aimin Hao School of Automation and Computer Science, Beihang University, Beijing, China

Baocun Hou Beijing Simulation Center, Beijing, China

Xiaolin Hu Department of Computer Science, Georgia State University, Atlanta, GA, USA

Korina Katsaliaki School of Economics, Business Administration & Legal Studies, International Hellenic University, Thessaloniki, Greece

Bo Hu Li School of Automation and Computer Science, Beihang University, Beijing, China

Jun Li Sugon Corp., Beijing, China

Tan Li Beijing Simulation Center, Beijing, China

Gregory R. Madey Department of Computer Science and Engineering, University of Notre Dame, Notre Dame, IN, USA

Gary R. Mayer Department of Computer Science, Southern Illinois University, Edwardsville, IL, USA

Il-Chul Moon Department of Industrial and Systems Engineering, KAIST, Daejeon, South Korea

Navonil Mustafee Centre for Innovation and Service Research, University of Exeter Business School, Exeter, UK

Aziz Naamane Aix Marseille Université, CNRS, ENSAM, Université de Toulon, LSIS UMR 7296, Marseille, 13397, France

Halit Oguztüzün Department of Computer Engineering, Middle East Technical University, Ankara, Turkey

Gürkan Özhan Department of Computer Engineering, Middle East Technical University, Ankara, Turkey

Duzheng Qin Beijing Simulation Center, Beijing, China

Hessam S. Sarjoughian Department of Computer Science and Engineering, Arizona State University, Tempe, AZ, USA

Andreas Tolk SimIS Inc., Portsmouth, VA, USA

Isaac I. Ullah School of Human Evolution and Social Change, Arizona State University, Tempe, AZ, USA

Gabriel Wainer Department of Systems and Computer Engineering, Carleton University, Ottawa, ON, Canada

Ming Yang Control and Simulation Center, Harbin Institute of Technology, Harbin, Heilongjiang, China

Levent Yilmaz Department of Computer Science and Software Engineering, Department of Industrial and Systems Engineering, Auburn University, Auburn, AL, USA

Bernard P. Zeigler RTSync Corp., Arizona Center for Integrative Modeling and Simulation, Sierra Vista, AZ, USA

Lin Zhang School of Automation and Computer Science, Beihang University, Beijing, China

Yu Zhang Department of Computer Science, College of St. Benedict, St. John's University, Collegeville, MN, USA

Qinping Zhao School of Automation and Computer Science, Beihang University, Beijing, China

Part I
Simulation Methodologies

Chapter 1
Toward Agent-Supported and Agent-Monitored Model-Driven Simulation Engineering

Levent Yilmaz

1.1 Introduction

The concepts that expound on the synergy of artificial intelligence and simulation date back to late 1970s when Professor Ören (1977) has introduced strategies that delineate the use of computer assistance in model-based activities and later when he promoted in (Ören 1978) the principled use of cybernetics for developing simulation software. In 1980s, Ören and his colleagues, Elzas and Zeigler, edited two volumes (Elzas et al. 1986, 1989) that explicated various facets of the synergy of modeling and simulation, artificial intelligence, and knowledge-based systems. These two volumes became highly influential in putting forward a pathway toward convergence of modeling and simulation (M&S) and artificial intelligence methodologies.

Concomitantly, the emergence of the software agent concepts (Shoham 1993) that embody and encapsulate knowledge-based deliberation, reasoning, autonomy, interaction, learning, and adaptation mechanisms led to the adoption of agents as model design metaphors in simulation modeling. However, this limited view of the use of agents in M&S was critiqued in Ören (2000a, b). In his panel statement (Ören 2000b) at the 2000 Winter Simulation Conference, Dr. Ören has pointed out a broader vision for the use of agents in M&S. Besides using M&S for modeling agent systems in the form of agent-based models, by which agent concepts are leveraged to create abstractions of the system of interest, Dr. Ören has suggested that agents can also be used to assist or support model behavior generation as part of the simulators or provide cognitive support as front-end or back-end interface to a simulation system. He called this expanded view as *agent-directed simulation* (ADS).

L. Yilmaz (✉)
Department of Computer Science and Software Engineering, Department of Industrial and Systems Engineering, Auburn University, Auburn, AL, USA
e-mail: yilmaz@auburn.edu

L. Yilmaz (ed.), *Concepts and Methodologies for Modeling and Simulation*, Simulation Foundations, Methods and Applications,
DOI 10.1007/978-3-319-15096-3_1

My collaboration with Dr. Ören on ADS started a few weeks before I joined to Auburn University as a tenure-track assistant professor in 2003. We met at the 2003 Summer Computer Simulation Conference that was held in Montreal, Canada. Following our initial discussions on the emerging trends and critical drivers of the use of software agent technologies in M&S, we developed a framework to structure and delineate the role of agents throughout the whole life cycle of M&S. A year later, this framework served as the basis for the first track of sessions on ADS that Dr. Ören and I organized as part of the 2004 Summer Computer Simulation Conference. Since the organization of this inaugural event, the ADS track of sessions has been and, as of 2014, continues to be one of the key features of the Summer Computer Simulation Conference series. The success of the ADS track in 2004 has led to the first Annual ADS Symposium that was held as part of the 2005 Spring Simulation Multiconference, formerly known as the Advanced Simulation Technologies Conference. Since 2005, the ADS Symposium continues to be organized annually by a growing organization committee. In 2007, Dr. Greg Madey and Dr. Maarten Sierhuis have joined the organization committee; this is followed by Dr. Yu Zhang joining in 2009. Since then, ADS has also been organized as a track of sessions as part of the European Modeling and Simulation Symposium and the SimulTech Conference series.

Our technical collaboration with Dr. Ören, along with other invited contributions, has led to the publication of the first book (Yilmaz and Ören 2009) on ADS. The Agent-Directed Simulation and Systems Engineering book is now viewed as the only book to present the synergy between modeling and simulation, systems engineering, and agent technologies while also expanding the notion of agent simulation to also deal with agent-monitored simulation and agent-supported simulation. This journey in ADS has resulted in numerous publications that range from advanced modeling and simulation methodologies such as multisimulation to concepts and demonstrations of agents with personality, emotions, anticipations, and understanding capabilities with application areas in engineering, defense, and human and social dynamics.

The objective of this chapter is to illustrate how agents can be used to facilitate development of next-generation simulators and assist in conducting simulation experiments. First, we introduce software agents and then characterize the three dimensions of the ADS framework that help explore the use of agents for simulation and the use of simulation for agents. This is followed by the introduction multisimulation and clarification of how agent-monitored and agent-assisted mechanisms facilitate its design and implementation. To emphasize the role of agents besides in model development and simulator (i.e., model behavior generator) design, we highlight the issues and challenges in goal-directed experimentation and present an agent-assisted and model-driven experiment management strategy to effectively address these challenges.

1.2 Agent-Directed Simulation

Agents are autonomous software modules with perception and social ability to perform goal-directed knowledge processing over time, on behalf of humans or other agents in software and physical environments. When agents operate in physical environments, they can be used in the implementation of intelligent (smart) machines and intelligent (smart) systems and they can interact with their environment by sensors and effectors.

The core knowledge-processing abilities of agents include goal-processing, goal-directed knowledge processing, reasoning, motivation, planning, and decision-making. The factors that may affect decision-making abilities of agents (in simulating human behavior) are personality, emotions, and cultural backgrounds. Abilities to make agents intelligent include anticipation (pro-activeness), understanding (avoiding misunderstanding), learning, and communication in natural and body language. Abilities to make agents trustworthy as well as assuring the sustainability of agent societies include being rational, responsible, and accountable. These characteristics lead to rationality, skillfulness, and morality (e.g., ethical agent, moral agent).

Synergy of simulation and agents is called agent-directed simulation (ADS). As shown in Fig. 1.1, it consists of contributions of simulation for agents and contributions of agents for simulation.

Agent simulation involves the use of simulation conceptual frameworks and technologies to simulate the behavioral dynamics of agent systems by specifying and implementing the behavior of autonomous agents that function in parallel to achieve objectives via goal-directed behavior. In agent-based model specifications, agents possess high-level interaction mechanisms independent of the problem being solved. Communication protocols and mechanisms for interaction via task

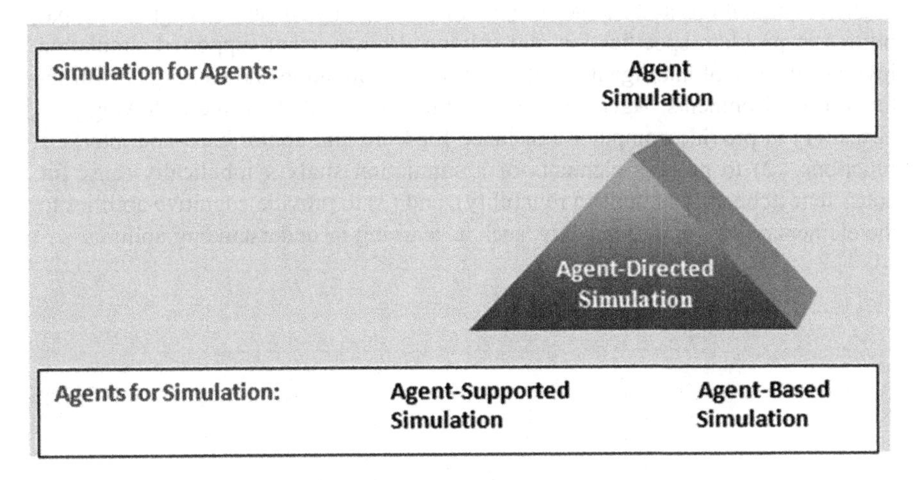

Fig. 1.1 Agent-directed simulation

allocation, coordination of actions, and conflict resolution at varying levels of sophistication are primary elements of agent simulations. Simulating agent systems requires understanding the basic principles, organizational mechanisms, and technologies underlying such systems.

The principal aspects underlying such systems include the issues of action, cognitive aspects in decision-making, interaction, and adaptation. Organizational mechanisms for agent systems include means for interaction. That is, communication, collaboration, and coordination of tasks within an agent system require flexible protocols to facilitate realization of cooperative or competitive behavior in agent societies Agent-based modeling in which agents are used as design metaphors to conceptualize and specify agent systems is becoming the most common methodology in agent simulation. On the other hand, simulation(s) can be used to realize deliberation architecture of agents in an agent system by simulating the cognitive processes in decision-making. While embedding simulations within the deliberation architecture of agents is not necessarily a novel concept on its own, it is still an unexplored territory in MAS design and implementation.

Agent-based simulation (or *agent-monitored simulation*) is the use of agent technology to monitor and generate model behavior. This is similar to the use of AI techniques for the generation of model behavior (e.g., qualitative simulation and knowledge-based simulation). Development of novel and advanced simulation methodologies such as *multisimulation* suggests the use of intelligent agents as simulator coordinators, where run-time decisions for model staging and updating take place to facilitate dynamic composability. The perception feature of agents makes them pertinent for monitoring tasks. Also, agent-based simulation is useful for having complex experiments and deliberative knowledge processing such as planning, deciding, and reasoning. Agents are also promoted and demonstrated as critical enablers to improve composability and interoperability of simulation models and software, in general.

Agent-supported simulation deals with the use of agents as a support facility to augment simulations and enable computer assistance by enhancing cognitive capabilities in problem specification and solving. Hence, agent-supported simulation involves the use of intelligent agents to improve simulation and gaming infrastructures or environments. Agent-supported simulation is used for the following purposes: (1) to provide computer assistance for front-end and/or back-end interface functions, (2) to process elements of a simulation study symbolically (e.g., for consistency checks and built-in reliability), and (3) to provide cognitive abilities to the elements of a simulation study, such as learning or understanding abilities.

1.3 Agent-Monitored Simulator for Exploratory Multisimulation

We define *multisimulation* as simulation of several aspects of reality in a study. It includes simulation with multimodels, simulation with multi-aspect models, and simulation with multistage models. Simulation with multimodels allows computational experimentation with several aspects of reality; however, each aspect and the transition from one aspect to another one are considered separately.

Simulation with multi-aspect models (or multi-aspect simulation) allows computational experimentation with more than one aspect of reality simultaneously. This type of multisimulation is a novel way to perceive and experiment with several aspects of reality as well as exploring conditions affecting transitions. While exploring the transitions, one can also analyze the effects of encouraging and hindering transition conditions. Simulation with multistage models allows branching of a simulation study into several simulation studies, each branch allowing to experiment with a new model under similar or novel scenarios. Each different strategy component characterizes a distinct aspect.

Multisimulation can be used to branch out multiple simulations, where each simulation uses a specific component configured with an exclusively selected strategy component. Similarly, multiple distinct stages of the problem can be qualified at a given point in time during the simulation by virtue of the evaluation of an updating constraint. In such a case, multisimulation enables branching multiple distinct simulations each one, which generates the behavior of distinct plausible stage within the problem domain.

Multisimulation with multimodels, multi-aspect models, or multistage models needs mechanisms to decide when and under what conditions to replace existing models with a successor or alternative. Staging considers branching to other simulation studies in response to a scenario or a phase change during experimentation. Graphs of model families facilitate derivation of feasible sequence of models that can be invoked or staged. More specifically, a graph of model families is used to specify alternative staging decisions. Each node in the graph depicts a model, whereas edges denote transition or switching from one model to another. Figure 1.2 depicts the components of the abstract architecture of a possible multisimulation engine.

A meta-simulator is an agent that generates staged composition of models by traversing the model stage graph and coordinates their simulation and staging within distinct simulation frames. Each frame simulates a distinct subset of models derived from the model stage graph. Note, however, that not all staged compositions are feasible or useful. Hence, the meta-simulator needs to consult with the model recommender before model staging to determine if emergent trigger or transition condition in the simulation is consistent with the precondition of the model to be staged. More than one model in a family can qualify for staging; in such cases, separate simulation frames need to be instantiated to accommodate and explore plausible scenarios.

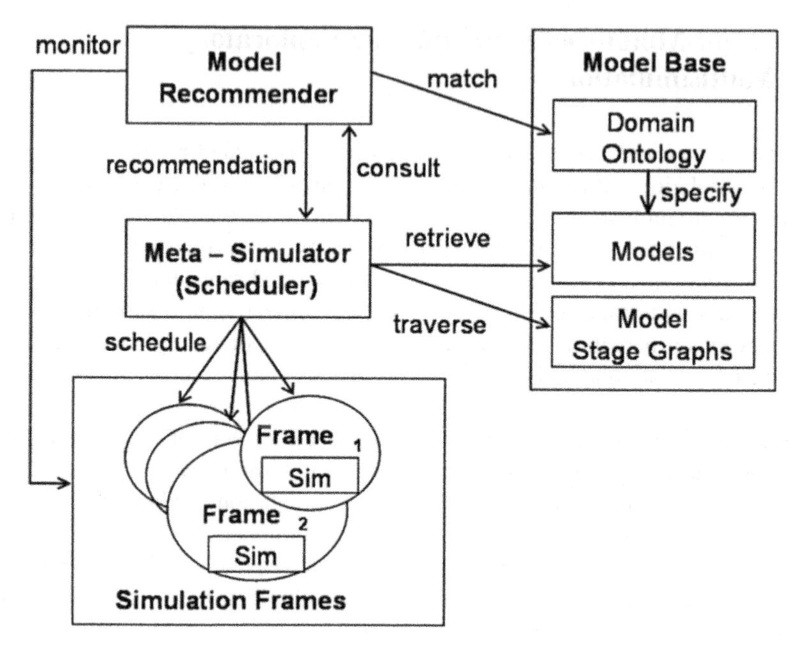

Fig. 1.2 Components of a multisimulation engine

Given a collection of models (or more generally, a family of models), a stage graph can be generated automatically by an optimistic approach that connects every available node (model) to every other node within the domain of problem. The edges in a model stage graph denote plausible transitions between models as the problem shifts from one stage to another. One can consider each model as a separate conflict management protocol (i.e., compromise over actions, compromise over outcomes, negotiation, and mediation) or a phase in the conflict process (i.e., escalation, resolution), where a phase (i.e., resolution) can constitute alternative models (i.e., mediation, negotiation, third-party intervention).

The subsets of staged models can be identified by traversing and enumerating the graph in some order (i.e., depth first). Infeasible paths may be due to an unreachable node, or it may result due to conflicts between the transition condition and precondition of the target model. Infeasible paths due to incompatible sequences of models are common. Each edge indicates that there is some legitimate solution; yet, it does not imply that every solution containing the edge is legitimate. As argued above, each model in a family of models is associated with a precondition. A precondition denotes the conditions required for a model to be instantiated. Hence, the feasibility of staging a successor model depends on the satisfiability of its precondition (relevance) by the condition of the transition and the post-condition of the predecessor model. As a result, not all enumerated staged sequences of model components are feasible.

Model recommendation in multisimulation can simply be considered as agent-supported exploration of the model-staging space that can be computed by a reachability analysis of the graph. There are two modes for the usage: (1) offline enumeration of paths using the graph and performing a staged simulation of each model in sequence one after the other, unless a model-staging operation becomes infeasible due to conflict between the transition condition and the precondition of the successor model, and (2) run-time generation of potential feasible paths as the simulation unfolds. In both cases, an online model recommender plays a key role to qualify a successor model.

The first case requires derivation of sequence of models using a traversal algorithm. The edges relate families of models. Therefore, the actual concrete models, the preconditions of which satisfy the transition condition, need to be qualified, since transition to some of these model components may be infeasible due to conflict between a candidate model and inferred situation. Identifying such infeasible sequences is computationally intractable; otherwise, it would have been possible to determine if the conjunction of two predicates is a tautology by using a polynomial time algorithm.

Experience in the component-based simulation paradigm, however, indicates that for most model components, preconditions are simple. Hence, it is possible to eliminate some models that violate the transition condition. For the remaining possible transitions, it is possible to select one of the three strategies: (1) omit all difficult qualification conditions, (2) decide on an edge-by-edge basis which specific models of a model family to include, and (3) include all difficult edges. Omitting all difficult associations between transitions and model preconditions is conservative. This strategy excludes all infeasible models. The cost is the exclusion of some feasible edges. Hand-selecting those associations between transition conditions and models facilitates inclusion of feasible models. Nonetheless, the costs involved with this level of accuracy are the potential human error and effort needed to filter out infeasible models. Choosing to include all difficult associations is liberal, in that it ensures inclusion of all feasible models. The cost is the inclusion of some infeasible models, hence the inclusion of some undesirable staged compositions that enforce models to be simulated even when their qualification conditions are violated. Nevertheless, it is possible to screen out such models using an online model recommender.

The second more ambitious yet flexible approach is to delay the enumeration process until a model is qualified at run time. Run-time generation of feasible staging using the graph of model families requires monitoring and evaluation of transition conditions as the simulation unfolds. An agent-planning layer connected to simulator would be capable of identifying, qualifying, and, if necessary, selecting and instantiating a model based on the specified preferences and options. Furthermore, in the case of an impasse or lack of knowledge on preferences among qualifying model switch strategies, a planning layer can guide exploring alternative contexts (games) in some order. The scheduler agent follows the recommendations made by the planner agent to instantiate distinct simulation frames.

Focus points maintain candidate models and associated simulations. A focus point manages branch points in the simulation frame stack. Suppose that a goal instance (i.e., stage transition condition) is at the top of the stack. If only a single model qualifies for exploration, then it is pushed onto the stack. Yet, if more than one model matches the condition, a simulation focus point is generated to manage newly created simulation branching (discontinuity) points. Each one of these simulation focus points has its own context. When a path is exhausted, the closest focus point selects the next available model to instantiate the simulation frame or return to the context that generated the focus point. As simulation games are explored, a network of focus points is generated. Determining which focus point should be active at any given time is the responsibility of the meta-scheduler. When more than one model is qualified, then scheduler needs to decide which one to instantiate. Control rules can inform its decision. Three steps involve in deploying a new simulation frame in such cases: matching, activation, and preference. The matching steps should both syntactically and semantically satisfy the request. The activation step involves running a dynamic set of rules that further test the applicability of models with respect to contextual constraints. Finally, the preference steps involve running a different set of rules to impose an activation ordering among the active frames.

1.4 Agent-Supported and Model-Driven Simulation Experiment Management

Model-driven engineering (MDE) has emerged as a practical and unified methodology to manage complex simulation systems development by bringing model-centric thinking to the fore. The use of platform-independent domain models along with explicit transformation models facilitates deployment of simulations across a variety of platforms. While the utility of MDE principles in simulation development is now well recognized, its benefits for experimentation have not yet received attention. However, simulation is the act of using simulators and models to perform goal-directed experimentation, and hence, model-driven experimentation needs to be an integral part of the overall MDE-based simulation development process. The provision of machine-interpretable experiment models grounded on the statistical Design of Experiments (DOE) methodology will not only enable computer assistance for selecting proper experiment designs but also facilitate reliable evaluation of results. On the other hand, ad hoc scenario configuration files that are often used for conducting experiments are not only difficult to maintain but also lack the capability to define the concepts, relations, and constraints associated with the DOE methodology.

We present our strategy, which brings together MDE (Stahl and Volter 2006), DOE, optimization, and Intelligent Agent Technology (Yilmaz and Ören 2009) to systematically design experiments and execute optimization algorithms by using abstract, but formal, experiment models defined in terms of a Domain-Specific Language (DSL). Model transformation (Rashid et al. 2011) methods facilitate synthesizing experiments and optimization meta-heuristics from high-level specifications and then allow their orchestration using software agents that interpret the design and conduct the experiments and optimization algorithms.

1.4.1 Domain-Specific Experiment Specification Based on the DOE Metamodel

In the context of MDE, it is mandatory to clearly specify the structure of the problem domain. In our case, it is the experiment design domain. The metamodel provides a domain vocabulary and grammar in the form of an abstract syntax, along with static semantics serving as constraints over the design. One of the first things an analyst must do to design an experiment is to identify the factors. An experiment may have many factors, each of which might be assigned a variety or range of values, called the levels of the factor in DOE. Factors are classified into types, including quantitative/qualitative, discrete/continuous, and controllable/uncontrollable. Simulations come in many flavors. There are deterministic, stochastic, and dynamic (terminating or nonterminating) simulations. A design is a matrix where every column refers to a factor and each row describes a particular combination of factor levels.

Each unique combination of factor levels is called a design point. We extend this characterization of DOE to identify key concepts, relationships, and attributes. Then, we develop a metamodel in the form of experiment domain ontology and encode it as a grammar for defining the Domain-Specific Language. The DOE ontology (Somohano-Teran et al. 2014) serves as the base model specifying the commonalities across DOE experiments. The differences and optional/alternative designs will be captured using a feature model that will be used to extend and configure the base model with selected experiment design options (e.g., factorial design vs. fractional factorial design).

To operationalize the metamodel, we use the *Xtext* environment, which is a platform for developing Domain-Specific Languages. The *Xtext* platform facilitates generation of the parser and a specific editor with syntax coloring and highlights for the developed DSL. The metamodel serves as the grammar of the language allowing users to define an experiment instance in terms of the vocabulary of the experiment domain ontology (Somohano-Teran et al. 2014).

1.4.2 Feature Models for Experiment Variability Modeling

An experiment design can have various mandatory, alternative, and optional features. Features are prominent attributes that facilitate modeling variants of experiments to support different objectives. For instance, the type of the experiment design (e.g., factorial, fractional factorial), the optimization strategy (e.g., evolutionary strategy vs. simulated annealing), and the analysis method (e.g., ANOVA vs. MANOVA) are potential features that collectively define plausible configurations of an experiment. A feature model is a model that defines the features and their dependencies in the form of a feature diagram. In our approach, features are interpreted as views into the DOE ontology, and selected features with their associated concept structures are woven into the base experiment model to derive the experiment specification. Each feature is defined in terms of a set of ontological constructs (e.g., concepts, associations, attributes) and is integrated into the experiment specification to configure the conceptual (i.e., data) and experiment workflow components (e.g., batch-run specification, variance reduction method, warm-up period, run length in terminating simulations, number of replications, pseudorandom number generator) of an experiment design.

1.4.3 Agent-Assisted Experiment Model Generators

The major elements of our tool suite for experiment synthesis are highlighted in Fig. 1.3. The DOE methodology and extant research in simulation experiment design (Kleijnen 2005; Sanchez et al. 2014) provide the basis for the formulation of experiment domain requirements. The resultant concept statement is used toward developing the DOE metamodel, which is defined in terms of the Ecore meta-metamodel over the Eclipse Modeling Framework.

The DOE ontology defines the vocabulary and grammar, i.e., the abstract syntax for building the experiment domain space model. To support the instantiation of the experiment models conforming to the DOE metamodel, a suitable Domain-Specific Language is needed. This involves the definition of the concrete syntax and the development of an appropriate editor. Tools that support the creation of DSLs are readily available, and we used the popular Eclipse-based Xtext environment. The feature model provides additional variants to bring additive variability to the base model defined by the DSL. The experiment model defined by the DSL is configured with the aspects specified in the feature model. Weaving an aspect element means that all properties of the element, including its associated components, are woven into the base model. Aspect-oriented modeling methods (Rashid et al. 2011) provided the requisite strategies to achieve this objective.

The Experiment Design Agent, shown in Fig. 1.2, evaluates the generated experiment design matrix and improves the effectiveness and efficiency of the design across the parameter space of the simulation application. A trade-off analysis between the number of design points and the number of replicates per design point is carried out in relation to the type of experiment being conducted. Consider, for instance, two options: one with many replicates per design point and another with more design points with fewer replicates.

The first option enables explicit estimation of response variances that can vary across scenarios. If the primary objective is to find robust system designs, then some replication at every design point is essential. If the goal is to understand the system

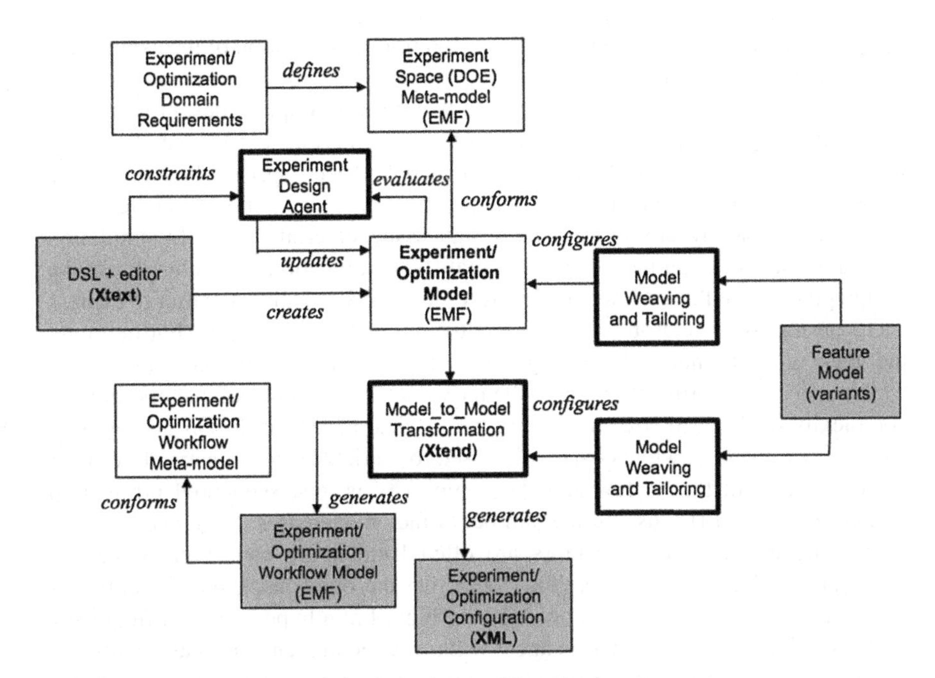

Fig. 1.3 Overview of experiment/optimization model synthesis

behavior, this requires understanding the variance, again mandating replication. However, if the goal is that of comparing systems and a constant variance can be assumed, then this constant can be estimated using classic ordinary least squares regression. Replication is then of less concern, and the second option (exploring more design points/scenarios) can be a better way to spend scarce computer resources. Even beyond the trade-off is the consideration of how many design points and replicates according to the users are needed to conserve or expend computational resources. Is a quick answer needed or are limited computational resources at hand? Or, is time not of the essence and computational resources are not severely constrained? These environments will dictate different appropriate DOE, and the agent interacts with the user to determine the best combination.

The Experiment Design Agent, if necessary, makes recommendations for updating the experiment design by evaluating it across a number of design criteria. A major design attribute is the number of scenarios required to enable estimation of metamodel parameters using saturated designs such as the fractional factorial design. To choose among different designs, *orthogonality* can be used to simplify computations while observing constraints on factor level combinations (e.g., in a queuing system, if arrival rates far exceed service rates, this will result in unacceptable steady-state average waiting times). The experiment design also determines the standard errors for the estimated parameters. The DOE literature uses several criteria to assure that the sum of these standard errors is minimal, and the design evaluation agent in updating the design prior to execution of the experiment plan can use such criteria.

Space-filling design is yet another area that allows sampling design points not only at the edges of the hypercube that defines the experiment space but also in the interior. A space-filling design with good and balanced space-filling properties helps users avoid making many assumptions about the nature of the response surface and hence results in a robust design (Sanchez et al. 2014). In simulation experiments, restricting factor values to realistic combinations may complicate the design process. To this end, we need features in the DSL to allow the user to express such constraints for interpretation and evaluation by the design agent. Furthermore, the agent should embody knowledge about the conditions and constraints under which specific experiment design schema work. The number of factors and the complexity of the response surface are two critical criteria to assess and classify designs. For instance, while coarse grids used for variable screening are effective if the number of factors is few and the complexity is low, sequential bifurcation (Sanchez et al. 2014) is used if the number of factors increases. Moderate levels of complexity and number of factors are often handled by composite or central composite designs (Law and Kelton 2000). On the other hand, if the number of factors and response surface complexity are high, Latin hypercube and frequency designs are often used. The design agent will make recommendations and configure the design matrix to better fit the characteristics of the problem and experiment space.

Following the instantiation of the experiment model under the Eclipse Modeling Framework (EMF), model-to-model transformation is utilized via the *Xtend* platform to synthesize both the experiment design matrix (in machine-interpretable XML format) and the experiment workflow model that includes specification of the computational activities for the selected experiment type. For instance, an evolutionary strategy optimization feature requires components (e.g., to support the variation, interaction, selection processes) that are different than elements of a strategy based on particle swarm optimization that seeks optimal design parameters. In this domain-specific architecture, the abstract experiment and feature models represent the experiment domain, whereas the generated workflow model along with the concrete experiment configuration represents the experiment workflow space. The workflow space model is used toward generating experiment infrastructure modules, which are initialized and coordinated by the Experiment Orchestration Agent to conduct the experiment according to the plan specified in the experiment configuration.

Figure 1.4 highlights the major components needed for the synthesis and execution of the experiment. By using the Eclipse-based *Xpand* Model-to-Code generation facility, we define a template for each workflow type to instantiate an application skeleton comprised of generic software components that implement the workflow. The orchestration agent initializes the experiment infrastructure modules and defines the simulation settings such as run length, warm-up period, variance reduction method setup, batch-run specifications, design point (scenario) initialization, and distribution of scenario executions.

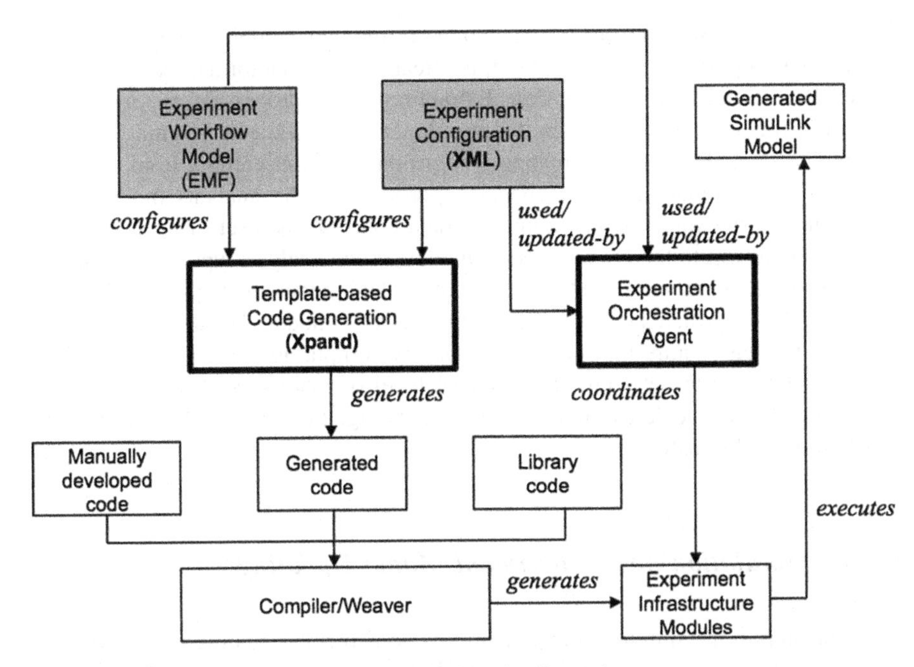

Fig. 1.4 Overview of experiment code synthesis and orchestration

The orchestration agent incorporates experiment design adaptation capabilities so that factors that are not significant in explaining the differences in the dependent variables are reclassified as control variables, and, if necessary, design schema will adapt as experimentation moves from variable screening to factor analysis and then to optimization.

1.4.4 Analysis Model Derivation

The aggregation of results for effective analysis and communication is a critical step. The mapping of the raw simulation data to abstract models that conform to specific configurations that lend themselves to effective communication and analysis benefits from model-driven engineering and transformation principles. For identifying robust solutions, a good analogy is exploratory analysis. Three-dimensional rotatable plots, contour plots, and trellis plots provide features that support exploration. Hence, data models that capture the ontology of these visualization artifacts are effective in conveying the results. Also, regression trees and Bayesian networks are effective ways of communicating which factors are most influential on the performance measures.

The process involved in transforming the raw simulation data to a format that can be consumed by external analysis tools has two main steps: (1) syntactic mapping and (2) semantic mapping. Syntactic mapping requires developing an injector that imports the contents of output stream data into a format that conforms to the data model of the simulation. Extraction is the process of exporting the content of the analysis model to an output file using the dedicated format used by the analysis tool. The semantic mapping aligns the concepts coming from the data model of the simulation to the concepts pertinent to the analysis model. The transformation reexpresses the simulation data concepts as a set of analysis concepts that conform to the grammar (metamodel) of the analysis type. For instance, the data elements that conform to the simulation data model are mapped to the leaf nodes of the Bayesian Net in a Bayesian Analysis model. On the other hand, ANOVA models require translating raw data by computing metrics that are used by the F-statistics to determine significance level of factors. Such mappings are encoded as part of the Atlas Transformation Language that we have been using for model-to-model transformation.

1.4.5 Simulation and Experiment Model Updating

The final step in the simulation experiment cycle is model updating in light of the analysis and evaluation performed by the Experiment Orchestration Agent. We consider two types of updates: (1) experiment model update and (2) simulation model update.

Adaptation of an experiment occurs at multiple levels. Based on sequential experiment results, specific factors are identified as significant, while others are classified as control variables. The reduction in the number of pertinent factors triggers a more detailed analysis of the levels of relevant factors. Such changes in the direction of exploration of the parameter space do not require an update in the experiment schema. However, experiment schema (metamodel) adaptation may be necessary when the observed response surface complexity and the change in the number of factors trigger, for example, an update from a central composite design to a Latin hypercube design. Schema adaptation can be followed by a complete schema revision, requiring a new experiment model consistent with the evolving purpose of the experiment. The initiation of an optimization process immediately after the discovery of the most pertinent factors calls for instantiation of a new schema that facilitates implementation of the most appropriate optimization protocol. Therefore, the experimentation effort consists of not only data collection and the number and length of simulation runs but also the effort required to generate the experimental designs and manage the runs.

The second type of update involves the coevolution of the simulation model with the experiment design. Figure 1.5 illustrates the related components that connect the technical spaces of the simulation and experiment models. Besides the updates needed within the parametric experiment design space, the results of the experiment may suggest revisions or refinements to the behavioral mechanisms within the simulation model's technical space.

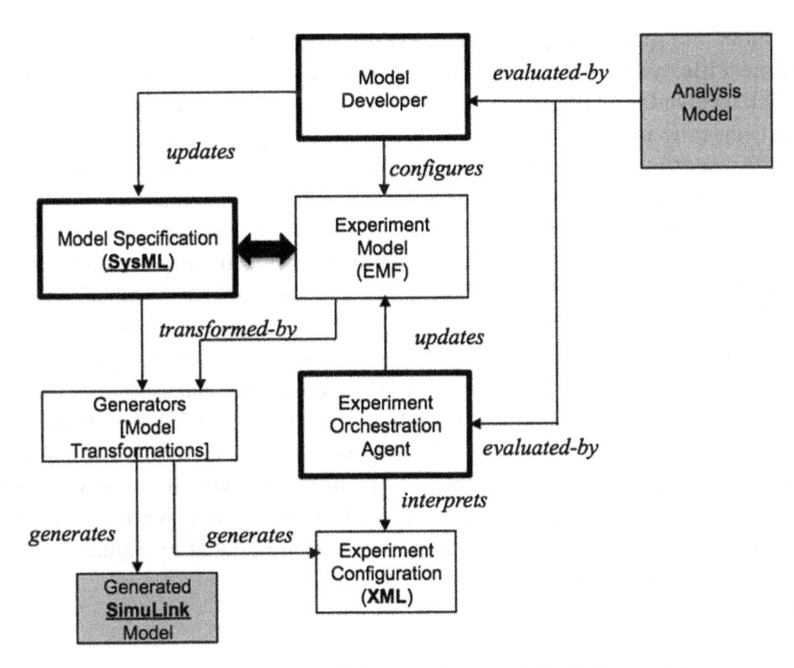

Fig. 1.5 Model updating – coevolution of the experiment and simulation models

A related benefit of connecting the experiment and simulation models is the design and transformation of a simulation model in a way that facilitates creating a list of potential factors and subsequently modifying their levels. At the same time, the analysis model and its derivation presents a challenge in conveying the results within the context of high-dimensional response surfaces.

1.5 Conclusions

Simulation involves goal-directed experimentation with dynamic models for a variety of purposes, including scientific discovery, training, education, and entertainment. Within the M&S life cycle, software agents can play various roles besides serving as model design metaphors. In this chapter, we examined how agent-monitored multisimulation concept can help explore alternative scenarios, as well as evolving and changing context in symbiotic adaptive simulations. In a similar vein, agent-assisted experiment management can support scientific discovery using proven and effective heuristics in experiment design and evaluation.

The presented strategy defines simulation experiments in terms of a life cycle with distinct phases, each with specific requirements and opportunities for computational assistance. The features of experiments and the governing processes are defined in terms of explicit experiment models, which can be made available at run time and be interpreted by intelligent software agents. By closing the loop via an experiment life cycle, we propose online adaptation of experiment models based on feedback received from previous iterations. As such, experiment models are used as explicit run-time introspective models to assist the experimentation process. Furthermore, experiments are defined at multiple levels of abstraction, starting with a feature model that uses the vocabulary of the experiment domain expert.

Agent-assisted mapping of the feature model onto a more specific experiment domain ontology, followed by model transformation, results in executable scripts for batch-mode experimentation. The proposed approach is founded on three critical pillars: (1) model-driven engineering, (2) agent-assisted experiment management, and (3) design of reproducible experiments. The synergistic interactions between these elements are leveraged to improve computational assistance for each phase of the life cycle. Ontology modeling is used to specify the structure of the experiment model and to allow the use of model transformers to shift the focus in experiment design to high-level features that are then compiled into executable experiment scripts. Agents are used to prune the experiment ontology space to relevant DOE concepts and attributes that are related to features selected by the experiment designer.

References

Elzas MS, Ören TI, Zeigler BP (eds) (1986) Modelling and simulation methodology in the artificial intelligence Era. North-Holland, Amsterdam, 423 p

Elzas MS, Ören TI, Zeigler BP (eds) (1989) Modelling and simulation methodology: knowledge systems' paradigms. North-Holland, Amsterdam, 487 p

Kleijnen JPC (2005) An overview of the design and analysis of simulation experiments for sensitivity analysis. Eur J Oper Res 164(2):287–300

Law MA, Kelton DW (2000) Simulation modeling and analysis. McGraw Hill, Boston

Ören TI (1977) Concepts for advanced computer assisted models. ACM Simuletter 9(1):8–9

Ören TI (1978) Rationale for large scale systems simulation software based on cybernetics and general systems theories. In: Ören TI (ed) Cybernetics and modelling and simulation large scale systems. International Association for Cybernetics, Namur, pp 151–179

Ören TI (2000a) Use of intelligent agents for military simulation, Lecture notes of simulation for military planning. International Seminars and Symposia Centre, Brussels

Ören TI (2000b) Agent-directed simulation – challenges to meet defense and civilian requirements. In: Joines JA et al (eds) Proceedings of the 2000 winter simulation conference. Orlando, ACM, New York, pp 1757–1762

Rashid A, Royer J-C, Rummler A (2011) Aspect-oriented model-driven software product lines. Cambridge Press, Cambridge/New York

Sanchez MS, Sanchez PJ, Wan H (2014) Simulation experiments: better insights by design. In: Proceedings of the 2014 summer computer simulation conference, ACM, New York, pp 79–86

Shoham Y (1993) Agent-oriented programming. Artif Intell 60(1):51–92

Somohano-Teran A, Dayibas O, Yilmaz L, Smith AE (2014) Toward a model-driven engineering framework for reproducible simulation experiment lifecycle management. In: Proceedings of the 2014 IEEE/ACM winter simulation conference, Savannah, 7–10 Dec 2014, pp 2726–2737

Stahl T, Volter M (2006) Model-driven software development. Wiley, Chichester/Hoboken

Yilmaz L, Ören TI (2009) Agent-directed simulation and systems engineering. Wiley, Berlin

Chapter 2
Service-Oriented Model Engineering and Simulation for System of Systems Engineering

Bernard P. Zeigler and Lin Zhang

2.1 Introduction

The model is the foundation for simulation activities (Ören et al. 1982), especially in regard to system of systems (SoS, a composition of systems which component systems have legacy properties). A valid model and correct simulator are necessary for obtaining simulation results that serve the intended use of the model (Zeigler et al. 2000). Verification, validation, and accreditation (VV&A) is the primary means of establishing the credibility of the simulation results (Pace 2004). VV&A is usually considered after model construction and involves calibration and/or validation of the established model in order to determine whether it is credible. This post-construction determination has important implications to discover model problems and defects, but it cannot solve the problem of how to get a correct model in the first place. Especially for complex systems, due to the complexity and uncertainty of the system, the modeling process can be very complicated, which makes VV&A of a model extremely difficult. Even if defects are found via VV&A, revision of the model will be very difficult and costly.

More importantly, for a system of systems, to construct a valid model is just the first step since a model of a SoS generally experiences a long term of evolution and management. As a result, the key issue for a complex system model is to guarantee the credibility of the full model life cycle with minimum cost.

To meet the challenges in development and management of the SoS model, this chapter introduces the concept of the *model engineering* (Zhang 2011; Zhang

B.P. Zeigler (✉)
RTSync Corp., Arizona Center for Integrative Modeling and Simulation,
Sierra Vista, AZ, USA
e-mail: zeigler@rtsync.com

L. Zhang
School of Automation and Computer Science, Beihang University, Beijing, China

L. Yilmaz (ed.), *Concepts and Methodologies for Modeling and Simulation*,
Simulation Foundations, Methods and Applications,
DOI 10.1007/978-3-319-15096-3_2

et al. 2014a), which aims at setting up a systematic, normalized, and quantifiable engineering methodology by exploring basic principles in model construction, management, and maintenance to manage the data, processes, and organizations/people involved in the full life cycle of a model to guarantee the credibility of its life cycle. Meanwhile, in recent years, service-oriented technology has been widely used in software intensive systems, as well as model construction and management of complex system simulation. Therefore, we will show how modeling engineering can take advantage of service-oriented technology to provide an efficient way of building and managing the model of a system of systems.

2.2 Some Related History

It helps to recount some history relevant to *service-oriented model engineering* (SOME) as appropriate to a volume dedicated to Tuncer Ören's 80th birthday. As early as 1973, Ören was expressing his normative views for modeling and simulation (M&S) methodologies (Ören 1973) and recently published a treatise on the synergies of simulation, agents, and systems engineering (Ören and Yilmaz 2012). Many of his views on these synergies are covered in the book with Yilmaz on *Agent-directed Simulation and Systems Engineering* (Yilmaz and Ören 2009).

As related by Ören and Zeigler (2012), Ören received his Ph.D. in systems engineering under the supervision of A.W. Wymore. His Ph.D thesis was greatly influenced by Wymore's axiomatic approach for his systems theory (Wymore 1967). Moreover, Ören's mechanical engineering background allowed him to appreciate the vital importance of developing software tools for M&S (Ören 1990). He has always been interested in learning, conceiving, and developing methodologies suitable for complex problems (especially for social problems) which are inherently nonlinear in nature. As part of his Ph.D. requirements, in the late 1960s, he developed a simulation model specification language called GEST (General System Theory implementer) (Ören 1971) based on Wymore's book (Wymore 1967). Part of his aim was to use a translator to generate a simulation program in a language which could be compiled or interpreted. Such translators were implemented later by his students. GEST's model specification language is based on Wymore's concept of systems composed of component systems and couplings that all components to exchange information through input and output ports. Since component systems and coupling recipes were already defined by Wymore, in set-theoretic notation, Ören concentrated on ease of robust specification and readability and avoided any set-theoretic representation.

Over the years, the scope of Ören's concerns has broadened to formulate a body of knowledge for M&S expressed in many publications and presented in detail in Ören (2005, 2014), where he describes a paradigm shift from use of the term M&S to the term simulation systems engineering (SSE): "In the early days, only very few were referring to M&S. Afterwards, to stress modeling process and the associated activities and environments, the term M&S is used by large number of simulationists. Currently, a very commendable shift of paradigm is being adopted to cover all aspects of simulation studies. This is to conceive M&S –within a larger perspective– as the Simulation Systems Engineering (SSE)." (Slight paraphrase of (Ören 2005))

In this article, we take a similarly broad perspective and probe the nature of model engineering in the context of systems engineering, particularly for systems of systems (SoS) and implemented with service orientation environments. To establish the background needed for this discussion, we briefly introduce the definition and theory of systems of systems as it relates to M&S, in particular to the discrete event system specification (DEVS) formalism for M&S. We then define, and examine in depth, model engineering for SoS which has deals with the full model life cycle. With these concepts as foundation, we analyze the services necessary to support model engineering and the requirements for design of a service-oriented model engineering and simulation environment. Consideration of the results of research in DEVS then enables us to give a more concrete characterization of such an environment. We close with a discussion of how model engineering and DEVS enable new frameworks for application areas and the opportunities for further research.

Some short definitions of terms we employ as initial concepts are drawn from Waite and Ören (2009):

- Body of Knowledge (BoK) – The set of justified true beliefs and competencies – explicit and implicit – that defines a discipline, practice, role, or field of endeavor
- Referent – n. Something referenced or singled out for attention, a designated object, real or imaginary, or any class of such objects
- Model – n. The representation of some referent
- Simulation – n. A mechanization of a model's evolution through time

Although definitions are still in flux, for our purposes, *service-oriented model engineering* is a form of model engineering that is based on a service approach to computation, and simulation systems engineering is an inclusive term that includes model engineering.

2.3 Theory of Systems of Systems

In systems theory as formulated by Wymore (Ören and Zeigler 2012), systems are defined mathematically and viewed as components to be coupled together to form a higher-level system.

As illustrated in Fig. 2.1, Wymore's (1967) systems theory mathematically characterizes:

- *Systems* as well-defined mathematical objects characterizing "black boxes" with structure and behavior.

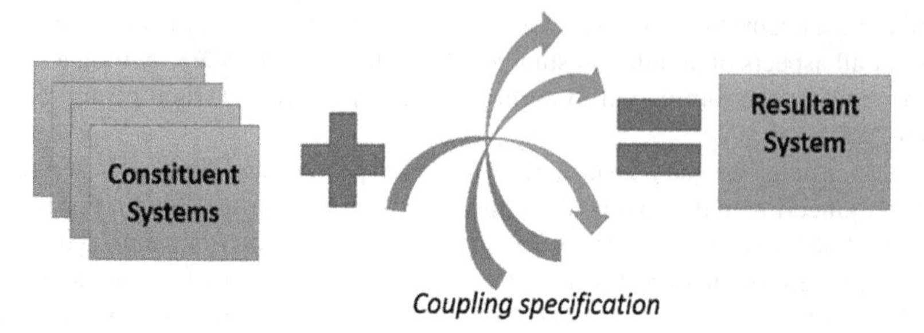

Fig. 2.1 Wymore's system composition

- *Composition of systems* – constituent systems and coupling specification result in a system, called the resultant, with structure and behavior emerging from their interaction.
- *Closure under coupling* – the resultant is a well-defined system just like the original components.

2.3.1 System of Systems

As illustrated in Fig. 2.2, a system of systems (SoS) is a composition of systems, where often component systems have legacy properties, e.g., autonomy, belonging, diversity, and emergence (Boardman and Sauser 2006). In this view, a SoS is a system with the distinction that its parts and relationships are gathered together under the forces of legacy (components bring their preexisting constraints as extant viable systems) and emergence (it is not totally predictable what properties and behavior will emerge). Here in Wymore's terms, *coupling* captures certain properties of relevance to coordination, e.g., connectivity, information flow, etc. *Structural and behavioral properties* provide the means to characterize the resulting SoS, such as fragmented, competitive, collaborative, coordinated, etc.

The main difference between SoS and general system composition is worth noting. SoS generally refers to systems composed of components that are already in existence and bring certain legacy properties "to the table" when placed into a new composition. This is in contrast to general system composition where components may be built from scratch for the distinct purpose of the new composition. This implies that in the SoS case, a key feature is that the compositions require integration and/or coordination to overcome the features and goals of the existing systems that don't align well with the new system goals. However, we remark also that the term SoS is still in flux and may sometimes mean complex systems whose components themselves are complex systems. Depending on the definition of "complexity" in this context, the two meanings may actually coincide.

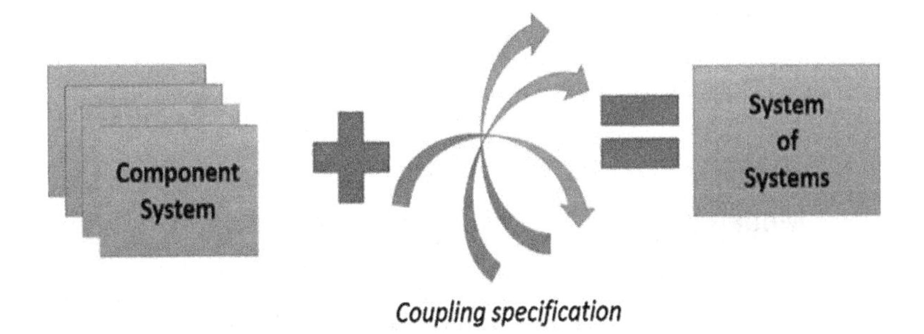

Fig. 2.2 System of systems

2.3.2 Discrete Event Systems Specification (DEVS) Formulation of SoS

The DEVS formalism (Zeigler et al. 2000), based on systems theory, provides a framework and a set of M&S tools to support systems concepts in application to SoS engineering (Mittal and Martin 2013). A DEVS model is a system-theoretic concept specifying inputs, states, and outputs, similar to a state machine. Critically different, however, is that it includes a time-advance function that enables it to represent discrete event systems, as well as hybrids with continuous components, in a straightforward platform-neutral manner. DEVS provides a robust formalism for designing systems using event-driven, state-based models in which timing information is explicitly and precisely defined. Hierarchy within DEVS is supported through the specification of atomic and coupled models. Atomic models specify behavior of individual components. Coupled models specify the instances and connections between atomic models and consist of ports, atomic model instances, and port connections (ports and connections are not shown here for simplicity). The input and output ports define a model's external interface, through which models (atomic or coupled) can be connected to other models.

As illustrated in Fig. 2.3, based on Wymore's systems theory, the DEVS formalism mathematically characterizes the following:

- DEVS Atomic and Coupled Models specify Wymore systems.
- *Composition of DEVS models* – component DEVS and coupling result in a Wymore system, called the resultant, with structure and behavior emerging from their interaction.
- *Closure under coupling* – the resultant is a well-defined DEVS just like the original components.
- *Hierarchical composition* – closure of coupling enables the resultant-coupled models to become components in larger compositions.

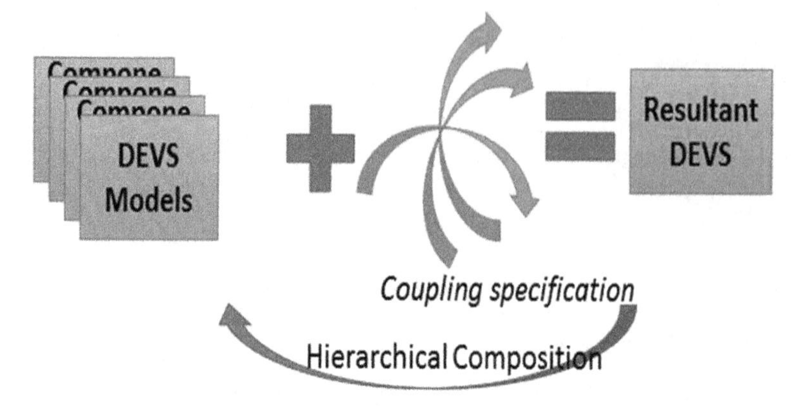

Fig. 2.3 DEVS formulation of systems of systems

2.4 Model Engineering for SoS

A model is an abstract expression of objects to study and embodies high intelligence of human beings in recognition of the world. With continuous development of science and technologies, the model is becoming more and more important. It refers not just to the process of modeling but also to the life cycle of model.

2.4.1 The Life Cycle of a SoS Model and Related Works

Generally, a model experiences requirement analysis, model design, model construction, model verification and validation (VV&A), model application, and model maintenance (Zhang 2011). These processes compose a complete life cycle of a model as is shown in Fig. 2.4.

How to build a right model is the core issue in simulation. A large number of research achievements on models have been obtained in the past dozens of years. These achievements are related to different phases in a model life cycle, e.g., modeling theory and method, VV&A, and model management.

The life cycle concept has not been emphasized enough in the simulation domain, and related research and applications are not sufficient (Balci 2012). Fishwick (1989)called the simulation model development process as simulation model engineering to emphasize engineering feature of the model development process, but no special explanation on its meaning was given, and no systematic method system was established.

In recent years, the international simulation community has become conscious of the unfavorable influences of missing foundational theory of M&S on the development of simulation curriculum. As a result, research on the M&S life cycle management is gradually attracting attention from academic circles. Radeski and

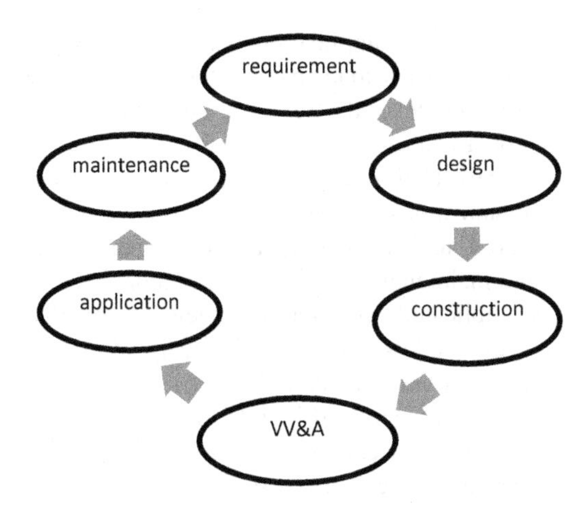

Fig. 2.4 The life cycle of a SoS model

Parr (2002) proposed the modeling simulation life cycle model framework, which defined organization mode and structure of the modeling simulation process, work products, quality assurance, and project management, and described the features and requirements of the life cycle phases such as development, use, maintenance, and reuse of the modeling simulation system. References (Fishwick 1990; Abdouni Khayari et al. 2010) achieved some valuable results in the model life cycle management and developed a model prototype management system, which provided valuable reference to model development for designers.

2.4.2 Challenges in Development and Management of SoS Models

As can be seen from the related work, current research on models generally focused on one phase in the model life cycle and is separate and diffuse. Although importance of the engineering idea is gradually recognized in applications of the full model life cycle, currently no complete theory and technology system and philosophy are available. So there are still lots of challenges in the life cycle of the model of SoS. As pointed in (Zhang et al. 2014a), some reasons for this situation are:

1. High complexity of the referent SoS – Firstly, a SoS is composed of various component systems, and the relationship between them is very complicated. Secondly, generally a complex system is dynamic, variable, and very uncertain. Thirdly, SoS generally performs emergent behaviors. These features make a SoS very complicated. The complexity of SoS leads to the complexity of the model itself.
2. The long life cycle of a SoS – With passage of time, the models should be continuously improved and changed. Different model versions are available. Each version may be applicable to different application phases. Different

versions and application phases of multiple models compose a complicated network. How to keep consistency and credibility of different parts and versions of the model is the key for model maintenance.

3. Model heterogeneity – A SoS is composed of many heterogeneous component models. Heterogeneity of models generally comes from different development organizations, different platforms and architectures, different development languages and databases, etc. Heterogeneity brings big challenges to integration and maintenance of the system models.
4. Complicated evolution of models – Generally, a SoS is in continuous evolution, so the models will be continuously adjusted and changed. Changes of different relations are very complicated in evolution due to system complexity, so the model elements and its relation should be completely tracked and managed to guarantee correctness of the model evolution.
5. Difficult model reuse – With growth of complexity of the systems to study, the roles and values of model reuse are very remarkable in model development and use of a SoS (Liu et al. 2008). Generally, a SoS includes multiple combined systems. A huge number of system models in past research and development practices have been accumulated. Correct and efficient reuse of models will reduce model development cost, greatly shorten development time, and effectively improve model credibility. Although some research on model reuse has been conducted, no efficient and practicable model reuse method is available now.
6. Massive processing data – Generally, a SoS entails a large amount of data to process, including the required modeling data, data generated in modeling, and data generated in the modeling process. Data processing includes data storage, inquiry, exchange, management, understanding, analysis, and mining, which bring many challenges.
7. The multidisciplinary collaborative model development – Collaborative model development is associated with different steps in the whole model life cycle. Collaboration is required on different phases, e.g., collaborative requirement analysis, collaborative design, and collaborative validation. All this work composes a huge engineering requirement and should be supported by appropriate management tools.
8. Higher requirements for system performance – Compared to a simple system, a SoS requires higher performance, e.g., higher requirements for reliability, security, credibility, cost, and energy saving. To guarantee that these performance requirements are met, special means should be required to analyze and process the models.

2.4.3 Meaning of Model Engineering

2.4.3.1 Concept of Model Engineering

Based on the state-of-the-art researches on the model, a systematic methodology was proposed to cope with challenges in model life cycle management of a SoS (Zhang 2011; Zhang et al. 2014a). The model development and management activities change from a spontaneous and random behavior to conscious, systematic, standardized, and manageable behavior by constructing a model engineering theory and methodology system in order to guarantee credibility of different model phases.

Zhang (2011; Zhang et al. 2014a) gave a definition of model engineering as follows:

Model engineering is defined as a general term for theories, methods, technologies, standards, and tools relevant to a systematic, standardized, quantifiable engineering methodology that guarantees the credibility of the full life cycle of a model with the minimum cost.

Here, *model engineering* involves the following meaning (Zhang 2011; Zhang et al. 2014a):

1. Model engineering regards the full life cycle of a model as its object of study, which studies and establishes a complete technology system at the methodology level in order to guide and support the full model life cycle process such as model construction, model management, and model use of a SoS.
2. Model engineering aims to ensure credibility of the full model life cycle; integrate different theories and methods of models; study and find the basic rules independent of specific fields in the model life cycle; establish systematic theories, methods, and technical systems; and develop corresponding standards and tools.
3. Model engineering manages the data, knowledge, activities, processes, and organizations/people involved in the full life cycle of a model and takes into account time period, cost, and other metrics of development and maintenance of a model.
4. Here, the model credibility is a comprehensive indicator and includes factors such as availability, accuracy, reliability, and quality of service (QoS).

2.4.3.2 Key Technologies of Model Engineering

As described in the above part, current research on the technologies related to the full model life cycle is preliminary and diffuse. For comprehensive and systematic application and implementation of model engineering, many key technologies should be studied (Zhang et al. 2014a). These technologies can be divided into six categories as shown in Fig. 2.5.

1. General technologies

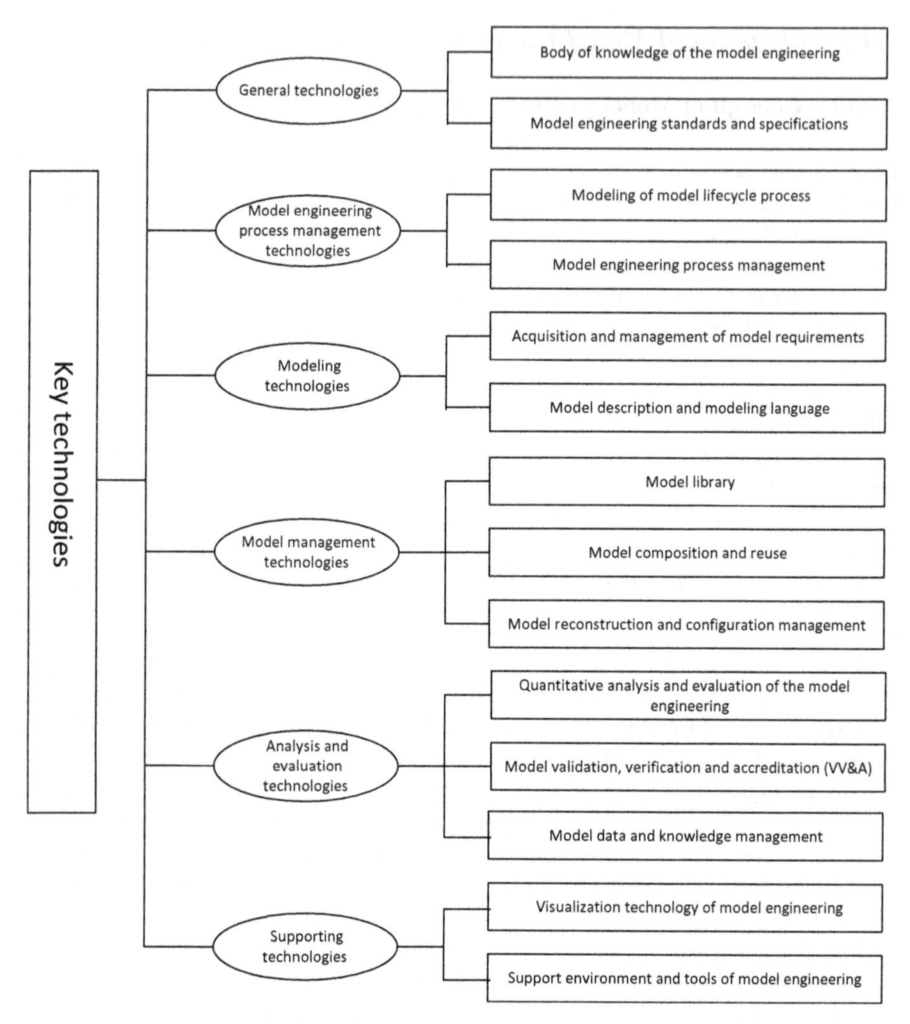

Fig. 2.5 Key technologies of model engineering

- Body of knowledge of the model engineering: The body of knowledge system (BoK) includes the concepts and terminologies involved in a specific research field. The model engineering BoK identifies the research scope of the model engineering and its boundary and relationship with other related subjects. Establishment of systematic and complete BoK requires long-term accumulation and extraction.
- Model engineering standards and specifications: Standards are the basis for the implementation of the model engineering. During the life cycle process of a model, each activity requires corresponding standards, including model development process, model description, model component interface,

model storage, model data exchange, model interoperation, model service, model maintenance, etc.

2. Model engineering process management technologies

   - Modeling of model life cycle process: The life cycle model of the model engineering aims to identify the structural framework of activities involved in model construction and management (Zeigler et al. 2000), which is the methodology to guide the model engineering, and ensure improvement of model quality and development efficiency and reduction of full model life cycle cost. Proper process models and corresponding implementation methods can be proposed by referring to the existing achievements in the system engineering and software engineering and other relevant fields and combining the model development features of SoS.
   - Model engineering process management: The data, knowledge, tools, persons/organizations, and technologies in the full model life cycle should be effectively managed with the model life cycle process model as the guide, with standards and specification as the basis, and with the project management methods and means as reference in order to get the dependable model with the minimum cost. The model maturity definition and control, performance management, flow monitoring and optimization, risk control, and cost control are important in model engineering process management.

3. Modeling technologies

   - Acquisition and management of model requirements: Accurate requirement acquisition is the key in modeling. Requirement acquisition and management is very challenging due to uncertainty and ambiguity of SoS. Requirement acquisition studies to extract, describe, parse, and validate requirement via automated or half-automated means. Requirement management studies how to reflect the changing requirements in the model construction and maintenance accurately and timely.
   - Model description and modeling language: Generally, a SoS contains multiple different systems with different properties such as qualitative systems, quantitative systems, continuous systems, discrete event systems, deterministic systems, uncertain systems, etc. One of the core issues in model development of SoS is how to take advantage of effective ways to describe the whole system. Therefore, it is required to study corresponding model description mechanism and structure and develop generic or specific description languages according to the characteristics of the various systems.

4. Model management technologies

   - Model library: The model library is the foundational platform to carry out model management and perform standardized encapsulation, storage, and query for the models (Ören and Zeigler 1979). The complicated applications such as model reuse, combination, and configuration management can be based on the model library. Traditional database technology, service-oriented technology, and cloud-computing technology can support construction and management of the model library.

- Model composition and reuse: The model composition and reuse is an important technology to improve model construction and maintenance efficiency and improve model credibility of SoS. It mainly studies how to use the existing model components to quickly and correctly compose complicated models according to the system requirements and includes standardized encapsulation of model components, intelligent model matching, model relation management, dynamic model combination, model consistency validation, and model service.
- Model reconstruction and configuration management: The requirements for model functions and performances change due to diversified requirements inside and environmental uncertainty, so the models should be quickly reconstructed or configured. The model reconstruction aims to adjust the internal structure without change of the main external functions of models, further optimize the model performance, and ease its understanding, maintenance, and transplant of models. Model configuration can adapt different requirements or change of models in function and performance by adjusting and optimizing internal components and parameters. For SoS model engineering, model reconstruction and configuration management are very important and challenging.

5. Analysis and evaluation technologies

- Quantitative analysis and evaluation of the model engineering: The quantitative analysis is one of main features of the model engineering. To ensure credibility of the full model life cycle, many steps should be analyzed, evaluated, and optimized in a quantitative manner, e.g., complexity analysis and evaluation of model development process, cost and benefit analysis and optimization, risk analysis and control, model availability and reliability analysis, and model service quality analysis.
- Model validation, verification, and accreditation (VV&A): The model VV&A technology is one important part in the model engineering. Although some rich research achievements have been achieved, they cannot meet the actual requirements of modeling simulation of SoS. Most research focuses on qualitative analysis, and quantitative and formalized analysis methods are lacking, so VV&A technology, especially VV&A quantitative analysis, and formalized analysis technology are still a main research focus in the model engineering.
- Model data and knowledge management: Many SoS models contain voluminous data to process. Some models are constructed based on massive data, or even exist in the form of data and their relations. Data management aims to effectively organize and use the data, especially massive data, and plays a key role in quantitative analysis of the model engineering. On the whole, the knowledge is divided into two classes in the model engineering. The class 1 indicates the knowledge in the model, e.g., some qualitative models include

massive knowledge rules. Another class indicates the knowledge on model development and management and generally includes experiences accumulated and extracted by developers and users in practices. Different knowledge should be managed and used in different manners to improve model quality and intelligence and automation of model construction and maintenance.

6. Supporting technologies
   - Visualization technology of model engineering: Visualization technology can be used on different phases of the model engineering, can realize transparent model development and management process, facilitate understanding and monitoring, and improve human–machine interaction efficiency. The visualization technology plays an important role in the model engineering.
   - Support environment and tools of model engineering: Implementation of the model engineering requires an integrated support environment and corresponding support tool to support different activities of model engineering, e.g., network collaboration, requirement management, process model construction and maintenance, model library management, qualitative and quantitative analysis and evaluation, data integration, knowledge management, model validation, and simulation experiment.

2.4.4 Body of Knowledge of Model Engineering

Model engineering is the resultant of fusion of many crossing subjects including the software engineering, system engineering, computer science and engineering, mathematics, system M&S, knowledge engineering, project management, quality management, and related application fields. Based on the body of knowledge in these disciplines, specific BoK of the model engineering is formed according to the requirements and features of the model engineering.

To establish the model engineering BoK, it is necessary to tease out the involved knowledge system in a systematic manner, extract features closely associated with the model-related activities from related fields, and summarize and condense those specific development technologies and management means of the model life cycle process. A preliminary BoK framework of the model engineering was given by Zhang (2011, 2014a). Two aspects are mainly considered in this process.

1. Identify the horizontal crossing relations between the model engineering and other closely associated subjects, properly tailor their overlapping parts, and make these overlapping parts reflect specific features of the model engineering.
2. Identify the modules in the model engineering system and vertical hierarchical relations and horizontal interface relations between modules, make the framework compose an organic whole, and serve for the full life cycle process of the model.

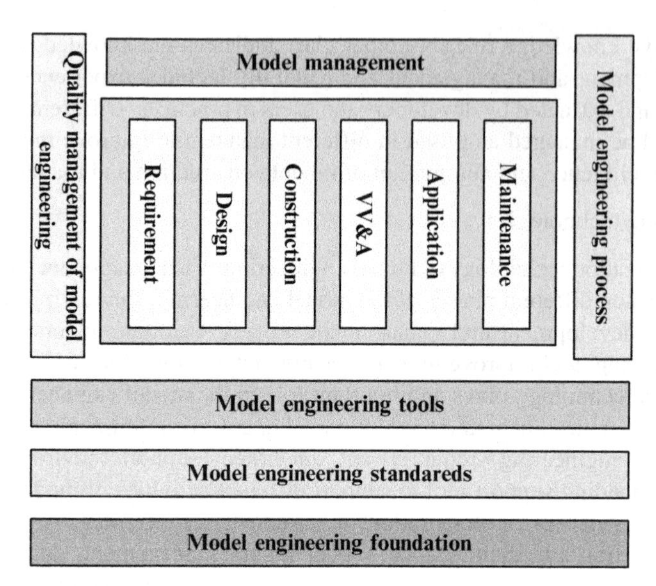

Fig. 2.6 The BoK framework of the model engineering

By introducing the model engineering BoK framework, we hope to reach the following targets:

1. Promote consistent opinion within academic circles on the meaning of model engineering
2. Identify the research scope of model engineering
3. Relate the position of model engineering to other subjects such as software engineering, system engineering, computer science, and mathematics and set their boundaries.

A BoK framework of model engineering is shown in Fig. 2.6 (Zhang 2011; Zhang et al. 2014a). The BoK framework of the model engineering is divided into five parts:

- Part 1: Foundation: including the basic concepts and terms, methodology, technical system, etc. It provides the basic guidance for the implementation of the model engineering and also is the foundation and guarantee of the model engineering independent of other subjects.
- Part 2: Model life cycle: it describes different phases of the full model life cycle at the technical level. This part modeling requirement, model design, model construction, model VV&A, model application, model maintenance.
- Part 3: Implementation and management of the model engineering: it includes the process management quality management of model engineering and model configuration management. All activities in the full model life cycle are managed and controlled implementation, process, and quality.

- Part 4: Model engineering tools: it provides the necessary software tools for the implementation and application of the full life cycle of the model engineering.
- Part 5: Related standards of model engineering: it includes rules, protocols, or specifications, which are necessary for implementation of model engineering and development of related tools.

The detailed contents of each part can be found in Zhang et al. (2014a).

2.5 Service-Oriented Model Engineering and Simulation Environment

The construction and management based on model engineering are shown in Fig. 2.7 (Zhang et al. 2014b). Taking modeling as an example, a simulation scenario is given, then multiple subtasks of the system are formed automatically according to the scenario, and automatic matching between tasks and processes is completed in the model engineering platform, so a new model is built. This just-built model can be added into the model library as a case; therefore, the model library is enriched. The use, management, and maintenance are completed by the full life cycle of model engineering.

2.5.1 Architecture of Service-Oriented Model Engineering and Simulation Environment

Service-oriented technology is one of the most powerful and popular technologies to the development, management, and integration of software intensive systems and has been widely applied to lots of different domains.

A service-oriented model engineering and simulation environment is a kind of software to support the implementation of model engineering (Zhang et al. 2014b) and simulation (Fig. 2.8). There are five layers including model component layer, model service layer, model management layer, simulation layer, and application layer. The functions of each layer are as follows:

1. *Model component layer*: there are various models, such as qualitative model, quantitative model, linear model, and nonlinear model. Meanwhile, these models are provided by different organizations and developers, which can lead to model heterogeneity, so model component layer is needed to classify and organize models.
2. *Model service layer*: this layer is a process of model normalization. It provides interface specifications among models and conducts unified service encapsulation and transformation (e.g., service–agent modeling (Si et al. 2009; Liu

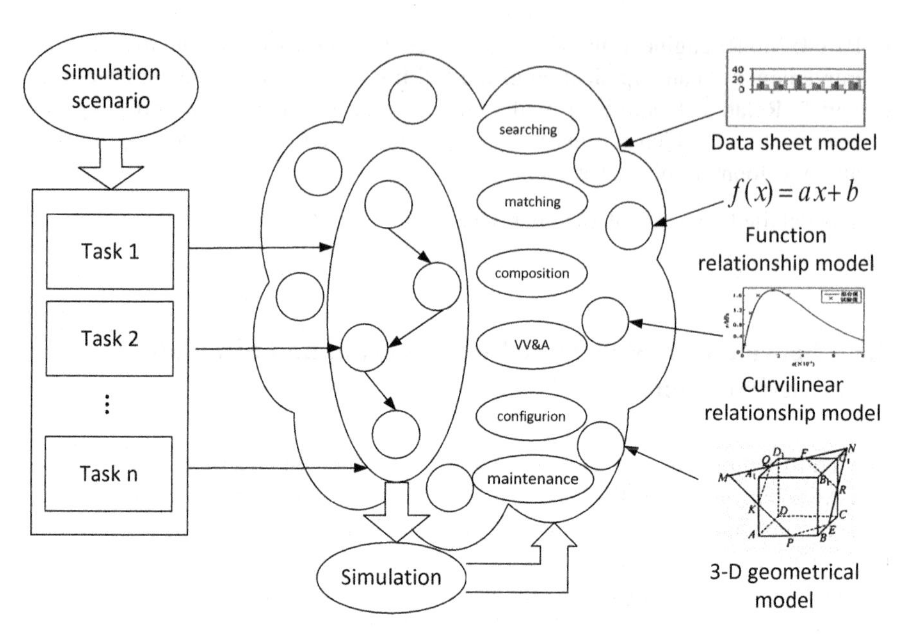

Fig. 2.7 Construction and management based on model engineering

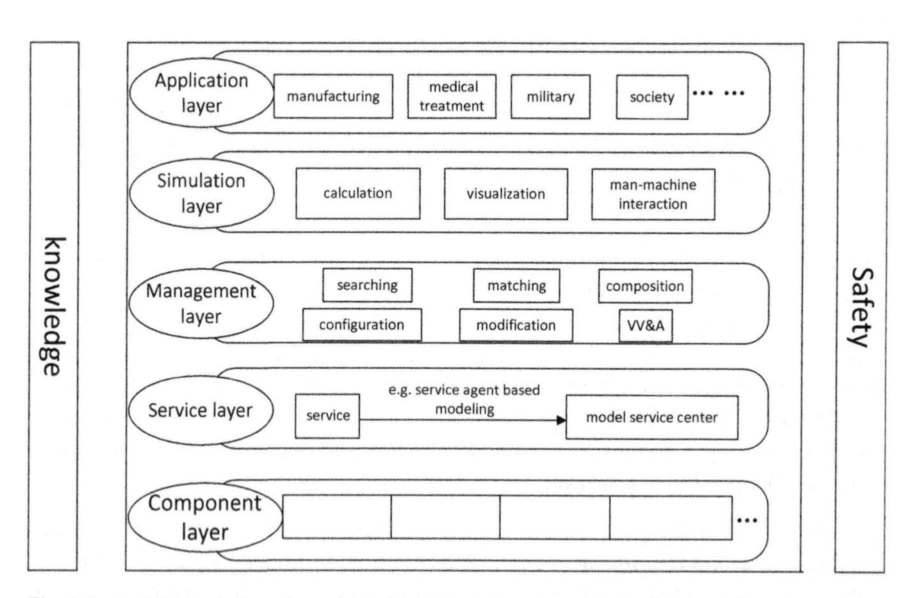

Fig. 2.8 Architecture of service-oriented model engineering and simulation environment

et al. 2014) is a kind of method), and then the standardized model is put into the model service center.

3. *Model management layer*: model management layer executes management and operation of models in the model service center, such as searching, matching,

composition, configuration, modification, VV&A, etc. Fast and accurately matching among models is guaranteed by management, so requirements of M&S can be satisfied.
4. *Simulation layer*: simulation layer mainly consists of simulating calculation, visualization, man–machine interaction. Simulating calculation means getting simulation results under the help of software and hardware. Visualization technology can realize transparent model development and management, as well as facilitate the process of understanding and monitoring; man–machine interaction can support different types of interactions.
5. *Application layer*: different kinds of applications can be carried out with the help of model engineering. These applications can include manufacturing, medical treatment, military, environment, society, etc. This reflects the value of model engineering itself and its contributions to society development.

2.5.2 Implementation of a Service–Agent-Based Model Engineering Supporting Environment

According to the idea of model engineering, the elements in a model library should have the characteristics of service oriented, intelligence, standardization, etc.

2.5.3 Encapsulation of Model Components

To achieve this purpose, we use service–agent (SA) that was proposed in Si et al. (2009) to encapsulate component models in the model library. Service–agent is a combination of service and agent, which can make service with agent characteristics (Si et al. 2009; Liu et al. 2014) and will be well suited to features of SoS. The structure of a SA is shown in Fig. 2.9.

A SA has several features (Liu et al. 2014): (1) A SA is an autonomous entity which observes and acts upon an environment and directs its activity toward achieving goals. (2) A SA pertains to SOA standard with XML-based protocols such as WSDL, SOAP, etc. (3) A SA has states, e.g., working state, prepared state, waiting state, and searching state.

After component models are encapsulated into SAs, the process of modeling of a complex system can be transferred into the process of composition of SAs in the model library. Theoretically, the composition of component models with SAs can be automatic and have the ability to be self-adapted to the uncertainty of SoS.

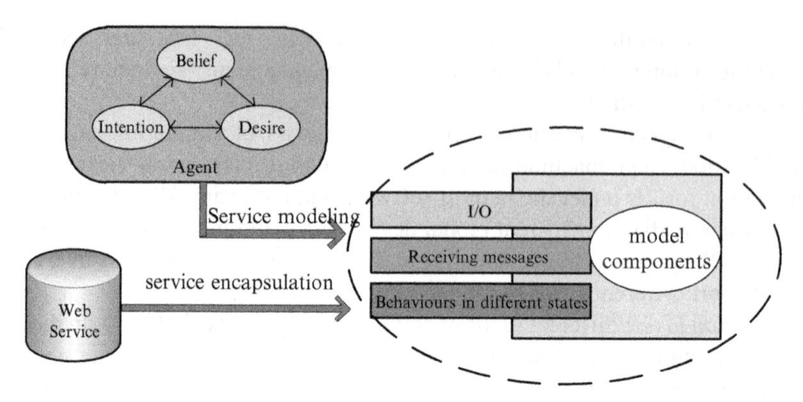

Fig. 2.9 Service–agent (SA) structure

2.5.3.1 Specifications of Service Agents

Specifications of SA are basis of communication, interaction, and composition among SAs. A set of SA specifications were given by Zhang et al. (2014b). The SA specifications are described as three parts: interface specifications, architectural specifications, and implementation specifications.

1. *Interface specifications*: SA external interface is either java component interface or web service, which can support integration of local area network (LAN) and wide area network (WAN). The purpose of user-oriented interface layer specification is to unify the descriptions of components and build standard calling interfaces for components. Component interface follows the principle of service orientation, which let users pay attention to functions provided by components, rather than the internal structure and state of components.
2. *Architectural specifications*: SA has the characteristics of environment awareness and special communication interface, which general simulation components do not have. These reflect the intelligence of SA. Architectural layer is designed to support the underlying services for interface layer, while conducts constraint and guidance as a component for realizing infrastructure by specialized technological standard for implementation layer.
3. *Implementation specifications*: One implementation for the SA is an agent based on JADE/JAVA platform. Compared with general agents, this endows SA with function operations in the form of web service and can support web integration. We discuss DEVS-based approaches to service-oriented model engineering in Section 2.6.

2.5.3.2 Cooperation Mechanism of SA

Cooperation mechanism of SA can be different according to different SA specifications and applications. A mechanism is given in Liu et al. (2014). In the mechanism, a SA is designed to have four states including the working state, prepared state, waiting state, and searching state. In working state, service agent provides service and receives output from its previous SA. In prepared state, all behaviors are blocked for a new message to come. Waiting state is for the workers. In this state, the worker receives other organization members' ID information and input/output matching information from the organizer. Searching state is for the organizer. The organizer adopts "first come first serve" strategy to choose workers.

A composition algorithm is given based on the above cooperation mechanism, and a software platform prototype for an abstracted SoS simulation problem is also developed (Liu et al. 2014). This prototype used JADE4.1 (Java Agent Development Framework) platform to perform model composition with the SA-based method. JADE is a MAS (multi-agent system) software development platform in JAVA language. Web services are published in a Tomcat7.0 container and are packaged by WSIG (JADE Web Service Integration Gateway) plug-in. The purpose of WSIG is to achieve the integration of MAS and WS (web service) architecture.

2.6 DEVS-Based Service-Oriented Model Engineering and Simulation Environment

Mittal (2014) describes model engineering for cyber complex adaptive systems, a very challenging class of SoS, by extending Model-Based Systems Engineering (MBSE) paradigms (Zeigler 1976; Zeigler et al. 2000; Mittal and Martin 2013). Applied to complex adaptive systems, model engineering must address distinct challenges posed in the M&S domain such as model composability and executability. These problems can be overcome with formalisms that distinguish models (which represent the essence of a SoS) from simulators (which are the platforms for executing the models to generate their behavior). To do so, we can employ the theoretical and conceptual frameworks such as the systems-based DEVS concepts presented earlier. The DEVS formalism provides a sound and practical foundation for the architecture of model engineering and simulation environment presented in Fig. 2.7. Some of the main reasons for basing the architecture on DEVS are the following:

- DEVS formalizes what a model is, what it must contain, and what it doesn't contain (e.g., experimentation and simulation control parameters are not contained in the model).
- DEVS represents a system of interest (SoI) using well-defined input and output interfaces. This is critical because composing models requires respecting such boundaries for the constituent referent SoIs.
- DEVS is *universal* and *unique* for discrete event system models in the sense that any system that accepts events as inputs over time and generates events as outputs over time is equivalent to a DEVS in a strong sense: i.e., its behavior and structure can be described by such a DEVS.

- A DEVS model is a system-theoretic concept specifying inputs, states, outputs, similar to a state machine. Critically different, however, is that it includes a time-advance function that enables it to represent discrete event systems, as well as hybrid systems with continuous components in a straightforward platform-neutral manner.
- DEVS-compliant simulators execute DEVS-compliant models correctly and efficiently. DEVS defines what's necessary to compose modularized models that will be executable by a compliant simulator.
- DEVS models can be executed on multiple different simulators, including those on desktops (for development) and those on high-performance platforms, such as multi-core processors.
- DEVS supports model continuity which allows simulation models to be executed in real time as software by replacing the underlying simulator engine.

Mittal (2014) stresses the fundamental difference between *software-based* discrete event simulation and *systems-based* discrete event simulation. While the former is strictly based on object-oriented software engineering paradigm (e.g., Schmidt 2006; Volter et al. 2006), the latter enforces Wymore's System Theory on the object-oriented discrete event simulation engine as shown in Section 2. Since cyber complex adaptive systems are multi-agent adaptive systems at the fundamental level, there are many agent-based modeling (ABM) tools available to represent them. Unfortunately, due to their software-based object orientation, the large majority of these tools do not conform to Wymore systems theory's closure under composition principle. In contrast, a DEVS-based agent has the notion of a system attached to it and is built on formal semantics that adheres to Wymore's systems theory. Such an approach makes it possible to develop a simulator, a simulation protocol, and a distributed high-performance engine for agent/system model's execution that ensures that closure of coupling is not violated. Moreover, DEVS formal specification allows it to interface with model-checking tools based on unified modeling language (UML) tools to supplement simulation with formal verification and validation, a critical feature of model engineering (Zeigler and Nutaro 2014).

Several M&S environments exist that support the DEVS-based methodology just described, including DEVS-Suite, CD++, DEVSim++, JAMES II, Python DEVS, and VLE (see the list at DEVS Standardization Group (2014) for descriptions). Mittal and Martin (2013) describe packaging all these functionalities in a netcentric DEVS Virtual Machine (VM) that provides and agent-execution environment to apply to cyber complex adaptive systems. The M&S environment MS4Me was developed as the first in a commercial line of DEVS products (ms4systems.com). It employs Xtext, an EBNF grammar, within the Eclipse Modeling Framework on the Rich Client Platform and the Graphical Modeling Project to provide a full-blown IDE specifically tailored to a DEVS development environment (Zeigler and Sarjoughian 2012).

2.6.1 System Entity Structure (SES)

The System Entity Structure (SES) formalism is an ontology representation of compositions, components, and coupling patterns that can be pruned to select a particular hierarchical model tree information structure. Automatic synthesis can then generate an executable simulation model through selection of model components in the model base (Zeigler 1984; Kim et al. 1990; Kim and Zeigler 1989). The SES is implementation agnostic and can be represented in various knowledge representation frameworks including standard relational data formalisms (Park et al. 1997; Kim and Kim 2006; Zeigler et al. 2013).

The SES/MB framework has been studied in various computational environments and applied to numerous industrial problems (Mittal et al. 2006; Cheon et al. 2008). Recently, the SES/MB framework has seen increasing application to M&S of system of systems (SoS). Commercial environments have been developed to enable more flexible representation of alternatives (including composition patterns and hierarchical components) and rule-based constraints for pruning enabling the development of suites of families of models (Zeigler and Sarjoughian 2012). Distributed simulations of complex federations in HLA can now be generated in addition to the original stand-alone simulations (Kim et al. 2013; Seo and Zeigler 2009).

The methodology has led to the Hierarchical Encapsulation and Abstraction Principle which allows combining the top-down paradigm for the constructive simulation with the bottom-up paradigm for the emergent simulation. Applications have been to simulation study of agent-based tactical and operational effectiveness of warfare and network vulnerability analysis (Chi et al. 2009; You et al. 2013; You and Chi 2009).

Table 2.1 summarizes the above review by examining the layers of service-oriented model engineering and simulation environment presented in Fig. 2.8 from the point of view of DEVS support.

DEVS enables new frameworks for application domains, especially those that feature continuous/and discrete systems that interact (sometimes called hybrid systems). Some examples such as production flows in the food industry, building energy design, quantum key distribution (QKD) systems, and agent-based transportation evacuation are presented in Table 2.2 in terms of their novel features and unique capability offered when compared to existing approaches. DEVS also supports tools for simulating such models. For example, a compiler that employs DEVS to execute models expressed in the well-known simulation language, Modelica.

Table 2.1 Layers of service-oriented model engineering and simulation environment

Layer	Description	DEVS support
Model component layer	Classifies and organizes models	Implementation agnostic, flexible representation of alternatives (including composition patterns and hierarchical components), and rule-based constraints for pruning enabling the development of suites of families of models
Model service layer	Provides model standardization with interface specifications, unified service encapsulation, and transformation	The DEVS formalism provides a formal basis for semantic and pragmatic interoperability among DEVS models using Service-Oriented Architecture and DEVS Namespace
Model management layer	Executes searching, matching, composition, configuration, modification, and VV&A	Pruning of the SES enables automatic synthesis of an executable simulation model through extraction and coupling of model components in the model center
Simulation layer	Provides simulation, visualization, man–machine interaction	The DEVS Abstract Simulator provides a standard distributed DEVS protocol for interoperation of DEVS simulators
Application layer	Enables model engineering applications to specific domains	Numerous applications have been done with DEVS-based M&S. Examples of the development of frameworks are given below

2.7 Conclusion

Inspired by Prof. Tuncer Ören's broad perspective on the M&S enterprise, we probed the nature of model engineering as spanning the life cycle of a model in the context of systems of systems engineering, particularly implemented with service orientation. We presented an architecture for service-oriented model engineering and simulation environments whose layers support the various activities of model engineering. Based on the results of DEVS research, we gave a more concrete characterization of such an environment and how model engineering and DEVS enable new frameworks for application areas. Unifying the various activities needed to produce credible models via the concept model engineering is only in its infancy.

Further research is needed to organize and deepen the body and knowledge and to probe each of the architectural layers we have identified, establish their cross connections, and add new ones as needed. The application context of service orientation is a good one to focus attention on the implementation of support for model engineering but not necessarily the only context in which support might be conceived. Similarly, DEVS theory and research have given much solid substance to the body of knowledge and practice of model engineering but more general

Table 2.2 DEVS-enabled frameworks

Application area	Novel feature	Unique capability
Components: processing units, conveyor belts	New framework for carrying out simulations of continuous-time stochastic processes	Keeping track of parameters related to the process and the flowing material (temperature, concentration of pollutant) is also considered. Since these parameters can change over time in a continuous manner, the possibility to transmit those laws as functions is introduced in the model
Development of DEVS models for building energy design	Allow different professions involved in the building design process to work independently to create an integrated model	Results indicate that the DEVS formalism is a promising way to improve poor interoperability between models of different domains involved in building performance simulations
Components: occupants, thermal network points, windows, HVACs, etc.		
Quantum key distribution (QKD) system with its components using DEVS	DEVS assures that the developed component models are composable and exhibit temporal behavior independent of the simulation environment	Enable users to assemble and simulate any collection of compatible components to represent complete QKD system architectures
Components: classical pulse generator, polarization modulator, electronically variable optical attenuator, etc.		
DEVS framework for transportation evacuation integrating event scheduling into an agent-based method	This framework has a unique hybrid simulation space that includes a flexible-structured network and eliminates time-step scheduling used in classic agent-based models	Hybrid space overcomes the cellular space limitation and provides flexibilities in simulating evacuation scenarios
Components: vehicles, agents		Model is significantly more efficient than popular multi-agent simulators. Keeps high model fidelity and the same agent cognitive capability, collision avoidance, and low agent-to-agent communication cost

theory of M&S should inform and unify such a body. For example, ultimately the elements, such as experimental frame, simulator, etc., and relations (modeling, simulation, applicability, etc.) of the theory of M&S must be brought in to fully consider the best practices for M&S and offer normative views on how to formulate the knowledge needed to make them better.

References

Abdouni Khayari RE, Musovic A, Lehmann A et al (2010) A model validation Study of Hierarchical and Distributed Web Caching Model. In: Proceedings of the 2010 spring simulation multi-conference, Orlando, 11–15 Apr, p 98

Balci O (2012) A life cycle for modeling and simulation. Simulation 88(7):870–883

Boardman J, Sauser B (2006) System of systems – the meaning of "of". IEEE/SMC international conference on system of systems engineering. IEEE, Los Angeles

Cheon S, Kim DH, Zeigler BP (2008) System entity structure for XML meta data modeling: application to the US climate normals. 17th international conference on software engineering and data engineering, Los Angeles

Chi SD, You YJ, Jung CH, Lee HS, Kim JI (2009) FAMES: fully agent-based modeling & emergent simulation SpringSim 2009. Proceedings of the 2009 spring simulation multiconference

DEVS Standardization Group (2014) http://cell-devs.sce.carleton.ca/devsgroup/?q=node/8

Fishwick PA (1989) Qualitative methodology in simulation model engineering. Simulation 52 (3):95–101

Fishwick PA (1990) Toward an integrated approach to simulation model engineering. Int J Gen Syst 17(1):1–19

Kim JK, Kim TG (2006) A plan-generation-evaluation framework for design space exploration of digital systems design. IEICE Trans Fund Electron Comm Comput Sci E89-A(3):772–781

Kim TG, Zeigler BP (1989) A knowledge-based environment for investigating multicomputer architectures. J Inform Soft Technol 31(10):512–520

Kim BS, Choi CB, Kim TG (2013) Multifaceted modeling and simulation framework for system of systems using HLA/RTI CNS 2013. Proceedings of the 16th communications & networking symposium, society for computer simulation international, San Diego

Kim TG, Lee C, Christensen CER, Zeigler BP (1990) System entity structuring and model base management. IEEE Trans Syst Man Cybern 20(5):1013–1024

Liu X, Tang Y, Zheng L (2008) Survey of complex system and complex system simulation. J Syst Simul 20(23):6303–6312 (in Chinese)

Liu J, Zhang L, Tao F (2014) Service-oriented model composition. In: Proceedings of SCS 2014 summer simulation multi-conference (SummerSim'14), Monterey, 6–10 July

Mittal S (2014) Model engineering for cyber complex adaptive systems. EMSS, Bordeaux

Mittal S, Martin JLR (2013) Netcentric system of systems engineering with DEVS unified process. CRC Press, Boca Raton, 712 p

Mittal S, Mak E, Nutaro JJ (2006) DEVS-based dynamic model reconfiguration and simulation control in the enhanced DoDAF design process. J Defense Model Simul 3(4):239–267

Ören TI (1971) GEST: a combined digital simulation language for large-scale systems. In: Proceedings of the Tokyo 1971 AICA (Association Internationale pour le CalculAnalogique) symposiumon simulation of complex systems, Tokyo, 3–7 Sept, pp B-1/1–B-1/4

Ören TI (1973) Application of system theoretic concepts to the simulation of large scale adaptive systems. In: Proceedings of the sixth Hawaii international conference on system sciences, Honolulu, 9–11 Jan

Ören TI (1990) Simulation methodology: top down approach. In: Sage AP (ed) Concise encyclopedia of information processing in systems and organizations. Pergamon Press, Oxford, pp 421–425

Ören TI (2005) Toward the body of knowledge of modeling and simulation (M&SBoK). In: Proceeding of I/ITSEC (Interservice/Industry Training, Simulation Conference), Orlando, 28 Nov–1 Dec, paper 2025, pp 1–19

Ören TI (2014) Publications, presentations and other activities of Dr. Tuncer Ören on modeling and simulation: normative views for advancement and advanced methodologies http://www.site.uottawa.ca/~oren/pubsList/MS-advanced.pdf
Ören TI, Yilmaz L (2012) Synergies of simulation, agents, and systems engineering. Expert Syst Appl 39(1):81–88
Ören TI, Zeigler BP (1979) Concepts for advanced simulation methodologies. Simulation 32 (3):69–82
Ören TI, Zeigler BP (2012) System theoretic foundations of modeling and simulation: a historic perspective and the legacy of A Wayne Wymore. Simulation 88(9):1033–1046
Ören TI, Zeigler BP, Elzas MS (eds) (1982) Simulation and model-based methodologies: an integrative view. Series: Nato ASI Subseries F, vol 10. NATO Advanced Institute Ottawa, Ontario, 26 July–6 Aug
Pace DK (2004) Modeling and simulation verification and validation challenges. Johns Hopkins APL Tech Dig 25(2):2004
Park HC, Lee WB, Kim TG (1997) RASES: a database supported framework for structured model base management. Simul Pract Theory 5(4):289–313
Radeski A, Parr S (2002) Towards a simulation component model for HLA. In: Proceedings of the 2002 fall simulation interoperability workshop (SISO Fall 2002). Paper ID 02F-SIW-079, Nov
Schmidt DC (2006) Model-driven engineering. IEEE Comput, Orlando, FL, 39(2):25–31. doi:10.1109/MC.2006.58
Seo C, Zeigler BP (2009) Interoperability between DEVS simulators using service oriented architecture and DEVS namespace. In: A joint symposium DEVS integrative M&S (DEVS) and high performance computing (HPC), Proceedings of the Spring Simulation Conference, San Diego, CA.
Si N, Zhang L, Tao F, Guo H (2009) Research on multi-agent system based service composition methodology in semantic SOA (in Chinese). In: Proceedings of the 5th conference on multi-agent system and control, Chongqing, 19–20 Sept
Volter M, Stahl T, Bettin J, Haase A, Helsen S (2006) Model-driven software development: technology, engineering, management. John, Chichester/Hoboken
Waite W, Oren TI (2009) Modeling & simulation body-of-knowledge index(M&S BOKIndex). http://sim-summit.org/WebinarBrief/BOK%20Webinar%20Brief%20Feb%202009%20v5%20Waite.pdf
Wymore AW (1967) A mathematical theory of systems engineering: the elements. John, New York
Yilmaz L, Ören TI (2009) Agent-directed simulation and systems engineering, 1st edn. Wiley Series in Systems Engineering and Management, –VCH, Weinheim
You YJ, Chi SD (2009) SIMVA: simulation-based network vulnerability analysis system SpringSim '09. Proceedings of the 2009 spring simulation multiconference
You YJ, Chi SD, Kim JI (2013) HEAP-based defense modeling and simulation methodology. IEICE Trans 96-D(3):655–662
Zeigler BP (1976) Theory of modeling and simulation, 1st edn. Wiley, New York
Zeigler BP (1984) Multifacetted modelling and discrete event simulation. Academic, London
Zeigler BP, Nutaro JJ (2014) Combining DEVS and model-checking: using systems morphisms for integrating simulation and analysis in model engineering. EMSS, Bordeaux
Zeigler BP, Sarjoughian HS (2012) Guide to modeling and simulation of systems of systems. Springer, Dordrecht, p 393
Zeigler BP, Kim TG, Praehofer H (2000) Theory of modeling and simulation: integrating discrete event and continuous complex dynamic systems, 2nd Revised edn. Academic Press Inc., Waltham, pp.11
Zeigler BP, Seo C, Kim D (2013) System entity structures for suites of simulation models. Int J Model Simul Sci Comput 4(3). doi:10.1142/S1793962313400060
Zhang L (2011) Model engineering for complex system simulation, The 58th CAST forum on new viewpoints and new doctrines, Li, 14–16 Oct

Zhang L, Shen YW, Zhang XS, Song X, Tao F, Liu Y (2014a) The model engineering for complex system simulation. In: The 26th European modeling & simulation symposium (Simulation in Industry), Bordeaux, 10–12 Sept

Zhang L, Li F, Song X, Liu YK (2014b) A supporting environment of model engineering for complex systems. In: The 11th Chinese intelligent system conference (CISC'14), Beijing, 18–19 Oct

Chapter 3
Research on High-Performance Modeling and Simulation for Complex Systems

Bo Hu Li, Xudong Chai, Tan Li, Baocun Hou, Duzheng Qin, Qinping Zhao, Lin Zhang, Aimin Hao, Jun Li, and Ming Yang

3.1 Introduction

A complex system is a kind of system that system composition is complex, system mechanism is complex, the interactions and energy exchanges between subsystems or between the system and its surroundings are complex, moreover, the overall properties of system are emergent, nonlinear, self-organized, chaotic and gaming, etc. Typical complex systems include complex engineering systems, complex society systems, complex biological systems, complex environment systems, complex military systems, complex network systems, etc. The research and application of complex systems are of great significance to both science and social economy.

High-performance modeling and simulation for complex systems (HPMSCS) refers to a kind of modeling and simulation technology, which integrates high-performance computing technology with modern modeling & simulation technology. And the objective is to optimize the overall performance of modeling, simulation execution, and result analysis for complex systems. Our primary research and practice indicate that High-Performance Modeling and Simulation for Complex Systems is becoming a new hotspot in the M&S community.

B.H. Li (✉) • Q. Zhao • L. Zhang • A. Hao
School of Automation and Computer Science, Beihang University, Beijing, China
e-mail: bohuli@buaa.edu.cn

X. Chai • T. Li • B. Hou • D. Qin
Beijing Simulation Center, Beijing, China

J. Li
Sugon Corp., Beijing, China

M. Yang
Control and Simulation Center, Harbin Institute of Technology, Harbin, Heilongjiang, China

L. Yilmaz (ed.), *Concepts and Methodologies for Modeling and Simulation*, Simulation Foundations, Methods and Applications,
DOI 10.1007/978-3-319-15096-3_3

The rapid development of HPMSCS is motivated by the application demands to support two types of users, which are the high-end M&S users of complex systems and massive users to acquire high-performance simulation cloud service on demand and fulfill three types of simulation, mathematical, man-in-loop, and hardware-in-loop/embedded simulation. Those application demands raise great challenges to traditional M&S technology.

For high-end M&S users of complex systems, the challenges include (1) high computing power; (2) high performance, high bandwidth, and low latency synchronization/communication network; (3) parallel I/O system with high-performance, high-throughput, high-scalability for the hardware-in-the-loop/embedded simulation; (4) user-friendly development environment for complex system modeling and simulation; (5) joint simulation of multiscale, multidisciplinary heterogeneous systems; (6) "parallel in three levels"; (7) big data processing; (8) verification, validation, and accreditation (VV&A); (9) intelligent analysis and evaluation of the simulation results; (10) low power consumption; (11) high reliability; (12) security.

For massive users to acquire high-performance simulation cloud service on demand, the challenges include (1) making high-performance simulation resources virtualized and servitized; (2) providing dynamic composition of different high-performance simulation services based on users' demands; (3) presenting a virtualized, high efficient, cooperative simulation environment to users; (4) providing an user-centered, distributed, cooperative, and interactive modeling and simulation development paradigm.

For years of research and practice, many key technologies of HPMSCS have made important development. And communications and cooperation with an international authority like Professor Tuncer Ören helped a lot in the progress of HPMSCS in China. This chapter will discuss details about our research on HPMSCS with respect to Ören's contribution in HPMSCS.

3.1.1 Ören's Contribution in HPMSCS

It was a great honor to take the opportunity to participate in the tribute volume for Professor Tuncer Ören since Ören has been one of the most respected friends and partners to both the Chinese simulation community and me. We have communicated and coworked closely on simulation technologies for years. Many concepts and ideas from Professor Tuncer Ören became inspirations for our research, especially for our research on High-Performance Modeling and Simulation for Complex Systems, which the following sections of this chapter are about to introduce. In this background section, I would like to discuss some important inspirations from Ören for our research on modeling, simulation, application, and the body of knowledge (BOK) of HPMSCS.

We started our research on simulation language for continuous and discrete hybrid systems early in 1986 (Li and Tong 1986). At that time, we noticed a series of Ören's papers (Ören 1978; Ören and Zeigler 1979), in which Ören proposed the idea of the separation of model session and experiment session. In traditional simulation language, the description codes of the model equations are interweaved with the simulation algorithm codes, which bring inconvenience to simulation engineers when they try to implement different simulation experiments on the same simulation model. The early proposal of separation from Ören enlightened us to develop the three-session language structure in our ICSL, which are initial session, model session, and experiment session (Li and Tong 1986). The three-session language structure is an early practice of the idea of object-oriented simulation.

Ören started the prior research on the artificial intelligence in simulation (Ören 1985) as well as the Knowledge-Based Modeling and Simulation System (Ören and Aytaç 1985). We tracked that "Most commonly used knowledge-based systems are rule-based systems" (Ören 1985) and started our research on rule-based systems, which we called quantitative and qualitative hybrid systems according to the application demands in the research on complex systems like C4ISR and human-in-loop simulations. Focused on the rule-based systems, we developed some new modeling methods like FuzzyCDG to graphically describe the rule-based systems (Li et al. 2011b) and new algorithms to solve the simulation of quantitative and qualitative hybrid systems (Fan et al. 2009).

Early in the year 2000, Ören proposed the concept of agent-directed simulation (ADS) to use agents as a framework to solve problems in complex simulation systems (Ören 2000a, b). We discussed ADS with Ören in several meetings from 2000 to 2008 and started our research on agents in late 2008 to solve some problems in the simulation of emergency in systems like the traffic and human groups (Li et al. 2011b). We combined agents with cellular automation to simulate human reactions in fire and traffic jams, where some interesting and useful results were found (Li et al. 2011b).

We started research on verification, validation, and accreditation (VV&A) in the late 1990s (Li et al. 2001), which set up a series of quality assurance issues on the simulation of Complex Product Virtual Prototype, based on the issues of Ören's "Type 1-Quality assurance in modeling and simulation" (Ören 1984). VV&A has now become an important QA methodology in the research and development of large-scale simulation in China.

With years of research on HPMSCS, the relative modeling, simulation, and application technologies have made great progress in the simulation community. Inspired by Ören's work in the BOK of modeling and simulation (Ören 2006), we started to set up the body of knowledge of HPMSCS to promote the research work of this rising and maturing discipline (Li et al. 2010d). Furthermore, based on Ören's 9,000 technical terms in modeling and simulation (Ören, T.I. et al. ISBN: 2-9524747-0-2.), we cooperated with Professor Ören to compile a Chinese–English and English–Chinese Dictionary of Modeling and Simulation Terms, which was published in China in June 2012 (Li et al. 2012a). As Ören said in the preface of that dictionary, "Words are labels to represent concepts"; the efforts we made in preparing the dictionary with Ören will have far-reaching influence in Chinese simulation research.

Professor Ören's work on basic simulation theory, AI and knowledge-based systems, ADS, QA, BOK, and Terminology of modeling and simulation has made great contributions to the development of this discipline and helped a lot in simulation research in China. In the following part of this chapter, the research of our team on HPMSCS inspired by Ören will be presented in detail.

3.2 Our Team's Contributions to HPMSCS

There are many key technologies related to HPMSCS. These can be classified into three areas: high-performance simulation modeling theory and method, high-performance modeling and simulation system theory and technology, and high-performance simulation application engineering technology.

3.2.1 Body of Knowledge of HPMSCS

Each direction has plenty of research hotspots, as shown in Fig. 3.1. Taking the BOK of HPMSCS as a framework and blueprint (Li et al. 2010d), our team has made a couple of breakthroughs around the three related directions that enrich the BOK and the HPMSCS discipline.

3.2.2 Complex System Simulation Modeling Method

Research of simulation modeling methods for qualitative and quantitative mixed systems basically includes three aspects:

1. Qualitative and quantitative Unified Modeling Method, including the system top-level description that is responsible for the top-level description of the static structure and dynamic behavior of the system and the domain-oriented description that is responsible for the description of kinds of domain models (including quantitative and qualitative models). The research fruit of our team is QR (quantitative-rule)–QA (quantitative-agent) modeling method (Fan et al. 2009).
2. Modeling the interface of quantitative and qualitative interaction. The interface converts the quantitative and qualitative interaction data into specific structure and format needed by the qualitative model and the quantitative model. The interface converts the quantitative and qualitative interaction data into specific structure and format needed by the qualitative model and the quantitative model.
3. Qualitative and quantitative time advance mechanism. The research fruit of our team is a QR (quantitative-rule)–QA (quantitative-agent) mixed time advancing

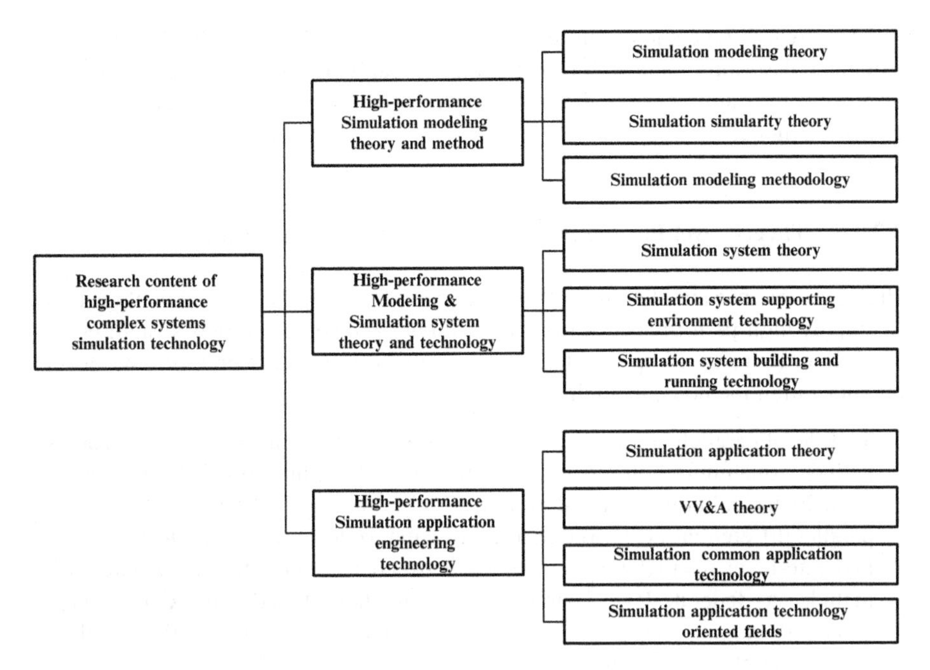

Fig. 3.1 Body of knowledge of HPMSCS

method, which realizes qualitative/quantitative mixed, hierarchical computing control and management of heterogeneous models (Fan et al. 2009)

3.2.2.1 Simulation Modeling Method Based on Metamodeling Framework

Simulation modeling method based on metamodeling framework mainly researches on the metamodel-based unified simulation modeling of the complex system features like multidisciplinary, heterogeneous, and emergency. The research outcome of our team is Meta Modeling Framework (M2F) (Li et al. 2010b), which proposes a hierarchical metamodel architecture, separating the continuous, discrete, qualitative mixed heterogeneous system models at the abstract level theoretically, so as to achieve unified top-level modeling for complex systems.

3.2.2.2 Simulation Modeling Method for Variable Structure System

Research of simulation modeling methods for variable structure systems mainly focuses on the dynamic variability of the simulation model content, interface, and connection, to support the complete modeling of the variable structure system.

The research outcome of our team is CVSDEVS (Yang and Li 2013), a DEVS-extended description norm for the complex variable structure system, which improves the ability of DEVS in describing the variability pattern and execution mode.

3.2.2.3 Three-Level Paralleling High-Performance Algorithm

High-performance simulation algorithm for complex systems is a kind of algorithm to employ high-performance simulation computers to solve complex system problems. In order to speed up simulation, our team focuses on research of three-level parallelization methods, including

1. Task-level parallelization methods for large-scale problems. Research outcomes include quantum multiagent evolutionary algorithm (QMAEA) (Zhang et al. 2009); cultural genetic algorithm (CGA) (Wu et al. 2010); and multigroup parallel differential evolution algorithm fusing azalea search (MPDEA);
2. Federate-level parallelization methods between federates. Research outcomes include a federate-level parallelization method based on RTI (Zhang et al. 2010a) and an event list–based federate-level parallelization method (based on optimistic methods)
3. Model/thread-level parallelization methods based on solving of complex models. Research outcomes include parallel algorithm of constant differential equations based on SMPS with load balance of right functions; GA-BHTR: genetic algorithm based on transitive reduction and binary heap maintenance (Qiao et al. 2010).

3.2.3 Complex Systems for Environment Technology

3.2.3.1 High Performance Cloud Simulation

High-performance cloud simulation, which was proposed by our project team in 2009 (Li et al. 2009), is a new network-based (including internet, internet of things, telecommunication network, broadcasting network, mobile network, etc.) and service-oriented intelligent, agile, and green simulation paradigm and means.

High-performance cloud simulation is an extension and development of cloud computing, which provides IaaS (Infrastructure as a Service), PaaS (Platform as a Service) and SaaS (Software as a Service). It enriches and expands "the sharing content of resources/capabilities, service patterns and supporting technologies" of cloud computing (Fig. 3.2).

Research of cloud simulation will be divided into the common technical research and application. Research on common techniques including portal layer technology (problem-solving environment technology, visualization of portal technology, pervasive portal technology, and so on), cloud simulation service layer technology

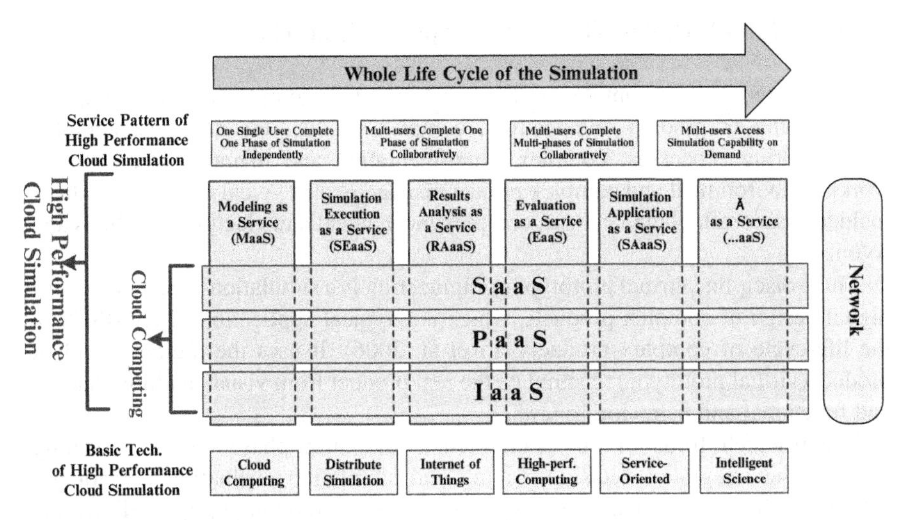

Fig. 3.2 Extension and development of cloud computing

(simulation service discovery, composition, interaction, management, and fault-tolerant migration, resource layer technology (Zhang et al. 2010a, b) of all kinds of simulation resource virtualization, service technology, system security (security mechanism and management technology), and standard technology (general and application fields of standards).

Application Research of service is mainly research on the natural sciences and engineering, social science, management science, life science and military fields such as digital/man in the loop/semi physical simulation service application. Research fruits in the key technology of cloud simulation include heterogeneous simulation resource and ability of virtualization technology (computer hardware, software, interoperability of heterogeneous/reusable simulation platform (such as RTI), virtual model, simulator and knowledge/content, many kinds of heterogeneous simulation resource and ability, high efficiency) to build the technology of virtual cloud simulation environment (simulation ability and resource registration and discovery, according to the optimal scheduling of simulation tasks automatically construct simulation virtual environment and multiuser simulation resources) (Zhang et al. 2010a, b), high-performance RTI (shared memory technology, based on structure, structure model, threads based on traditional RTI performance, improve the real-time RTI, scalability, and throughput and suitable for SMP server and multicore machine (Zhang et al. 2010a, b)) and efficient fault-tolerant virtual cloud simulation environment migration technology (to achieve during the simulation run environment/system error detection, evaluation and simulation of intermediate results' fast storage and loading, and the replacement of environment found and migration technology of based on virtualization technology).

3.2.3.2 Multi-discipline Virtual Prototyping Engineering

Complex products are complex systems featured with complex customer requirements, complex product composition, complex product technologies, complex manufacturing processes, complex experimentation and maintenance, complex working environment and complex project management. Typical complex products include spacecraft, plane, and car, complex mechanical and electrical products and so on.

Multi-discipline virtual prototyping engineering is a simulation-based method of digital design of complex products, which is a typical application of HPMSCS in the life-cycle of complex products (Li et al. 2006). It uses the digital model of product (virtual prototype) to simulate the real product from visual, auditory, tactile and functional and behavioral views.

The research fruits of our team include the multi-phase unified modeling method, the integrated decision-making and simulation evaluation technology, the comprehensive management and prediction methodology, as well as the multi-discipline virtual prototyping engineering platform COSIM (Li et al. 2006).

3.2.3.3 Complex System Modeling and Simulation Language

High-performance simulation language for complex systems is a highly efficient problem-oriented software system for modeling and simulation of complex systems. It allows users to focus on a complex system itself and can greatly facilitate software development and debugging efforts of system modeling and simulation and high-performance computing. Other domain-oriented simulation language (such as biological system simulation problem oriented language, multidisciplinary virtual product simulation oriented language and so on) can be developed based on this language. Key technologies of simulation language include

1. Simulation language framework. Generally speaking, the simulation language framework consists of one modeling environment (model and experiment description language, translation/compiler, utility), libraries (model libraries, algorithm libraries, and function libraries), and one simulation run and control engine and result processing software (Li et al. 2011b).
2. The descriptive language of model and experiment. The simulation language uses the component-based extensible language architecture, which consists of the initialization module, the model module, and the experiment module. The initialization module includes the initial value of parameter settings and syntax and statement of algorithm settings. The model module mainly depicts static and dynamic description statements and syntax, including the continuous, discrete, qualitative, or mixed system. The experiment module describes various statements of experimental operation (such as running with initial values, stop, drawing, etc.) and experiment processes. The language usually adopts the

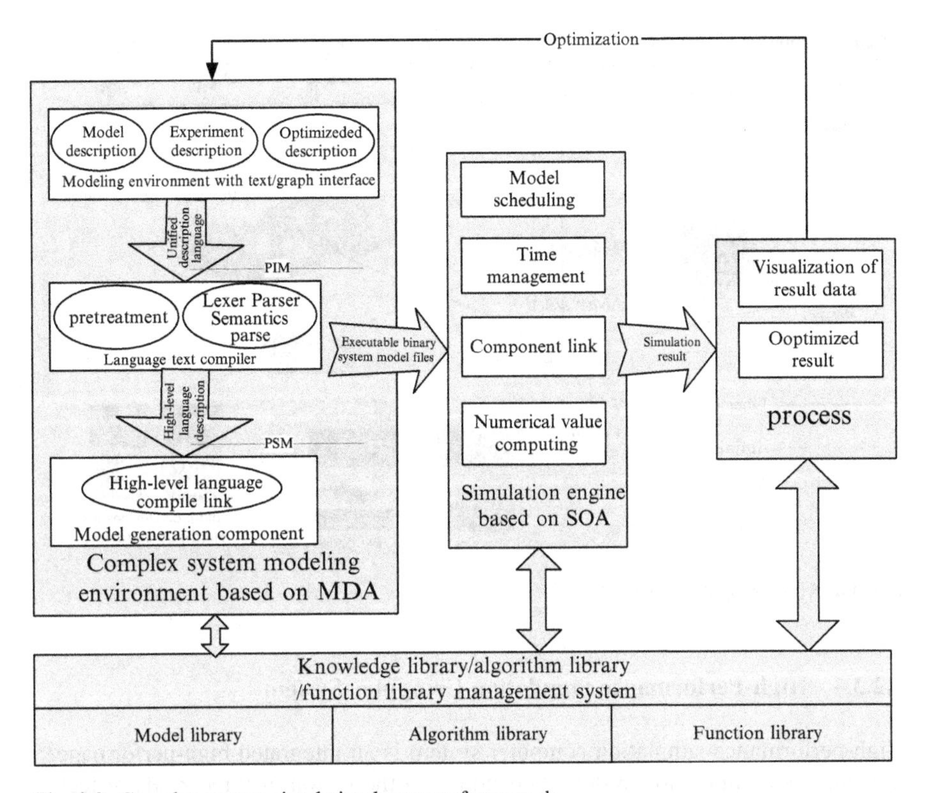

Fig. 3.3 Complex system simulation language framework

similar descriptive statement of simulated problems itself (continuous, discrete, qualitative, mixed system) to describe the model (such as using $dx/dt = f(x)$ in continuous system models).

3. Complex system simulation modeling technology. As shown in the section "Complex system simulation modeling method." The effective dynamic parallelized compiler technology. The research contents include lexical parsing of the simulation language, parallelism analysis of the complex system problem, and dynamic parallel compilation of objective codes (Li et al. 2011b). The core of this technology is to automatically decompose and parallelize the problem described by simulation language and to automatically link to the corresponding function library, model library, and algorithm library. The method that our team takes is to achieve text conversion based on the operations of C++ file stream and string class and parallel computing based on OpenMP/TBB guidance statement and VC, through the analysis of the intrinsic properties of the target machine architecture (vector/parallel multiprocessors) as well as the characteristic of the problem (dependency relationship/loop equivalent transformation, etc.) (Fig. 3.3).

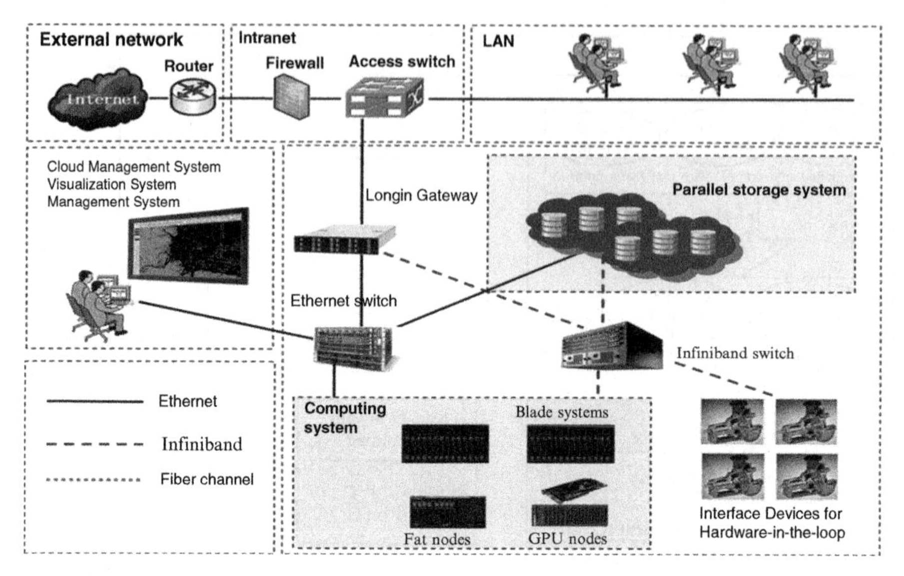

Fig. 3.4 System topology

3.2.3.4 High-Performance Simulation Computer System

High-performance simulation computer system is an integrated high-performance modeling and simulation system, depending on the fusion of three kinds of technologies, including the newly emerging information technologies (e.g., Cloud Computing, Internet of Things, Big Data, Service-Oriented Computing), modern modeling and simulation technology, and high-performance computer system technology to support two types of users (high-end users of complex system modeling and simulation, massive users to acquire high-performance simulation cloud services on demand) to complete three types of simulation (mathematical, man-in-the-loop, hardware-in-the-loop/embedded simulation). The objective is to optimize the overall performance of modeling, simulation execution, and result analysis.

The research works of our team include

System architecture of high-performance simulation computer – System architecture research contains system topology (as shown in Fig. 3.4), system function composition (as shown in Fig. 3.5), hardware system architecture (as shown in Fig. 3.6), and software system architecture (as shown in Fig. 3.7). The features include integrated architecture for two types of users (high-end users of complex system modeling and simulation and massive on-demand users of high-performance simulation cloud services); integrated architecture for three types of simulation (mathematical simulation, man-in-the-loop simulation, and hardware-in-the-loop embedded simulation).

High-efficiency simulation system architecture composed of Hardware System Layer, Parallel OS Layer, Parallel Complier System Layer, Simulation

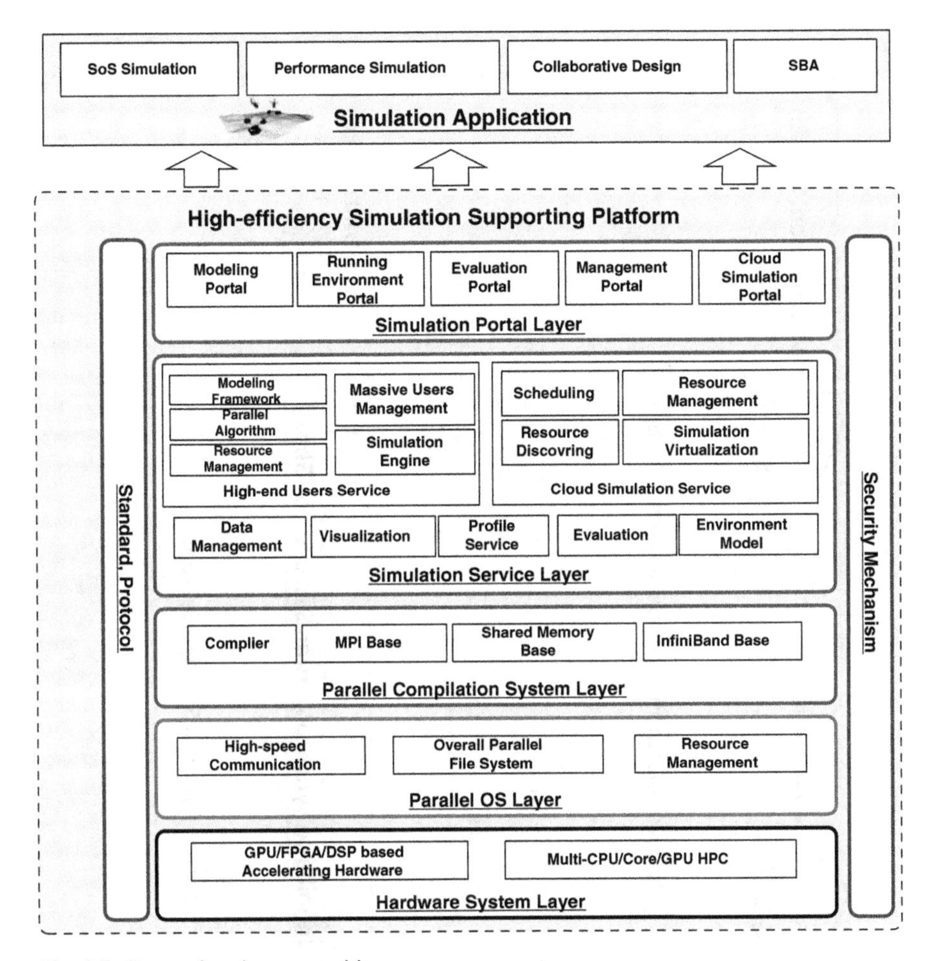

Fig. 3.5 System function composition

Service Layer, and Portal Layer. And relevant standards, protocols and security mechanisms are adopted on all these layers shown in Fig. 3.3.

Currently, high-efficiency simulation hardware, which mainly consists of FPGA, GPU, dedicated accelerator, and general commercial multicore/many-core processors is developing toward high-efficiency and energy-saving direction by combining generalization and specialization, local customization, and targeted optimization method. High-efficiency simulation hardware technology is used to address specific complex simulation demand by optimized hardware design, as shown in Fig. 3.5.

Software systems support the parallel operating system for real-time simulation, providing high-end simulation and cloud simulation as well as the problem-oriented simulation language and model/algorithms bases.

High-performance simulation hardware technology – A kind of design and implementation technology of hardware fitting to the characteristics and needs of

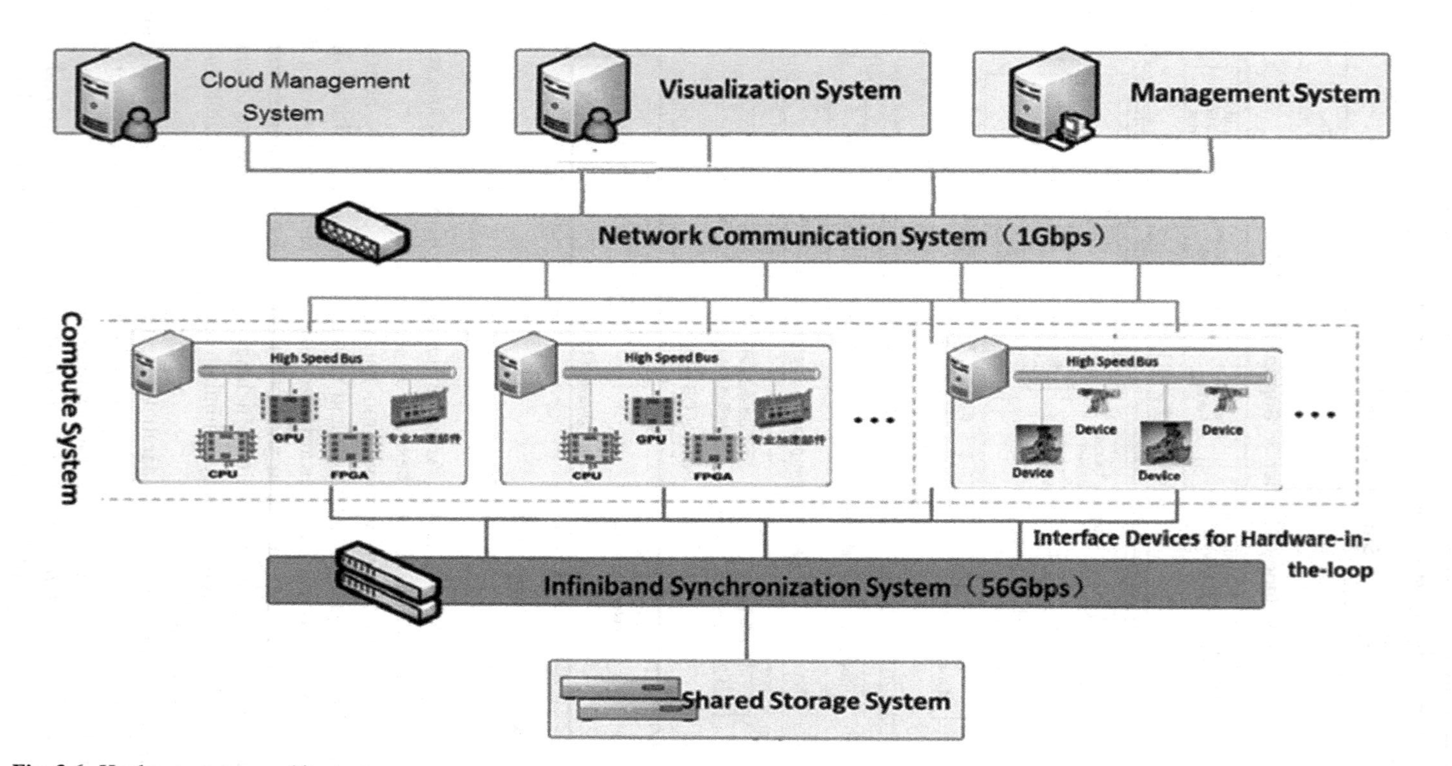

Fig. 3.6 Hardware system architecture

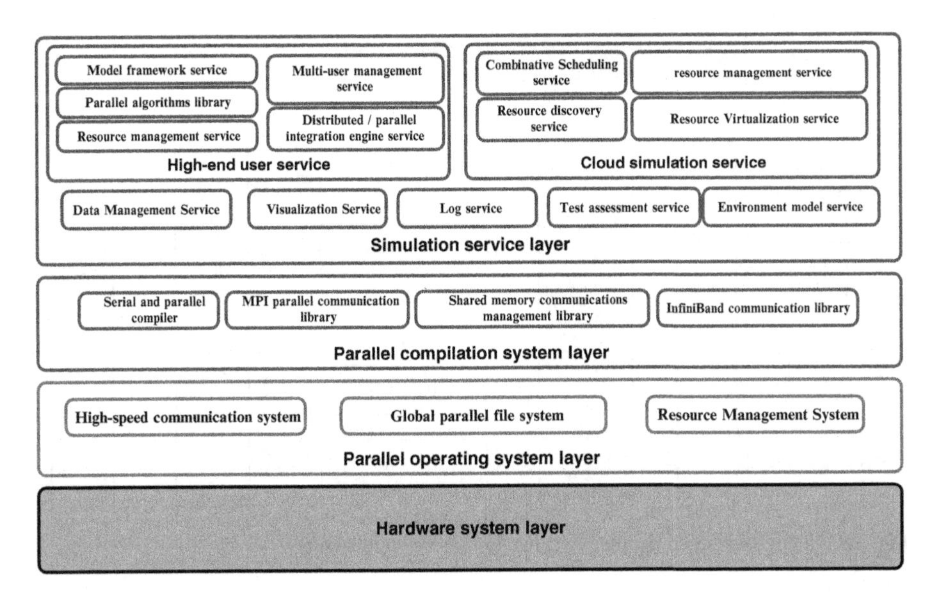

Fig. 3.7 Software architecture

complex system simulation; currently, high-performance simulation hardware is developing toward the high efficient direction that FPGA, GPU, dedicated accelerator, and general multicore/many-core processors play the main role, with a combination of generalization and customization, local customization, and corresponding optimization. The research fruits of our team on high-performance simulation hardware technology include, (a) CPU + GPU-based heterogeneous high-performance computing systems, (b) Application-oriented high-bandwidth, low-latency interconnection network (Yao Yiping 2012), (c) Simulation-oriented hardware acceleration components based on CPU + GPU, and (d) Interface subsystem for hardware-in-the-loop/embedded simulation.

High-performance simulation software technology—High-performance simulation software supporting technology is a kind of design and implementation technology of supporting software fitting to the characteristics and needs of complex system simulation; for now, high-performance simulation software is evolving toward the direction of componentization and automatic parallelization; there are four research contents, including (a) Parallel operating system technology, (b) High-performance parallelization compiler technology, (c) High-performance parallel compiler technology, and (d) HLA/RTI on high-performance computers (Zhang et al. 2010a, b).

3.2.3.5 High-Performance Visualization Technology

High-performance simulation visualization technology mainly researches on GPU group–based parallel visualization system architecture; data organization and scheduling technology of large-scale virtual scene; two-level parallel rendering

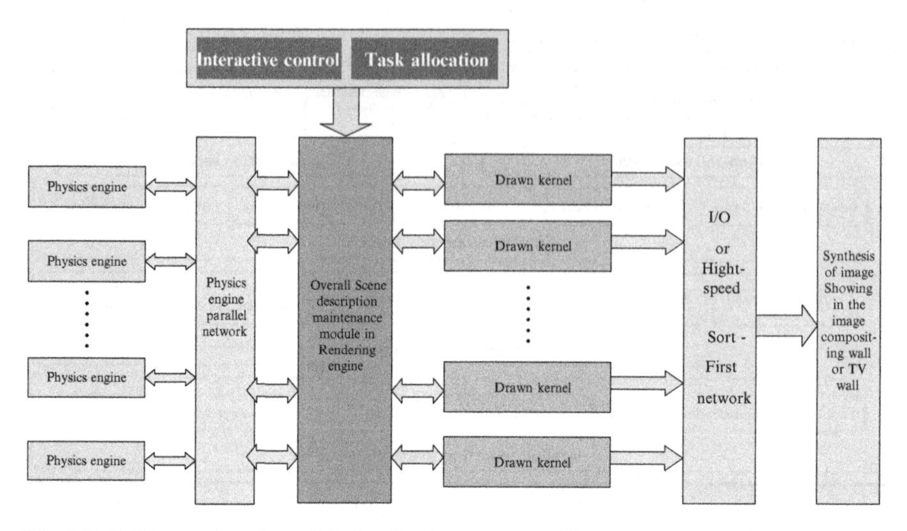

Fig. 3.8 GPU-group based parallel visualization system architecture

based on multimachine and multicore technology; efficient visualization technology of inconstancy objects in battlefield environment; real-time dynamic global illumination (Li et al. 2012b) (Fig. 3.8).

3.2.4 *Simulation Application Engineering Technology of Complex Systems*

3.2.4.1 VV&A of Complex System M&S

High-performance simulation VV&A technology for complex systems includes verification: ensuring that the system models express the user's research targets; validation: ensuring that a simulator processes a satisfactory range of accuracy; accreditation: ensuring that the simulation results satisfy a user's specified criteria. Research includes VV&A of life cycle; VV&A of the whole system; hierarchical VV&A; VV&A of the entire personnel; VV&A of the full range of management (Wang et al. 1999).

The research work in our team to carry out system modeling and simulation of VV&A and reliability evaluation, independently research into system of systems combat simulation system's credibility evaluation tool set, significantly improving the whole life cycle of the credibility of simulation system (Yang et al. 2003), and will further study on VV&A model technology, VV&A process lifecycle collaboration technology of multi stage as well as the physical experiments and simulation integration VV&A technology.

3.2.4.2 Management, Analysis, and Evaluation Technologies of Simulation Results

The complex system simulation experiments and results management, analysis and assessment technology for complex system simulation, a series of functional simulation of data acquisition, data management, visualization and analysis processing and intelligent evaluation, as well as the simulation results analysis and evaluation and optimization provide full support for the application of personnel.

The main research contents include experimental data acquisition technology, mass data management technology (storage, query, analysis, and mining), experimental data analysis and processing technology, the simulation experiment data visualization technology, and intelligent simulation and evaluation technology (Li et al. 2001).

The research fruit in complex system simulation and evaluation of the team can be taken as an example as an effective assessment tool for complex system simulation application, which provides a dynamic simulation data acquisition to evaluate the simulation, application of a variety of evaluation algorithms are evaluated, and the simulation process of playback analysis and other functions, the evaluation of the prototype system, the modeling and Simulation of complex systems in engineering implementation. Its characteristic lies in, modeling, support technology to develop complex evaluation model integration, deployment and operation, openness, scalability, reusability and flexibility; support dynamic data acquisition, evaluation model of loading and the simulation playback; support to data driven mode effectively connecting the algorithm model, to automate the evaluation process (Li et al. 2006).

3.3 Application Examples

3.3.1 Application of Complex Systems with High-Performance Simulation Language

In the modeling and simulation of complex systems that realize co-simulation and optimization system based on parallel simulation, prototype of complex system modeling and simulation language has been applied to implement the complicated electromagnetic environment simulation calculation, the qualitative and quantitative process, using modeling and simulation language text and graphical environment to realize the efficient continuous discrete combination of qualitative and quantitative system (Li et al. 2011b) (Fig. 3.9).

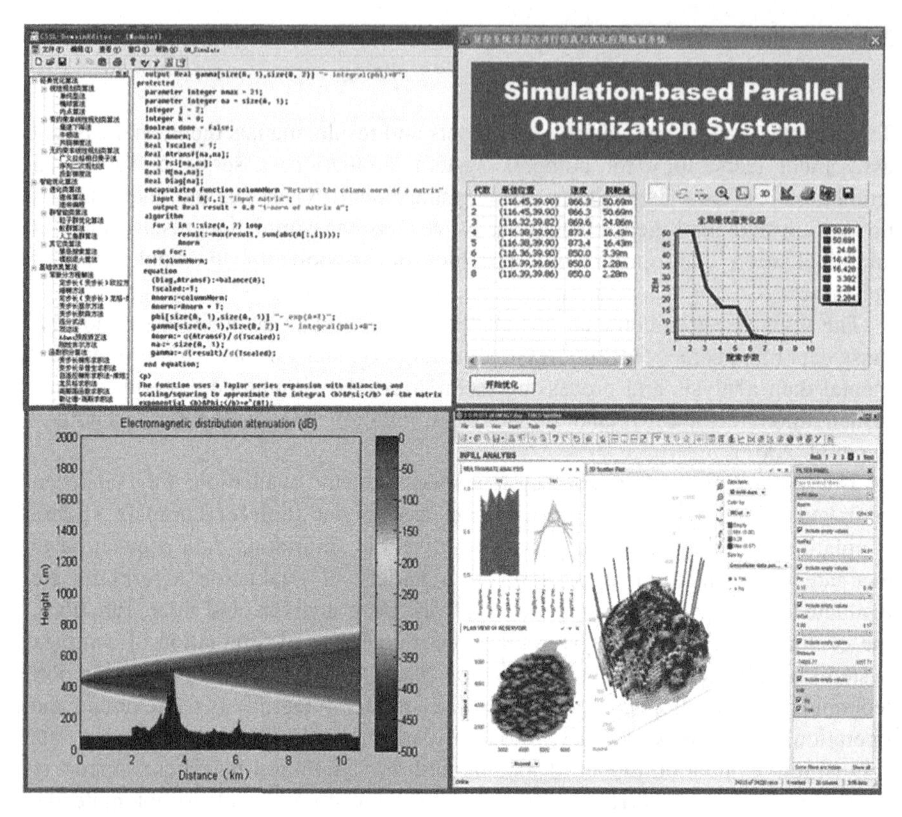

Fig. 3.9 Application of M&S language for complex system

3.3.2 Application of Complex Systems with High-Performance Cloud Simulation

Based on high-performance cloud simulation technology, the team developed a cloud simulation system prototype, which supports four kinds of cloud simulation application mode, including personalized virtual desktop mode, batch mode, collaborative simulation model as well as the ability to trade mode. Cloud simulation prototype system developed a cloud simulation platform and constructed the cloud simulation service system where the resources and capabilities are virtualized as services to users. High-performance cloud simulation application and demonstration of the landing gear system collaborative simulation is shown as (Zhang et al. 2010a, b) (Fig. 3.10).

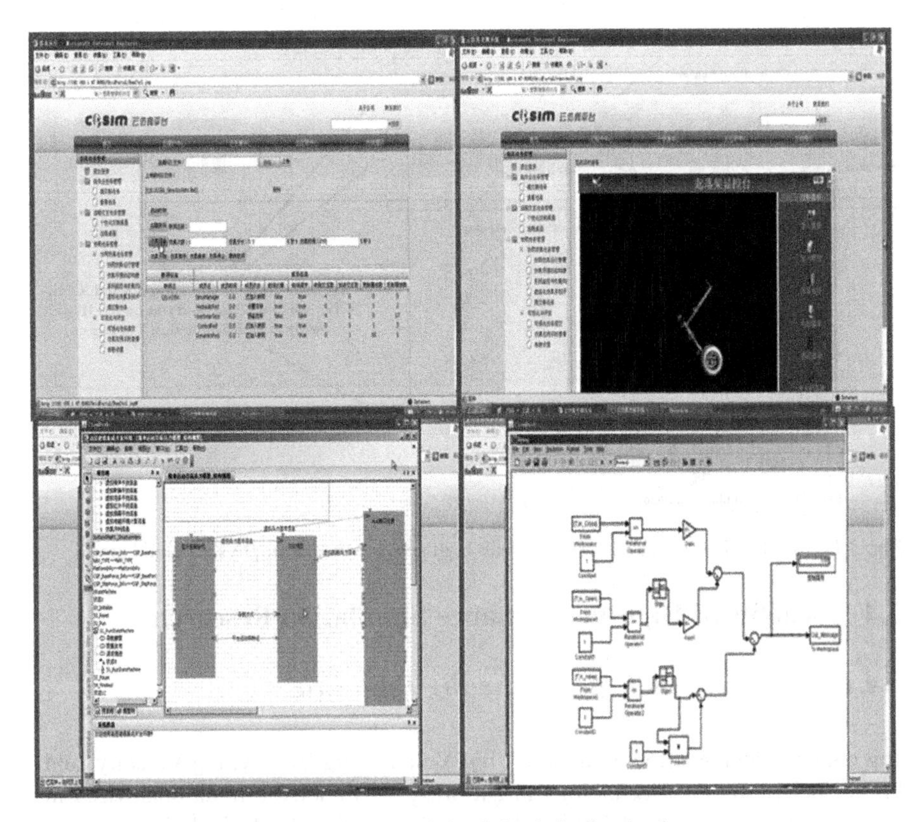

Fig. 3.10 Application of high performance "cloud simulation" technology

3.3.3 Application of Complex Systems with High-Performance Semiphysical Simulation

Based on the high-performance simulation computer, we set up a semiphysical simulation system which includes control system models running on the computer and physical devices. The environment model and production models are run on physical simulators, communicating with the digital control models through InfiniteBand and PCI-E Cable. The structure of the application of complex systems with high-performance semiphysical simulation is shown in Fig. 3.11.

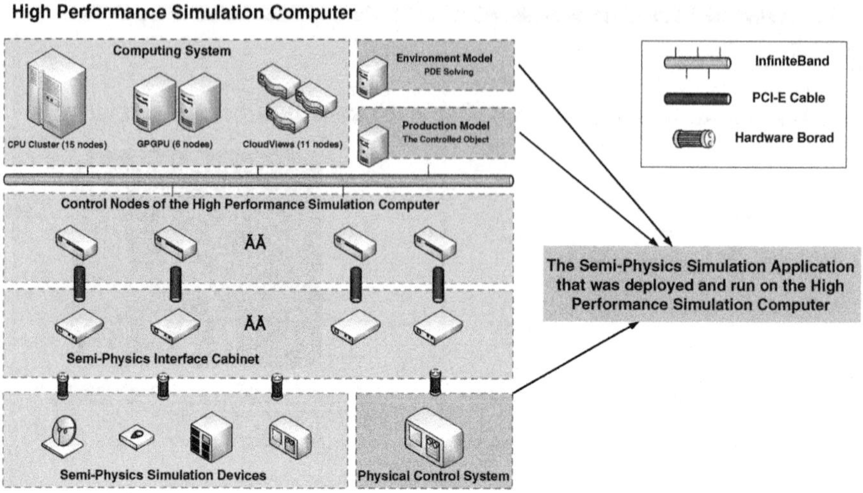

Fig. 3.11 Application of Complex system with high performance semi-physical simulation

3.4 Benefits of High-Performance Simulation Research

3.4.1 M&S Technologies in the Big Data Era

Big data are data featured with "4 V", big Volume, big Variety, big Veracity, and big Velocity. Big data technology might bring revolutions to traditional M&S in many aspects (Academy of Chinese Science Association 2014), including

1. The revolutions in simulation mind and research pattern, such as the knowledge mining in Big Data, the further fusion of reductionism and holism, and the intelligent reasoning of the data and simulation results
2. The revolutions in modeling, such as the extension from traditional mechanism-based modeling to Big Data–based modeling
3. The revolutions in simulation supporting systems, such as the establishment of the unified smart cloud simulation system based on pervasive network and Big Data middleware; the introduction of Big Data management and mining technology into current simulation algorithms and systems
4. The revolutions in simulation application engineering, such as the Big Data–based VV&A, the intelligent simulation result analysis technology based on Big Data, etc.

3.4.2 Pervasive Simulation Technology

Pervasive simulation technology aims at realizing the pervasive simulation pattern, in which the combination and spontaneous interaction of physical space and simulation space is everywhere and people can get simulation service transparently wherever and whenever (Li et al. 2012).

Relative research contents include (1) the architecture of pervasive simulation based on web-based distributed simulation technology, grid computing technology, and pervasive computing technology; (2) the development of pervasive simulation-oriented software platform and middleware; (3) the new intercommunication channel between user and simulation computing service; (4) the new simulation application model for pervasive computing mode; (5) the new simulation services to satisfy the requirements of pervasive simulation; (6) the collaborative management and integration technology of the information space and the physical space; (7) the self-organization, self-adaptability, and high toleration of pervasive simulation based on pervasive computing.

Some key technologies have been studied primarily by our research teams (Tang and Li 2007), including (1) application mode and architecture of pervasive collaborative design; (2) implicit simulation service invocation framework; (3) context-aware technology in pervasive collaborative design; (4) service-self migrating technology based on mobile agents.

3.4.3 *Embedded Simulation Technology*

Embedded simulation is to embed the simulation system into the real system. Embedded simulation can enable operators to see the virtual world (virtual world) and the real system through the interaction of various subsystems to complete real-time performance monitoring, information visualization, operation, management, decision support, training, test, and evaluation functions (Abate et al. 1998). Promising research directions of embedded simulation technology include

1. Training based on embedded simulation. Embedded simulate the operation environment and targets in real physic equipment to provide vivid training experience to the operator.
2. Decision and control technology based on embedded simulation. Collect the real-time data in the environment and run the simulation real time to find the right decision to adjust dynamically (Asia Simulation Company 2009).

3.4.4 *Intelligent Simulation Technology for Complex Systems*

Intelligent simulation technology combines simulation technology with intelligent science, which is composed of brain science, cognitive science, and artificial intelligence to research on the basic theory and realization of human intelligence

(Shi 2014). The intelligent science enables the intelligent recognition, fusion, computing, control, and analysis of the "Human – Machine – Things – Information" in simulation.

Promising research topics of intelligent simulation technology for complex systems include the intelligence-based simulation modeling methods, the intelligence-based high-performance simulation supporting platform, and the intelligence-based high-performance simulation application engineering.

3.4.5 Mobile Internet–Based Simulation Technology

Mobile Internet technology refers to technologies that enable the mobile client to access the Internet via mobile communication networks (Features, key technologies and application of Mobile Internet. http://CrazyCoder.cn/ 2010). Mobile Internet–based simulation technology will be an important technology to realize the on-demand socialized simulation.

In the future, mobile Internet technologies like SOA, WEB X.0, Widget/Mashup, P2P/P4P, and SaaS/Cloud Computing and Protocols like MIP/SIP/RTP/RTSP will be fused with simulation systems to support the on-demand socialized simulation.

3.4.6 Cyberspace Simulation Technology

The concept of "Cyberspace" was primarily proposed by US-NSF in their paper (NSF Advisory Committee 2013), which refers to the virtual space in computers and networks. And the cyberspace becomes the fifth space besides Land-space, Ocean-space, Aero-space, and Outer-space (Hu 2013). Important research topics in cyberspace simulation technology include

1. Cyberspace modeling and algorithm, including the multiscale, multipattern simulation modeling, advanced discrete method of PDE, the simulation-based large-scale system optimization, etc.
2. Cyberspace high-performance simulation, including the simulation programming and runtime technology for huge systems; petabyte/exabyte-level Big Data simulation and analysis
3. Cyberspace high-performance simulation application, including the VV&R (verification, validation, and reproducibility) for huge cyberspace systems; data-sensitive simulation visualization

Acknowledgments Sincere thanks go to the team members, including Yiping Yao, Weijin Wang, Xiao Song, Fei Tao, Ni Li, Shuai Fan, Han Zhang, Yabin Zhang, Zhihui Zhang, Tingyu Lin, Chen Yang, Chi Xing, Yingying Xia, etc.., as well as project team members in the related projects funded by 973, 863 (Chinese National Science and Technology Projects) etc.

References

Abate CW, Bahr HA, Brabbs JM (1998) Embedded simulation for the army after next

Academy of Chinese Science Association (2014) Challenges and thoughts of modeling and simulation in BIG DATA era. Chinese Science Technology Press, Xi'an, China

Asia Simulation Company (2009) Embedded Simulation based Scenario Control System

Fan S, Li BH, Chai X et al (2009) Research on complex system oualitative and quantitative modeling technology. J Syst Simul 16(10):2166–2173

Features, key technologies and application of Mobile Internet (2010) http://CrazyCoder.cn/

Li BH, Tong C (1986) A powerful simulation language ICSL. IMACS Transactions on Scientific Computation. pp 364–371

Li BH, Chai X et al (2001) Virtual prototyping engineering for complex product [A]. The proceedings of SeoulSim 2001

Li BH, Chai X et al (2004) Some focusing points in development of modern modeling and simulation technology. J Syst Simul 16(9):1871–1878

Li BH, Chai X, Hou B et al (2006) Research and application on CoSim (Collaborative Simulation) Grid [C]. The proceeding of MS-MTSA'06, Toronto, Canada

Li BH, Chai X, Hou B et al (2009) Networked modeling & simulation platform based on concept of cloud computing—cloud simulation platform [J]. J Syst Simul 21(17):5292–5299

Li T, Li BH, Chai X (2010a) An M2F-based modeling approach to the emerge mechanism in complex systems. J Beijing Univ Aeronaut Astronaut

Li T, Chai X, Hou B et al (2010b) Research and application on meta modeling framework of complex product multidiscipline virtual prototype. The proceeding of ICMSC 2010, Hongkong, China

Li BH, Chai X, Hou B et al (2010c) Some key technologies in complex system modeling and simulation. Xiangshan science conference, 5, Beijing, China

Li BH, Li T et al (2010d) The body of knowledge of complex system modeling and simulation. The proceedings of Xiangshan science meeting, Beijing, China

Li BH, Chai X, Yao Y et al (2011a) New advances in cloud simulation. Asia simulation conference, Seoul

Li T, Li BH, Chai X (2011b) Research and application on complex system modeling & simulation language architecture. Technical Report. School of Automation and Computer Science, Beijing University

Li T, Li BH, Chai X (2012) Research and application on complex system modeling & simulation language architecture. J Beijing University Aeronaut Astronaut 38(9):1240–1244

Li BH, Ören T, Zhao X, Wu Q, Chen Z, Xiao T, Gong G (2012a) Chinese-English English-Chinese modeling and simulation dictionary. Chinese Science Press, Beijing, China

Li BH, Zhao X et al (2012b) The primary design report of high performance simulation computer. Beijing Simulation Center, Beijing, China

Ören TI (1978) A personal view on the future of simulation languages (Keynote Paper). In: In: Proceedings of the 1978 UKSC conference on computer simulation. IPC Science and Technology Press, Chester, 4–6 Apr 1978, pp 294–306

Ören TI (1984) Quality assurance in modelling and simulation: a taxonomy. In: Ören TI, Zeigler BP, Elzas MS (eds) Simulation and model-based methodologies: an integrative view. Springer, Heidelberg, pp 477–517

Ören, T.I. (1985). Intelligence in Simulation - Editorial. ACM Simuletter, 16:1, 2

Ören TI, Aytaç KZ (1985) Architecture of MAGEST: a knowledge-based modelling and simulation system. In: Javor A (ed) Simulation in research and development. North-Holland, Amsterdam, pp 99–109

Ören TI (1987) Quality assurance paradigms for artificial intelligence in modelling and simulation. Simulation 48(4):149–151

Ören TI (2000a – Invited Paper) Agent-directed simulation – challenges to meet defense and civilian requirements. In: Joines JA et al (eds) Proceedings of the 2000 winter simulation conference. Orlando, 10–13 Dec 2000, pp 1757–1762

Ören TI (2000b – Invited Presentation) Use of intelligent agents for military simulation. Lecture notes of simulation for military planning, 21–22 Sept 2000. International Seminars and Symposia Centre, Brussels
Ören TI (2005 – Invited Tutorial) Toward the body of knowledge of modeling and simulation (M&SBOK). In: Proceedings of I/ITSEC (interservice/industry training, simulation conference), Orlando, 28 Nov–Dec 1; paper 2025, pp 1–19
Ören TI (2006) Body of knowledge of modeling and simulation (M&SBOK): pragmatic aspects. In: Proceedings of EMSS 2006 – 2nd European modeling and simulation symposium, Barcelona, 4–6 Oct 2006, (Presentation)
Ören TI, Zeigler BP (1979) Concepts for advanced simulation methodologies. Simulation 32 (3):69–82
President's Information Technology Advisory Committee (2005) Computational science: Ensuring America's competitiveness, Washington D.C., USA
Qiao K, Tao F, Zhang L, Li Z (2010) Partner selection in virtual manufacturing based on a genetic algorithm maintained by binary heap and transitive reduction[C]. 2010 I.E. international conference on intelligent computing and integrated systems (ICISS 2010), Guilin, China
Shi Z (2014) Intelligent science technology. Chinese Computer Education
Shuai F, Xudong C, Tan L (2010) Knowledge handling methods in qualitative & quantitative fault diagnosis platform. Comput Integr Manuf Syst 16(10):2166–2173
Tao F, Zhang L, Zhang ZH, Nee AYC (2010) A quantum multi-agent evolutionary algorithm for selection of partners in a virtual enterprise. CIRP Annals-Manufacturing Technology, Edinburgh, Scotland
Wang Z, Yang M, Zhang B (1999) The present and future of simulation verification, validation and acceptance (VV & A). J Syst Simul 11(5):321–340
Wu Y, Zhang L, Tao F et al (2010) Study on high effective simulation of complex electromagnetic environment. 7th EUROSIM congress on modeling and simulation, Edinburgh, Scotland
Xiaofeng L, Li BH (1996) An algorithm for automatic parallel allocation and scheduling of a simulation task SMPS. J Syst Simul 13(7):1733–1737
Yang M, Zhang B, Ma P, Wang Z et al (2003) Five key issues of the development of simulation systems VV&A [J]. J Syst Simul 15(11):1506–1508
Yang C, Li BH, Chai X, Chi P (2013a) An efficient dynamic load balancing method for simulation of variable structure systems. In: Proceedings of 8th EUROSIM congress on modelling and simulation, London, England, pp 525–531
Yang C, Li BH, Chai X (2013b) Ivy: a parallel simulator for variable structure systems under multi-core environment. Int J Serv Comput Orient Manuf 1(1):103–123
Yilmaz L, Ören T (2010–Invited paper) Intelligent agent technologies for advancing simulation-based systems engineering via agent-directed simulation. SCS M&S Magazine. SCS M&S Magazine, July
Yiping Y (2012) High performance parallel simulation technology and its applications in SoS research. University of National Defense Science Technology, Changsha, China
Zhang L, Wu Y, Huntsinger RC et al (2009) Key issues in electromagnetic environment modeling & simulation. International summer computer simulation conference, Kyoto, Japan
Zhang Z, Li BH, Chai X et al (2010a) Research on high performance RTI architecture based on shared memory. The proceedings of ICCEE 2010, Beijing, China
Zhang Y, Li BH, Chai X et al (2010b) Research on virtualization-based simulation environment dynamically building technology for cloud simulation. The proceedings of ISAI 2010, Beijing, China
Zhang Y, Li BH, Chai X et al (2011) Research on the virtualization-based cloud simulation resource migration technology. J Syst Simul 23(6):268–1272
Zhen T, Li BH, Chai X (2007) Application of context-sensitivity in Pervasive Simulation Grid. J Comput Integr Manuf Syst 14(8):1550–1558

Chapter 4
Dynamic Data-Driven Simulation: Connecting Real-Time Data with Simulation

Xiaolin Hu

4.1 Introduction

Data plays an essential role in almost every aspect of computer modeling and simulation. The importance of data in modeling and simulation was discussed in many of Tuncer Ören's works. In a well-cited early work (Ören and Zeigler 1979), Ören and Zeigler pointed out that one of the shortcomings of conventional simulation techniques was the lack of needed tools for managing data and models. Data was regarded as an important factor in the conceptual framework of simulation.

In another work devoted to the topic of "Impact of Data on Simulation" (Ören 2001), Ören systematically studied the relationship between data and simulation and elaborated the multiple ways data can impact simulations. According to the article, "Data can occur in several phases of a simulation study: In formulating a model, in formulating an environment (static or dynamic) where the model resides, in providing input to excite the model, and as the behavior of the model". Furthermore, "In modeling, data is needed for parameter fitting and parameter calibration. Afterwards, data is needed for validating the model and experimental conditions." In this article, Ören also differentiated two types of simulations: *standalone simulation*, where the use of simulation is independent of the real system, and *online simulation*, where a simulation runs concurrently with a real system.

Using online simulation to support real-time decision-making was identified as an important application of simulation: "Simulation has the potential of surpassing its own abilities of being an off-line decision making tool to be also an on-line decision support tool for complex and important problems." (Ören 2000). Ören's differentiation of two types of simulations and discussions of how data can impact simulations provide a conceptual framework to categorize the many existing works

X. Hu (✉)
Department of Computer Science, Georgia State University, Atlanta, GA, USA
e-mail: xhu@cs.gsu.edu

L. Yilmaz (ed.), *Concepts and Methodologies for Modeling and Simulation*,
Simulation Foundations, Methods and Applications,
DOI 10.1007/978-3-319-15096-3_4

of using data in simulation. Most of these existing works were developed from the perspective of modeling, e.g., using data to support model design, model calibration, and model validation. In this book chapter, we focus on using data from the perspective of simulation, i.e., assimilating real-time sensor data into a running simulation model in the context of online simulation.

4.1.1 The Need of Dynamic Data-Driven Simulation

Computer simulations have long been used for studying and predicting behaviors of complex systems such as wildfires, urban traffic, and infectious disease spread. The accuracy of these simulations depends on many factors, including data used in the simulations and fidelity of the simulation models. Considering wildfire spread simulation as an example, the simulation relies on terrain data, vegetation data, and weather data in the wildfire area. Due to the dynamic and stochastic nature of wildfire, it is impossible to obtain all this data with no error. For example, the weather data used in a simulation is typically obtained from local weather stations in a time-based manner (e.g., every 30 min). Before the next data arrives, the weather is considered unchanged in the simulation model. This is different from the reality, where the real weather constantly changes (e.g., due to the mutual influences between wildfires and the weather). Besides data errors, the wildfire behavior model introduces errors too because it is an abstraction of the real world. Due to these errors, the predictions from a simulation model will inevitably be different from what is in a real wildfire. Without assimilating real-time data from the real wildfire and dynamically adjusting the simulation model, the difference between simulation and real wildfire is likely to grow continuously.

Incorporating real-time data into a running simulation model has the potential to significantly improve simulation results. Unfortunately, until recently, this idea has not received much attention in the simulation research community. While sophisticated simulation models have been developed for various applications, traditional simulations are largely decoupled from real systems by making little usage of real-time data from the systems under study. This was partially due to the limited availability and quality of real-time data in the past. With recent advances in sensor and network technologies, the availability and quality of such real-time data have greatly increased. As a result, a new paradigm of *Dynamic Data-Driven Simulation (DDDS)* is emerging where a simulation system is continually influenced by real-time data streams for better analysis and prediction of a system under study.

Figure 4.1 illustrates the idea of dynamic data-driven simulation based on the application of wildfire spread simulation. In the figure, the top part represents the wildfire simulation model; the bottom part represents the real wildfire. As the wildfire spreads, streams of real-time data are collected from fire sensors, e.g., ground temperature sensors deployed at various locations of the fire area. This real-time sensor data (referred to as *observation data*) reflects the dynamically changing system states (e.g., fire front location and fireline intensity, which are usually not

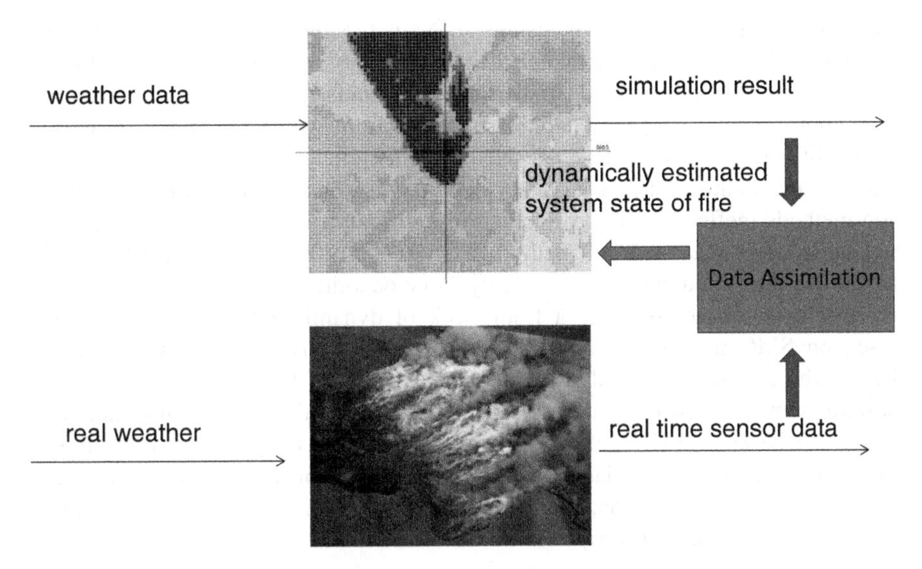

Fig. 4.1 Dynamic data driven simulation for wildfire spread prediction

directly observable and need to be estimated) of the wildfire. The real-time observation data is assimilated by the data assimilation component to improve simulation results. By coupling a simulation system with real-time data, dynamic data-driven simulation can greatly increase the power of simulation-based study. In the wildfire example, based on the real-time sensor data of a wildfire, a wildfire simulation can better estimate the current fire front and fire intensity and thus provide initial conditions for more accurate fire spread simulations in the future. The sensor data also carries "feedback" information for a simulation to calibrate its model parameters to reduce discrepancies between its simulation results and the observations. The dynamically calibrated model parameters can improve simulation results as time moves forward. These capabilities are very useful for supporting real-time decision-makings of wildfire containment.

The work of dynamic data-driven simulation is closely related to *Dynamic Data-Driven Application System* (*DDDAS*), which entails the ability to dynamically incorporate data into an executing application simulation and, in reverse, the ability of applications to dynamically steer measurement processes (Darema 2004). DDDAS advocates a conceptual framework including bidirectional influence between the application simulation and the measurement system. Dynamic data-driven simulation is also related to the work of *data assimilation* by incorporating data into a running model. Data assimilation is an analysis technique in which the observed data is assimilated into the model to produce a time sequence of estimated system states (Bouttier and Courtier 1999). It has achieved significant success in fields such as oil and gas pipeline models and atmospheric, climate, and ocean modeling. Important estimation techniques used in data assimilation include Kalman filter and its variance (Kalman 1960; Evensen 2007). Conventional estimation techniques, however, cannot effectively support dynamic data-driven simulation because many sophisticated simulation models (such as cellular automata-based wildfire spread simulation models and agent-based crowd behavior simulation models) lack the analytic structures from which functional forms of probability distributions can be derived. Furthermore, these simulation models usually have nonlinear, non-Gaussian behavior, which makes conventional estimation methods ineffective. To overcome these difficulties, in our work we employ *Sequential Monte Carlo* (*SMC*) methods (Doucet et al. 2001) to assimilate real-time sensor data into simulation models for dynamic data-driven simulation.

This book chapter presents a framework of dynamic data-driven simulation based on SMC methods. Section 4.2 provides an overview of SMC methods. Those who are not interested in the technical aspect of Bayesian inference may skip this section. Section 4.3 describes the overall dynamic data-driven simulation framework based on SMC methods. Section 4.4 describes the potential applications and challenges of dynamic data-driven simulation. Section 4.5 presents an illustrative example of dynamic data-driven simulation for wildfire spread simulation. Section 4.6 concludes this chapter.

4.2 Overview of Sequential Monte Carlo (SMC) Methods

SMC methods, also called *particle filters*, are sample-based methods that use Bayesian inference and stochastic sampling techniques to recursively estimate the state of dynamic systems from some given observations. A dynamic system is formulated as a discrete dynamic state-space model, which is composed of the system transition model of Eq. (4.1) and the measurement model of Eq. (4.2) (Jazwinski 1970). In the equations, t is the time step, s_t and m_t are the state variable and the measurement variable (the observation) respectively, the function f defines the evolution of the state variable, and the function g defines the mapping from the state variable to the measurement variable, where γ_t and ω_t are two independent random variables representing the state noise and the measurement noise.

$$s_{t+1} = f(s_t, t) + \gamma_t. \tag{4.1}$$

$$m_t = g(s_t, t) + \omega_t. \tag{4.2}$$

Due to its stochastic nature, the system transition function Eq. (4.1) is often referred to as the *state transition density* and is represented by a probability distribution $p(s_t|s_{t-1})$; the measurement function Eq. (4.2) is often referred to as the *measurement density* and is represented by a probability distribution $p(m_t|s_t)$. Typically, the system state is hidden and cannot be observed. Nor can it be directly computed from the measurement based on Eq. (4.2) (i.e., the inverse problem). Thus, there is a need to estimate the dynamic changing state based on measurements and the system's state evolution over time. Figure 4.2 shows the relationships

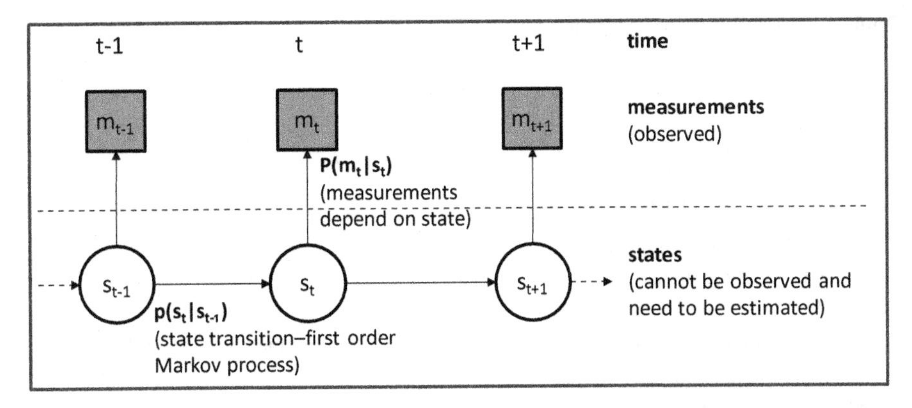

Fig. 4.2 States and measurements of a stochastic dynamic system

among a system's states, measurements, state transition function, and measurement function. In the figure, $p(s_t|s_{t-1})$ is the state transition model and is considered as a first-order Markov process; $p(m_t|s_t)$ is the measurement model that maps system states to measurements.

SMC methods are sample-based methods that approximate the sequence of probability distributions of interest using a large set of random samples named particles. These particles are propagated over time using importance sampling (IS) and resampling mechanisms. It has been shown that a large number of particles are able to converge to the true posterior even in non-Gaussian, nonlinear dynamic systems (Liu and Chen 1998). For systems with strongly nonlinear behavior, SMC methods thus are more effective than the widely used Kalman filter and its various extensions, which approximate beliefs by their second-order characteristics and some linear correction (updating) procedure. The recursive nature of SMC methods also makes it more suited for dynamic data-driven simulation that incorporates real-time data in a sequential manner.

As a sample-based variant of Bayes filters, SMC methods represent the belief Bel (s_t) by a set S_t which includes N weighted samples distributed according to Bel(s_t)

$$s_t = \left\{ \left\langle s_t^{(i)}, w_t^{(i)} \right\rangle | \, i = 1, \ldots, N \right\}$$

where each $s_t^{(i)}$ is a state, and the $w_t^{(i)}$ are non-negative numerical factors called importance weights, which sum up to one. A time update of the basic SMC algorithm (the bootstrap filter algorithm (Gordon et al. 1993), which chooses the proposal density to be the system transition density), is outlined in Table 4.1 below. It describes the basic algorithmic structure of SMC methods, from which various sampling and resampling techniques can be developed and incorporated.

This algorithm implements a sequential importance sampling and resampling (SISR) procedure. In each iteration, the algorithm receives a sample set s_{t-1}, representing the previous belief of the system state, and an observation m_t and

Table 4.1 Bootstrap filter algorithm

1. *Initialization*, $t = 0$.
   For $i = 1, \ldots, N$, sample $s_0^i \sim p(s_0)$ and set $t = 1$.
2. *Importance sampling step*
   For $i = 1, \ldots, N$, sample $\tilde{s}_t^{(i)} \sim p\left(s_t | s_{t-1}^{(i)}\right)$ and set $\tilde{s}_{0:t}^{(i)} = \left(s_{0:t-1}^{(i)}, \tilde{s}_t^{(i)}\right)$.
   For $i = 1, \ldots, N$, evaluate the importance weights
   $\tilde{w}_t^{(i)} = p\left(m_t | \tilde{s}_t^{(i)}\right)$
   Normalize the importance weights.
3. *Selection step (resampling step)*
   Resample with replacement N particles $\left(s_{0:t}^{(i)}; i = 1, \ldots, N\right)$ from the set $\left(\tilde{s}_{0:t}^{(i)}; i = 1, \ldots, N\right)$ according to the importance weights.
   Set $t \leftarrow t + 1$ and go to step 2.

then generates N samples representing the posterior belief. In the importance sampling step, each sample in s_{t-1} is used to predict the next state $\tilde{s}_t^{(i)}$. This is done by sampling from the prior distribution $p\left(s_t | s_{t-1}^{(i)}\right)$ that represents the system transition (the simulation model). The importance weight for the i-th sample is computed from $p\left(m_t | \tilde{s}_t^{(i)}\right)$. These weights are normalized after generating all N samples. In the resampling step, N offspring samples are drawn with probability proportional to the normalized sample weights. These samples represent the posterior belief of the system state and are used for the next iteration.

4.3 Dynamic Data-Driven Simulation Based on Sequential Monte Carlo Methods

DDDS uses *data assimilation* to dynamically incorporate real-time observation data into a running simulation model. Whenever new observation data arrives (or after every fixed period of time, e.g., every 30 min), data assimilation is carried out to assimilate the new observation data. The goal of data assimilation is to provide an updated estimate of the "current" system state, which is often hidden and cannot be observed. The estimated system state is then used to provide initial conditions to simulate/predict the system dynamics in the future. Since initial conditions have significant impacts on simulation results, a simulation starting from more accurate initial conditions will lead to improved simulation results.

Figure 4.3 shows how data assimilation works in dynamic data-driven simulation. In the figure, the current time step is denoted as t. Data assimilation is carried out using the observation data at time t to estimate the current system state. The estimated system states at time t are then used as initial conditions to simulate/

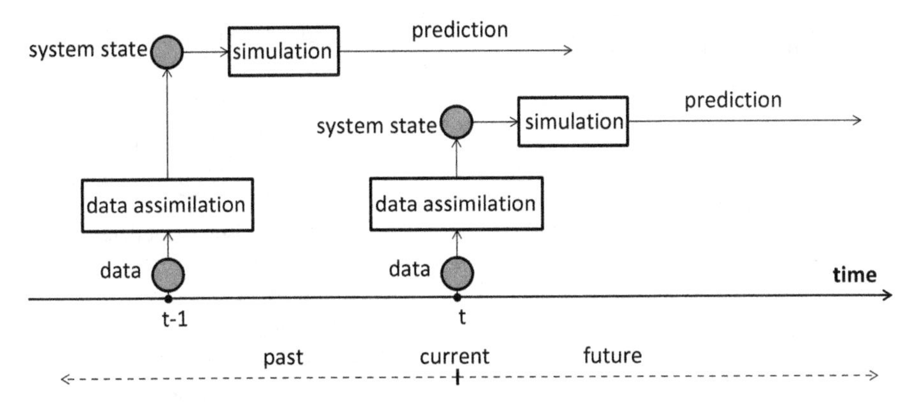

Fig. 4.3 Data assimilation in dynamic data driven simulation

predict how the system evolves in the future. Note that at a previous time step $t-1$ (in the past), data assimilation was also carried out using the data at $t-1$, and simulations were run to simulate/predict the system evolution after $t-1$. With new observation data arrived at time t, the data assimilation at time t enables new simulations for updated predictions of the future. As time moves forward, this process continues, and the simulation system keeps adjusting itself to improve simulation/prediction results. This compares to traditional simulations where the simulation is decoupled from the real system and the difference between simulation and real system is likely to grow larger and larger as time increases. We note that in Fig. 4.3, the time steps $t-1$ and t are used to indicate the stepwise nature of the process. The actual time interval between two consecutive steps is usually defined by how often sensor data is collected, for example, every 30 min.

The need of dynamic data-driven simulation asks for advanced data assimilation methods that can work with sophisticated simulation models. Sequential Monte Carlo (SMC) methods hold great promise in this area. A key advantage of SMC methods is their ability to represent arbitrary probability densities. This makes them effective methods for data assimilation for complex simulation models, which usually have nonlinear, non-Gaussian, unsteady behaviors. Meanwhile, SMC methods are nonparametric filters that do not rely on analytic functions of probability distributions. This feature is especially suited for sophisticated simulation models (such as cellular automata-based wildfire spread simulation models and agent-based traffic simulation models) that lack the analytic structures from which functional forms of probability distributions can be derived. Furthermore, SMC methods are recursive methods that are able to recursively adjust their estimations of system states when new observation data becomes available. This feature is suited for dynamic data-driven simulation where new sensor data arrives sequentially and the simulation system needs to be continuously updated. Due to these reasons, we choose SMC methods to carry out data assimilation in DDDS.

Figure 4.4 shows a DDDS framework based on SMC methods (specifically the bootstrap filter algorithm described in Section 2). Major components of the

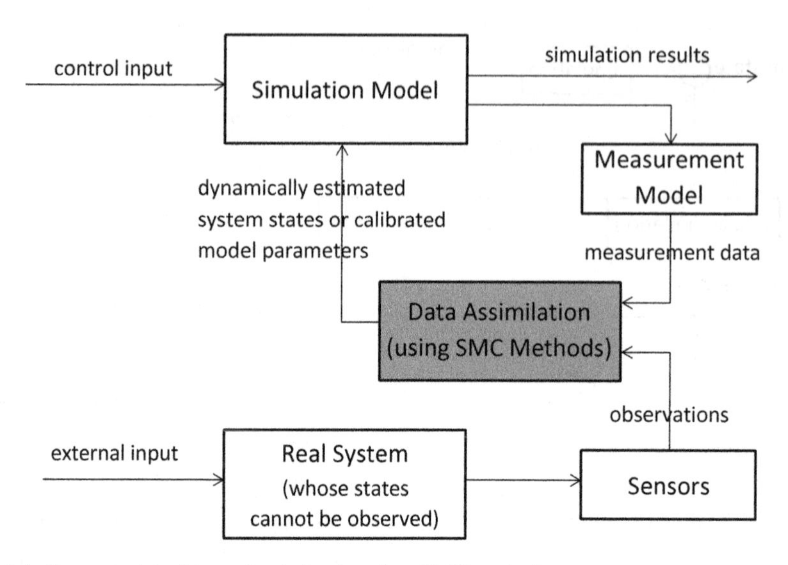

Fig. 4.4 Dynamic data driven simulation based on SMC methods

framework include a *real system* under study, a *simulation model* that simulates the dynamic behavior of the real system. The real system is influenced by external inputs, e.g., weather input that is external to the wildfire system in the wildfire simulation example. Similarly, the simulation model is influenced by control input that aims to capture the external inputs of the real system. As mentioned earlier, due to data errors and model errors, there exist differences between simulation results and the real system. Furthermore, the real system's states, which change over time, cannot be directly observed and are unknown to the simulation model. This makes a simulation start from an initial state different from the state of the real system, leading to inaccurate simulation results.

A major task in dynamic data-driven simulation is to dynamically estimate the "current" state of the real system and then feed the estimated states to the simulation model for follow-on simulations. This is achieved through *data assimilation* based on SMC methods, which assimilate real-time sensor data to infer the "current" system state. As will be discussed later, data assimilation also makes it possible to dynamically calibrate model parameters to reduce discrepancies between simulation and the real system. To support dynamic data-driven simulation, *sensors* are deployed, and real-time sensor data (referred to as *observations* in Fig. 4.4) is collected. Meanwhile, a *measurement model* is developed that maps system state to observations according to the sensors used. With this measurement model, one can compute the *measurement data* for a given simulated system state. The difference between measurement data (computed from simulated system state) and real observations (collected from the real system) carries information about how well the simulation is doing compared with the real system. This information is utilized by the data assimilation component to dynamically estimate the real system's state.

To support data assimilation based on the bootstrap filter algorithm, one needs to define the system state appropriately and develop the associated models and algorithms following the structure of the bootstrap filter algorithm. The major activities are described below.

- *Define the system state*. The system state evolves over time and needs to be estimated. Typically, the system state is a vector of state variables.
- *Specify the system transition model* (Eq. (4.1) in Section 2). Because the simulation model defines how the system state changes over time, the simulation model essentially is the system transition model. One issue may arise if the simulation model is a deterministic simulation model. In this case, stochastic behavior (e.g., by adding random noises) needs to be introduced in order to construct a stochastic state transition model required by SMC methods. An example can be found in Xue et al. (2012).
- *Define the measurement model* (Eq. (4.2) in Section 2). The measurement model maps system state to observation data collected by sensors. Thus the measurement model is sensor dependent—it should be constructed according to the specific sensors deployed in the real system. In the example to be shown in Section 5, the sensor data is collected from multiple ground temperature sensors deployed at different locations of a fire area. Accordingly, the measurement model is to compute a vector of temperature data (corresponding to the deployed sensors) from the system state, i.e., the state of the burning wildfire.
- *Develop the sampling, weight updating, and resampling methods* to be used by the bootstrap filter algorithm. Sampling is to run simulation to the next data assimilation time point (e.g., 30 min later) and obtain the new system state. Weight updating is based on the difference between the real observation data and the measurement data computed from the simulated system state and assigns weights accordingly. The resampling method is to sample new particles according to their normalized weights. Several standard resampling methods, e.g., multinomial resampling, systematic resampling (see Douc et al. (2005) for a comparison of the different resampling methods), have been developed and can be used.

With all the above components developed, the bootstrap filter algorithm can be run to carry out data assimilation. In this algorithm, the set of system states is represented by a set of particles, each of which represents a full system state. Figure 4.5 shows the procedure of data assimilation based on the bootstrap filter algorithm. In the figure, the rectangle boxes represent the major activities in one step of the algorithm, and the circles represent the data/variables. The data assimilation runs in a stepwise fashion. At time step t, the set of particles from time step $t-1$ (denoted as S_{t-1} in Fig. 4.5) are fed into the system transition model. Each sample in the sample set represents a specific system state. This whole sample set represents the prior belief of the system state. For each particle in S_{t-1}, the system transition model uses the state represented by the particle to produces a new sample by running simulation from time $t-1$ to time t. The resulting system state set is denoted as S'_t. To compute the importance weights of the particles, for each state in

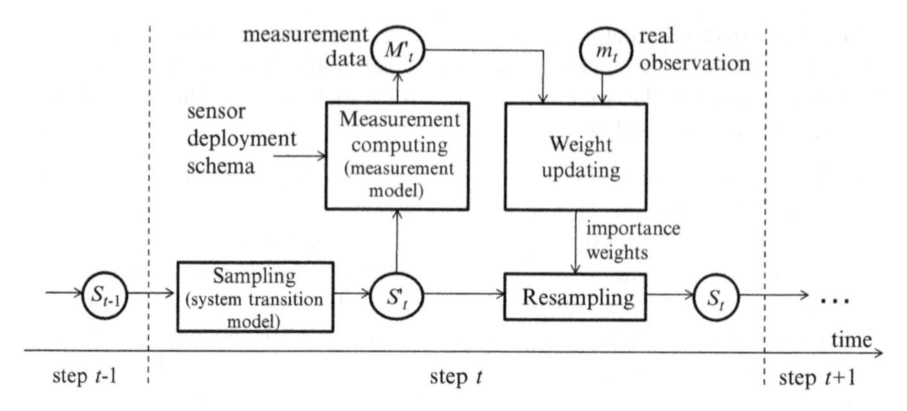

Fig. 4.5 Data assimilation based on the bootstrap filter algorithm

S'_t, its measurement data is computed according to the measurement model and the sensor deployment schema of the real sensors. The set of measurement data for all states in S'_t are denoted as M'_t. Then, considering each measurement data vector in M'_t as the mean vector, the probability density value of the real observation data vector m_t (the data collected from real sensors at time t) is calculated based on a multivariate Gaussian distribution. This density value is used to update the importance weight of the corresponding particle. After normalizing the weights of all particles, a resampling algorithm is applied to generate a new set of offspring samples according to the probabilities proportional to the normalized weights of particles. These samples (denoted as S_t) represent the posterior belief of the system state and are used for the next iteration.

4.4 Applications and Challenges

DDDS represents a new paradigm where a running simulation system is coupled with the real system by assimilating real-time sensor data. The capability of assimilating real-time sensor data to improve simulation results is especially important for supporting real-time decision-making. Within this context, below we list several applications that DDDS enables.

- *Dynamic state estimation*: Estimating the dynamically changing system states from observation data is a fundamental task in DDDS. The state estimation from real-time data allows a simulation to start from a state "closer" to the real system's state and thus leads to more accurate simulation results.
- *Online model parameter calibration*: Besides estimating system state that represents the system behavior, it is also desirable to dynamically calibrate model parameters that characterize the system structure. One can formulate the problem of online model parameter calibration as a joint state–parameter estimation and uncertainty assessment problem, which treats model parameters

as stochastic state variables that need to be estimated (see Moradkhani et al. (2005) as an example).

- *Dynamic data-driven event reconstruction*: We define dynamic data-driven event reconstruction as the process of estimating the occurrences and characteristics (e.g., when, where) of some events of interest. Such events are not explicitly modeled by the simulation model but can significantly affect the system behavior. In the wildfire example, while a wildfire is spreading, new fires may be ignited in the vicinity of fire front, resulting in multiple fires. Being able to estimate the occurrences of such new fires from real-time sensor data can greatly improve fire spread simulation results.

Due to the complexity of simulation models and the data assimilation algorithms, DDDS also faces several major challenges as described below.

- The first challenge is associated with the fact that many simulation models (such as cellular automata-based fire spread simulation and agent-based traffic simulation) do not have the analytic structures as the ones in numeric models (e.g., partial differential equations). The lack of analytic structures makes it infeasible to derive functional forms of probability distributions from which samples can be easily drawn and density can be calculated. As a result, many advanced particle filtering methods developed for numeric models cannot be applied to data assimilation for simulation models. New methods need to be developed in order to improve the effectiveness of SMC methods to work with simulation models with nonanalytic structures.
- The second challenge is associated with the high-dimensional state space of simulation models. High-dimensional filtering is a fundamental challenge for SMC methods (see Snyder et al. (2008) for detailed discussions on this). Simulation models often have a large number of state variables and thus a high-dimensional state space. The high-dimensionality asks for advanced sampling and resampling methods in order for SMC methods to achieve effective inference and quick convergence in data assimilation.
- Another important challenge is associated with the high computation cost of applying SMC methods to sophisticated simulation models. SMC methods have a demanding computation cost because of the large number of particles, each of which needs a full-scale simulation to sample new particles for the next time point. This issue of high computation cost is especially true for large-scale simulations such as simulations of large-scale wildfires. Developing advanced methods to reduce the demanding computation cost is crucial for supporting real-time decision-making.
- Research challenges also come from how sensor data is collected, which is an important part of data assimilation. In general, advanced sensors and dense senor deployment are needed in order to collect high-quality sensor data to be used in data assimilation. However, this will increase the cost of sensors and sensor deployment. An important research task is to study how to deploy sensors in effective manners and how to extract useful information from sensors for data assimilation.

4.5 An Illustrative Example: Dynamic Data-Driven Wildfire Spread Simulation

We present an example to illustrate how dynamic data-driven simulation works based on the application of wildfire spread simulation. This example is adapted from our previous work (Xue et al. 2012), where interested users can find more details. In this example, the wildfire spread simulation model is a discrete-event simulation model called DEVS-FIRE (Hu et al. 2012; Ntaimo et al. 2008). DEVS-FIRE is a two-dimensional cellular space model where the forest is modeled as a two-dimensional cell space. The cell space contains individual forest cells, each of which contains its own GIS data and weather data. Each cell in the cell space is represented as a DEVS (Zeigler et al. 2000) atomic model and is coupled with its eight neighbor cells according to the Moore neighborhood. Consequently, the forest cell space is a coupled model composed of multiple forest cell models. Fire spread is modeled as a propagation process where burning cells ignite their unburned neighbors. The rate of spread of a burning cell is calculated using Rothermel's fire behavior model (Rothermel 1972) and then decomposed into eight directions corresponding to the eight neighboring cells. More details about the DEVS-FIRE model can be found in Hu et al. (2012) and Ntaimo et al. (2008).

Ground temperature sensors are (sparsely) deployed at different locations of the fire area. These sensors collect data that reflects the local temperature of the sensor location. For example, when a sensor is far away from the burning fire front, its temperature data is the ambient temperature; as the fire front spreads and gets closer to the sensor location, its temperature data increases. Different sensors provide different temperature data due to their different locations in the area. This data changes dynamically as the fire spreads. In our example, we assume sensor data is collected every 20 min. Our goal is to assimilate the sensor data to estimate the dynamically changing system state of the fire, based on which to improve simulation results.

To apply SMC methods for data assimilation, the system state, measurement data, system transition model of state evolution, and the measurement model that maps system state to measurement data need to be defined. In this example, the DEVS-FIRE model is composed of many cells, each of which captures the state (e.g., unburned or burning, fireline intensity, etc.) of the corresponding local region in the fire area. Based on this implementation, we can define the overall system state as an n_c dimensional vector fire $\in \text{FIRE}^{n_c}$, where

$$\text{FIRE} = \{< \text{unburned}, 0 > ,\ < \text{burning},\ \text{FI} > ,\ < \text{burned},\ 0 >\},$$

n_c is the total number of cells in the whole cell space, and the second element in each tuple indicates the fireline intensity. We define the measurement variable m_t as an n_s (the number of sensors)-dimensional vector containing temperature values from all the sensors deployed in the fire area. With system state and measurement variable defined, we then formulate a nonlinear state space model as shown in the equation below.

$$\begin{cases} \text{fire}_{t+1} = \text{SF}(\text{fire}_t, t) + \gamma_t, \\ m_t = \text{MF}(\text{fire}_t, t) + \omega_t. \end{cases}$$

where fire_t and fire_{t+1} are system state variables of fire spread at time step t and time step $t+1$ respectively; SF is the system transition function. In this work, SF() is based on the DEVS-FIRE simulation model (a simulation model defines how the system state evolves over time):

$$\text{SF}(\text{fire}_t, t) = \text{DEVSFIRE} \quad (\text{fire}_t, \theta_t, \Delta t),$$

where Δt is the time duration of a data assimilation step (20 min in this example); parameter θ_t captures the slope, aspect, fuel, and weather data used in computing fire spread behavior; γ_t is the system transition noise that introduces stochastic elements to the system transition model; the function DEVSFIRE() represents the DEVS-FIRE simulation model. For the measurement model, m_t is the measurement variable; $\text{MF}(\text{fire}_t, t)$ is the measurement function mapping fire states to measurements, and ω_t is the measurement noise. Given a fire state at time step t, through SF (), a fire state at time step $t+1$ can be obtained; also, through MF(), the corresponding measurements can be calculated.

The measurement function MF() maps a fire state to a measurement vector (temperature data of deployed sensors). The number of sensors and their locations are predefined based on how real sensors are deployed in the fire area. Given a system state, for a particular temperature sensor, the measurement model is used to compute the sensor's temperature data based on the fire intensity and the spatial distance between the sensor and the fire front (see Xue et al. (2012) for more details). As an illustrative example, Fig. 4.6a shows a snapshot of a simulated fire's fire front. Figure 4.6b shows the temperature data collected by the temperature sensors, with different colors indicating different temperature values. In Fig. 4.6b, the blue dots in the outside area represent the locations of the ground temperature sensors. Note that as the fire front evolves, the collected temperature sensor data changes over time too.

4.5.1 Experimental Results

We use the identical-twin experiment, which is widely used in data assimilation research, to show how the data assimilation works and the data assimilation results. The purpose of identical-twin experiments is to study data assimilation in ideal situations and evaluate the proximity of the prediction to the true states in a controlled manner. In the identical-twin experiment, a simulation is first run, and the corresponding data is recorded. These simulation results are considered as

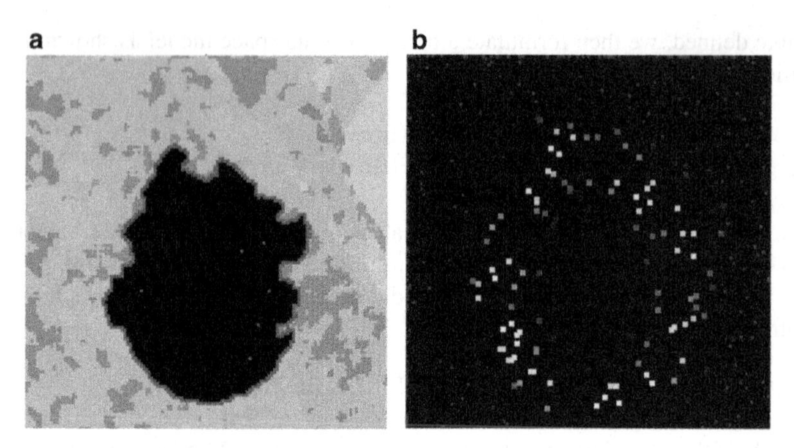

Fig. 4.6 Fire front and temperature sensor data. (**a**) Fire front of the a fire. (**b**) Temperature sensor data

"true"; therefore, the observation data (real-time sensor data) obtained here is regarded as the real observation data (because it comes from the "true" model). Consequently, we estimate the system states from the observation data using SMC methods and then check whether these estimated results are close to the "true" simulation results. In the following description, we use three terms, *real fire*, *filtered fire*, and *simulated fire*, to help us present the experimental results. A real fire is the simulation from which the real observation data is obtained. A simulated fire is the simulation based on some "error" data ("error" in the sense that the data is different from that used in the real fire), for example, imprecise weather data. This is to represent the fact that wildfire simulations usually rely on imperfect data as compared to real wildfires. Finally, a filtered fire is the data assimilation–enhanced simulation based on the same "error" data as in the simulated fire. In our experiments, we intend to show that a filtered fire gives more accurate simulation results by assimilating sensor data from the real fire even if it still uses the "error" data.

The differences between a real fire and a simulated fire are due to the imprecise data, such as wind speed, wind direction, GIS data, and fuel model, used in the simulation. In this example, we choose to use the imprecise wind speeds as the "error" data. Specifically, the real wind speed and direction are 8 (m/s) and 180° (from south to north) with random variances added every 10 min. The variances for the wind speeds are in the range of -2 to 2 (m/s), and the variances for the wind direction are in the range of -20 to 20 (degrees). We introduce errors to the wind speeds and make the wind directions to be exactly the same as the real wind directions. Two simulations are carried out. In the first simulation (referred to as case 1), the wind speed is randomly generated based on 6 (m/s) with variances added in the range of -2 to 2 (m/s). In the second simulation (referred to as case 2), the wind speed is randomly generated based on 10 (m/s) with variances added in the range of -2 to 2 (m/s). Figure 4.7a displays the real fire after 3 h of simulation. Figure 4.7b, c show the two simulated fires for the same simulation duration (3 h).

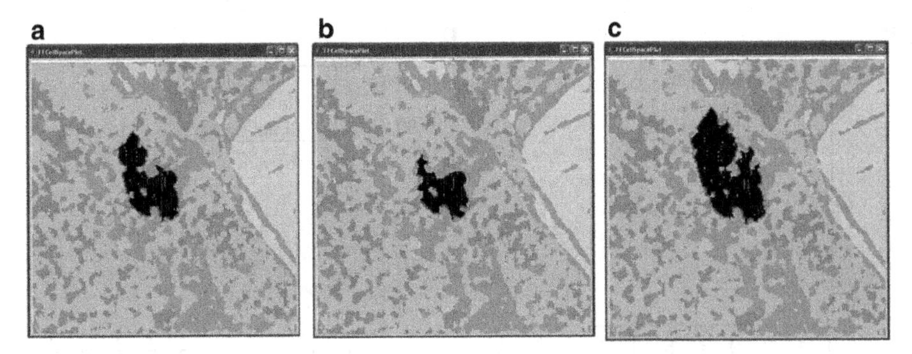

Fig. 4.7 Inaccurate fire spread simulations due to imprecise wind speed. (**a**) Real fire. (**b**) Simulated fire 1: wind speed is smaller than real wind speed. (**c**) Simulated fire 2: wind speed is larger than real wind speed

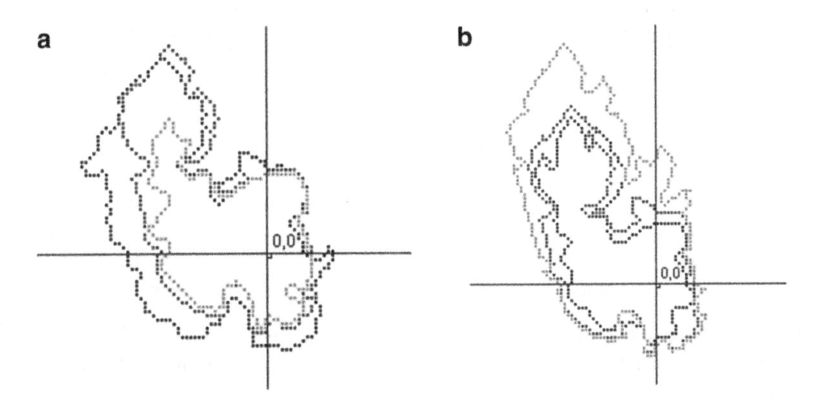

Fig. 4.8 Data assimilation results by assimilating temperature sensor data. (**a**) Data assimilation result for the fire spread simulation using smaller wind speed. (**b**) Data assimilation result for the fire spread simulation using larger wind speed

In the figures, the burning cells and the burned cells are displayed in red and black respectively. The other colors display different fuel types of the cells. From the figures, we know that the real fire and the simulated fires have large deviations due to the imprecise wind speeds. In the first simulation, the real fire spreads faster than the simulated fire because the real wind speeds are larger than the error wind speeds. In the second simulation, the real fire grows slower than the simulated fire since the real wind speeds are smaller than the error wind speeds.

By assimilating the temperature sensor data into DEVS-FIRE, the filtered fires were obtained. Figures 4.8a, b display the filtered fires (displayed in blue) after 3 h of simulation, compared with the real fire (displayed in red) and the simulated fires (displayed in green). Figure 4.8 shows that in both cases, the filtered fire shapes match the real fire shape better than the simulated fires do. This is particularly true

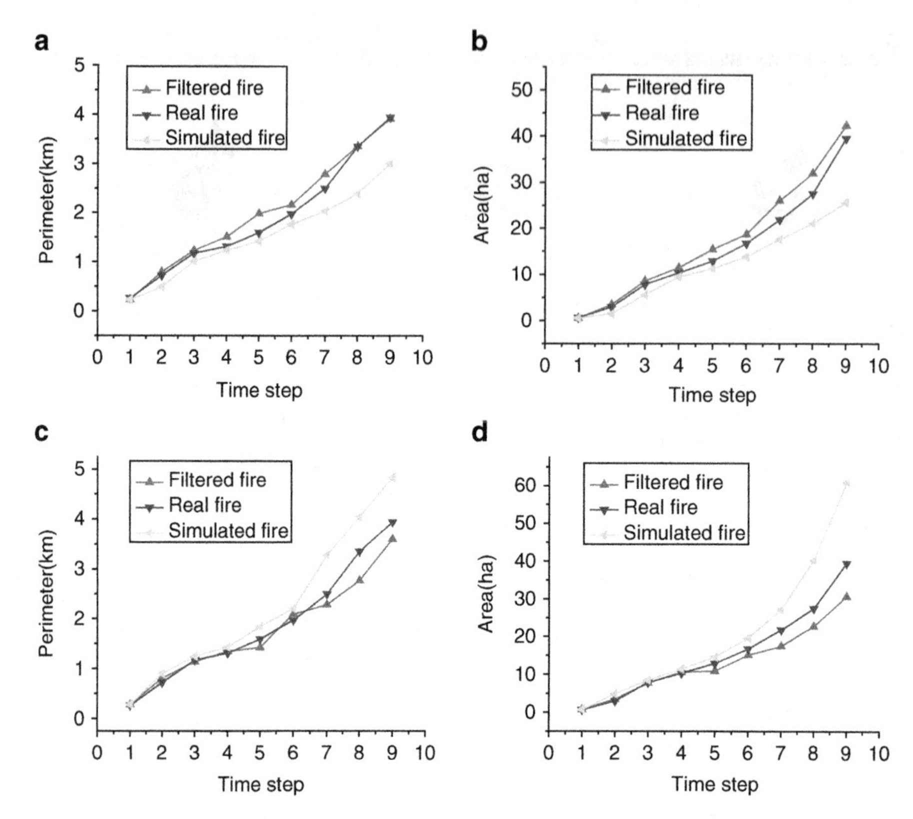

Fig. 4.9 Perimeters and areas of real fire, simulated fires, and filtered fires for case 1 and case 2. (**a**) Perimeters for case 1; (**b**) Burned areas for case 1; (**c**) Perimeters for case 2; (**d**) Burned areas for case 2

at the head area (the north direction) of the fire, where the fire spreads fast. For example, in Fig. 4.8a, the simulated fire is much smaller than the real fire at the head area due to the smaller wind speeds. However, using the same "error" wind speeds as the simulated fire did, through data assimilation the filtered fire was able to match the real fire well at the head area. Similar effects can be seen for the second case as shown in Fig. 4.8b. This demonstrates the effectiveness of the data assimilation method.

We note that the current data assimilation method still has a lot of room for improvement. For example, in Fig. 4.8a, although the filtered fire matches well with the real fire at the head area, it spreads faster than the real fire does on the west side. This asks for further development of the data assimilation method for improving the data assimilation results. More results of data assimilation can be found at (Xue et al. 2012).

To quantitatively show the data assimilation results, Figs. 4.9a, b display the perimeters and burned areas of the real fire, the simulated fire, and the filtered fires for case 1 from data assimilation step 1–9. Figures 4.9c, d show the perimeters and areas of the real fire, the simulated fire, and the filtered fires for case 2 from data assimilation step 1–9. In both cases, we carried out ten independent runs of data assimilation experiments and display the average results from the ten runs in the figures. These figures show that the differences between the real fire's perimeter and the filtered fires' perimeters are smaller than those for the simulated fires. The same trend holds true for the burned areas. We note that in the beginning, the differences among the real fire, the simulated fire, and the filtered fire are small because the fire sizes are small. As the fire size grows larger, the effect of data assimilation becomes more obvious. Since the filtered fires are much closer to the true system states, simulations starting from these fires (i.e., using the filtered fires as initial conditions) will lead to more accurate simulation/prediction results in the future.

4.6 Conclusions

Data is an essential part of computer modeling and simulation and will play even more important roles for modeling and simulation in the future. The importance of data in modeling and simulation was recognized by Ören in several of his works. In this book chapter, we present dynamic data-driven simulation as a new simulation paradigm where a simulation system is continually influenced by the real-time data for better analysis and prediction of a system under study. A framework of dynamic data-driven simulation based on Sequential Monte Carlo methods is presented. We describe the applications and challenges associated with dynamic data-driven simulation and provide an illustrative example of dynamic data-driven simulation based on the application of wildfire spread simulation.

References

Bouttier F, Courtier P (1999) Data assimilation concepts and methods. Presented at the Meteorological Training Course Lecture Series. European Centre for Medium-Range Weather Forecasts, Reading, England, pp 1–58

Darema F (2004) Dynamic data driven applications systems: a new paradigm for application simulations and measurements. Proc. International Conference on Computational Science. Springer-Verlag Berlin Heidelberg, pp. 662–669.

Douc R, Cappe' O, Moulines E (2005) Comparison of resampling schemes for particle filtering. In: Proceeding int'l symposium image and signal processing and analysis, Istanbul, Sept 2005

Doucet A, Freitas ND, Gordon N (eds) (2001) Sequential Monte Carlo methods in practice. Springer, New York

Evensen G (2007) Data assimilation. The Ensemble Kalman filter. Springer, Berlin/New York

Gordon NJ, Salmond DJ, Smith AFM (1993) Novel approach to nonlinear/non-Gaussian Bayesian state estimation, Radar and Signal Processing. IEE Proceed F 140(2):107–113

Hu X, Sun Y, Ntaimo L (2012) DEVS-FIRE: design and application of formal discrete event wildfire spread and suppression models. Simulation 88(3):259–279

Jazwinski AH (1970) Stochastic processes and filtering theory. Mathematics in science and engineering. Academic, New York

Kalman RE (1960) A new approach to linear filtering and prediction problems. J Basic Eng 82 (1):35–45

Liu J, Chen R (1998) Sequential Monte Carlo methods for dynamic systems. J Am Stat Assoc 93:1032–1044

Moradkhani H, Hsu K-L, Gupta H, Sorooshian S (2005) Uncertainty assessment of hydrologic model states and parameters: sequential data assimilation using the particle filter. Water Resour Res 41:W05012. doi:10.1029/2004WR003604

Ntaimo L, Hu X, Sun Y (2008) DEVS-FIRE: towards an integrated simulation environment for surface wildfire spread and containment. Simul Trans Soc Model Simul Int 84(4):137–155

Ören TI (2000) Responsibility, ethics, and simulation. Trans SCS, San Diego 17(4):165–170

Ören TI (2001 – Invited Paper) Impact of data on simulation: from early practices to federated and agent-directed simulations. In: Heemink A et al (eds) Proceedings of EUROSIM 2001, Delft, 26–29 June 2001

Ören TI, Zeigler BP (1979) Concepts for advanced simulation methodologies. Simulation 32 (3):69–82

Rothermel RC (1972) A mathematical model for predicting fire spread in wildland fuels. USDA For. Serv. Res. Pap. INT-115, Ogden

Snyder C, Bengtsson T, Bickel P, Anderson J, Anderson J (2008) Obstacles to high-dimensional particle filtering. Mon Weather Rev 136:4629–4640

Xue H, Gu F, Hu X (2012) Data assimilation using sequential Monte Carlo methods in wildfire spread simulation. ACM Trans Model Comput Simul (TOMACS) 22(4), Article No. 23

Zeigler BP, Kim TG, Praehofer H (2000) Theory of modeling and simulation: integrating discrete event and continuous complex dynamic systems, 2nd Revised edn. Academic Press Inc., Waltham

Part II
Modeling Methodologies

Chapter 5
Learning Something Right from Models That Are Wrong: Epistemology of Simulation

Andreas Tolk

5.1 Introduction

> Essentially, all models are wrong, but some are useful. (Box and Draper 1987, p. 424)

Epistemology is the branch of philosophy that deals with gaining knowledge. It is closely related to ontology, the branch that deals with questions such as "What is real?" and "What do we know?" When using modeling and simulation (M&S), we usually do so to either apply knowledge, in particular when we are using them for training and teaching, or to gain knowledge, for example, when doing analysis or conducting virtual experiments. But none of our models represents reality as it is. They are only valid within their limitations, which leads to the famous quote of Box that "all models are wrong." The question is therefore: how can we learn something from these models? What are the epistemological foundations for us simulationists?

Already in his early contributions, Tuncer Ören (1977) observes that simulation is ubiquitously applied across many domains as a tool to solve problems, but is it really scientific to use simulations when we are interested in real-world solutions? Can we as the modeling and simulation community justify the use of simulation as a tool that we provide to scientists, scholars, and decision-makers? Is the use of simulation a scientifically sound method? To be able to answer these questions, we have to revisit what we understand regarding the application of science to solve problems.

The basic elements of the scientific method are learned in elementary school—especially in conjunction with science fairs. It all starts with a question. The next step is to conduct background research. Is there already a theory that answers the

A. Tolk (✉)
SimIS Inc., Portsmouth, VA, USA
e-mail: andreas.tolk@gmail.com

L. Yilmaz (ed.), *Concepts and Methodologies for Modeling and Simulation*, Simulation Foundations, Methods and Applications,
DOI 10.1007/978-3-319-15096-3_5

question within the current body of knowledge? If not, we formulate a hypothesis that predicts the expected outcome. Using an experiment, we then evaluate the constructed hypothesis. We identify the independent and dependent parameters, capture expected causality and resulting observations, eliminate all disturbing factors and aggregate nonessential influences, and then run our experiment as often as required to attain statistical significance at the level desired. Resultant data is then analyzed in light of the following hypothesis: do the observed data points correlate with expected values within the target margins of error? If they do, our hypothesis is supported and ultimately may become part of the embedding theory. If they do not, the hypothesis is rejected as it could not be supported by the observations during the experiment.

What does this have to do with modeling and simulation (M&S)? Much more than we might expect. It begins with the process of modeling. *Modeling is a task-driven purposeful simplification and abstraction of a perception of reality* (Tolk 2013a). Let us have a closer look at the components of this definition:

Task-driven: a model is generated for a task, such as to answer a question within the domain of analysis or provide certain functionality, such as supporting training. Like the question that initiates the scientific method, the task drives the modeling process.

Purposeful: modeling is a creative act. The various activities are driven by the task and are done knowingly and purposefully to reach the goal to the greatest extent possible.

Simplification and abstraction: as in the experiment described above, elements that are not important and only distract from the main event are eliminated from the model. Furthermore, components that may have an effect but are considered secondary or less important can be combined as a form of data reduction techniques.

Reality: our work shall be rooted in empirical data, no matter if we assume a positivistic or postmodern worldview.

Perception: our perception is shaped by physical–cognitive aspects and constraints. The physical aspect defines what attributes of an object are observable with the sensory system of the observer or, more generally, the information about the object that can be obtained (this can include gaining insight from literature, discussions with colleagues, using instruments, etc.) Cognitive aspects are shaped by the education and the knowledge of the observer, their paradigms, and even knowledge of related tools associated with the tasks.

The result of the modeling phase is a conceptualization shaped by the task as well as by the physical–cognitive abilities of the modeler. This conceptualization is then captured using a modeling method and subsequently transformed into an executable simulation. This process is also shaped significantly by the resources and abilities of the programmers. Are the computers powerful enough to translate the concepts into something directly executable, or do memory and processor capacities require additional simplifications? Do the languages support all manipulations envisioned? Are there numerical challenges that influence the quality of the simulation?

This introduction demonstrates good reasons why there can be no "correct" model and therefore no "correct" simulation, as they always will be limited to their constraints and assumptions that are made in the process of their creation. However, we will see in the following section that this is not really surprising for the scientist or the simulationist and that we nonetheless learn from models and simulations.

To guide the reader on this path, we will start with an introduction to the philosophical fields of ontology and epistemology, leading to a short history of science, both from a simulationist view. These views shape our discussion on the use of models and resulting limits and constraints. The outcome of these analyses, many of them based on work initiated by Tuncer Ören, as will be demonstrated, will lead to a grand challenge for the simulation community to evolve within the community of scientists by including epistemological perspectives in curricula for simulationists as a pillar of our profession.

5.2 Ontology and Epistemology for Simulationists

> Theories cannot be proven to be generally correct, but we can state that they have not been falsified so far by new observations or insights!. (Popper 1935)

Our current curricula in the domains of science and engineering often neglect the ethical and philosophical foundations that are so important for each discipline. Interestingly, philosophers of science seem to be more interested in the foundations of M&S than simulationists themselves. Since 2004, the workshop on Epistemological Perspectives on Simulation (EPOS) convenes every second year (Frank and Troitzsch 2005; Squazzoni 2009; Grune-Yanoff and Weirich 2010). The following two sections summarize some of the fundamentals every simulationist should be aware of.

Ontology and epistemology pose the questions of what we know and how we can gain knowledge. To answer these questions, one should begin with a short overview of science and what is considered to be knowledge and what role M&S can play in this regard.

5.2.1 A History of Science: A Simulationist's View

In the introduction of this chapter, we looked at the scientific method. Interestingly, this method is relatively young, as described by Goldman (2006). In the same era, two scientists shaped our understanding of what science knows and how we can increase this knowledge.

Sir Francis Bacon lived from 1561 to 1626 in England. He is considered the father of the *inductive–empirical method*. He postulated that knowledge can only be derived by observation and data collection. Only that which can be observed should be the subject of science. With the development of more and more sophisticated sensor systems, the discussion arose as to what extent observations made by sensors are valid. In particular, when they were made under limiting constraints, such as the use of small probes under a microscope, their validity was often questioned. This discussion continues to this day in some domains of modern physics. Overall, Bacon's idea was that knowledge could be derived only from observation, data collection, and data analysis.

Parallel to these efforts, René Descartes (1596–1650), who worked in France in defining the *deductive–rational method*, assumed that a world order of physical laws can be described by mathematical models. These perfect models are the basis of knowledge. He followed the traditions of ancient Greek thinkers that led to the Platonic–Aristotelian view: that rational thought is linked via deduction to reality. For Descartes, observations and data were means to validate knowledge.

These two competing visions of the scientific methods, also known as Baconian empiricism and Cartesian rationalism, were unified in the work of Sir Isaac Newton (1642–1727). His work on *Mathematical Principles of Natural Philosophy* (original 1687 in Latin, Newton et al. 1955) was the standard literature for more than 200 years and is still the foundation of high school physics. Newton utilized mathematical models that allowed for coherent and consistent interpretation of observations: a conceptualization of experience used to generalize observations (knowledge generation, theory development) as well as having to be validated by empirical observations (theory testing). Newton's pragmatic approach employed both visions.

This new approach was not immediately embraced by all, as the discussions between Clark, a student of Newton, and the German philosopher and mathematician Gottfried W. Leibniz show (Alexander 1998). In particular, continental Europe clinched to the ideas of perfect laws reigning the universe. Newton's premise was that science does not *discover* truth about natural laws but that scientists *construct* interpretations of experiences. For him, a theory was the best model available that explains all observations and experiences without internal contradictions. If new observations are made that contradict the old theory, the theory is falsified and needs to be extended or replaced. In this view, science is understood as a series of models.

The work of Newton was continued by John Locke (1632–1704) and David Hume (1711–1776), who introduced the ideas of skepticism and challenges of inductive reasoning. On what evidence can we assume that our experiences are universal or even repeatable in the future? Immanuel Kant (1724–1804) reintroduced a kind of certainty and stability with his cosmological theory. His *Universal Natural History and Theory of the Heavens* (original 1755, Kant and Jaki 1981) established the rational mind as the objective element. Again, the model was the central role of knowledge: a rational conceptualization.

In the nineteenth century, scientific knowledge became more and more directly applicable. Technological innovation was driven by scientific insights; real-world problems were solved by applying scientific principles to engineer solutions. The application became in many domains more important than the theory. At the same time, the physics of the nineteenth century started to discover alternatives to principles assumed to be inviolable (Harman and Harman 1982). Discovery and successful applications to alternatives of Euclidian geometry, such as Fourier's equations, emphasized the need for models that clearly stated their assumptions and constraints. It also led to the possibility of having two models based on incompatible theories that were equally valid themselves. Assumptions and constraints needed to be formulated to capture these premises in order to unambiguously communicate results and ideas. In summary, the model and its assumptions became the best representation of the theory.

The twentieth-century views can be reflected when looking at the work of Bertrand Russell (1872–1970), Henri Poincaré (1854–1912), and Percy Bridgman (1882–1961). They agreed on the principles that scientific knowledge is not about reality but focuses more on common concepts, how to measure them, and how to express them. Mathematics, and in particular logic, was understood as the common language to express these concepts. Science was interpreted as a commonly conceptualized and experiential world expressed by models based on logic. Their work became the foundation of logical positivism, embracing the idea that only science can provide a rational account of the world. In their view, truth was driven by a physicalistic–deterministic account of reality.

This viewpoint was shaken by Albert Einstein (1879–1955) and Niels Bohr (1885–1962) with their theories of relativity and quantum theory. The work of Werner Heisenberg (1901–1976) and Erwin Rudolf Josef Alexander Schrödinger (1887–1961) added to the new challenges: scientific theories had become complex and often argued against empirical observations of daily life. They were no longer considered to have explanatory value. Accordingly, the view on science and its models changed again. It was no longer understood as a converging process getting closer and closer to reality but a discontinuous and even revolutionary process in which old theories were replaced by new ones in a nonconvergent but progressive process (Kuhn 1962). Again, science was a series of models continuing to replace each other.

Today, 400 years after the modern story of science began, we are in a "science war" between natural scientists and supporters of the idea that scientific knowledge is a mere social construction, also referred to as relativism. The role of modeling has not diminished. The definition of modeling as a task-driven purposeful simplification and abstraction of a perception of reality holds under all paradigms and allows for communication between scientists across the borders of paradigms and belief systems. As stated by Robert Rosen (1998), "I have been, and remain, entirely committed to the idea that modeling is the essence of science and the habitat of all epistemology."

The history of science from the simulationist's view is a series of models that are created to explain the world. We use models to explain what we see (inductive) as well as to forecast what happens based on our experience (deductive). When creating a model, we establish a theory by replacing empirically observed correlation with assumed causality. When we compare our model predictions with actual observations, we validate our theory as implemented in the model. Science came to the understanding that theories are always just "as good as they can be." Scientists never know what new observations will be made or correlation will be discovered that cannot be explained with the current best theory. They also understand that, as Popper (1935) shows in his work, theories cannot be proven to be generally correct. The best we can do is to observe that so far, no contradictions between theory and observation occurred. Therefore, asking if we can learn anything from models that are wrong is like asking: can we learn anything from science?

With the advent of more and more powerful computers, the simulation components of M&S started to influence science as well (Winsberg 2010). The use of simulation in science was only recently highlighted by Arieh Warshel. He delivered his Nobel Lecture on "Computer Simulations of Biological Functions: From Enzymes to Molecular Machines" on December 8, 2013, at Aula Magna, Stockholm University, where he was introduced by Professor Sven Lidin, Chairman of the Nobel Committee for Chemistry. In 2006, a National Science Foundation Blue Ribbon Panel even envisioned simulation as the third pillar of scientific work, standing as an equal partner beside theory and experimentation. In the next section, this claim will be evaluated in more detail.

5.2.2 Referential and Methodological Ontology

Philosophically, ontology is understood as the *study of being* or the study of what exists. Epistemology is the *study of how we come to know* or how we define knowledge, represent it, and communicate it with others. These two fields are often accompanied by teleology, which deals with design and purpose or application, and axiology, the study of values, normally in the ethical sense. Ören (2000) uses these views to make a strong point for philosophy and ethics within simulation.

Most simulationists, however, think of computer tools such as Protégé when they discuss ontology. Among the more known approaches are the Discrete-event M&S Ontology (DeMO) (Silver et al. 2011) and the Component-Oriented Simulation and Modeling Ontology (COSMO) (Teo and Szabo 2008). These ontologies describe how we model which methods we use to describe simulations, federations, and the like. Hofmann et al. (2011) refer to this type of ontology as methodological ontologies that address the question: *How shall we model*?

Gruber (1995) understands ontology from the information systems perspective as a formal specification of a conceptualization in communicable and reusable form within a community of interest. This conceptualization is a collection of concepts, their characteristic attributes, their relations, activities, and behavior as well as governing processes shared within the community of interest and captured in unambiguous form. Hofmann et al. (2011) refer to this type of ontology as referential ontologies that address the question: *What shall we model*!

Referential and methodological ontologies are two dimensions equally important to understanding M&S. Common referential ontologies such as the National Information Exchange Model (NIEM) establish a common set of concepts that can be used to exchange information between systems of the same community of interest. With the help of these referential ontologies, two systems can detect if they have an overlapping information space within the application domain. Only if two systems have some referential concepts in common can they then exchange information. However, this is only part of a solution. The methods and paradigms used in the participating systems must be mappable to each other as well. This can be evaluated by looking at the methodological ontologies. Referential and methodological ontologies are captured in formal languages. As discussed by Tolk et al. (2013), the mathematical branch of the Model Theory can therefore be applied to identify if two systems talk about the same concepts (referential) using equivalent means to do so (methodological). Both dimensions are equally important.

Coupling models to compose their functionality to provide the user with a broader set of information is a good practice. However, the simulationist must be aware of the dangers as well. Winsberg (2010, pp. 72–92) uses nanosciences as an illustration. In his example, scientists are interested in how cracks evolve and move through materials in order to predict, e.g., the stability of a bridge or a building. To address this problem, three different levels of resolution are necessary. In order to understand how cracks begin, sets of atoms governed by *quantum mechanics* are modeled in small regions. These regions are embedded into medium-scale regions that are governed by *molecular dynamics*. Finally, most of the material is neither cracking nor close to a developing crack and can be modeled using continuum mechanics based on *linear-elastic theory*. The challenge is that these three theories—quantum mechanics, molecular dynamics, and linear-elastic theory—cannot all be valid at the same time; each exists with excluding principles. Nevertheless, a simulationist can easily establish a heuristic to create handshakes between these simulation systems. But what is the value of such a composition? According to Winsberg (2010), there are still many gray zones when it comes to applying simulation in support of science. Actually, simulation may create in itself a new field of scientific studies addressing the simulatability of problems.

In summary, epistemology of M&S applications must address methodological constraints of the M&S domain as well as referential constraints of the application domain. The next section will look into the use of models in more detail.

5.3 How Do We Use Models and Simulations?

> Simulation is like a gem: it is multifaceted!. (Ören 1984)

How do simulationists see their models being used? Tuncer Ören is one of the leading experts to answer this question. The seminal paper by Ören and Zeigler (1979) on "Concepts for Advanced Simulation Methodologies" laid the foundation of a simulation taxonomy that is still used and referenced today worldwide. The 1989 proceedings edited by Elzas, Zeigler, and Tuncer Ören on "Modelling and Simulation Methodology: Knowledge Systems Paradigms" comprises ideas like model bases usable as knowledge repositories that are still on the research agenda of scholars today.

Simulations have been successfully applied in hundreds of application areas for a variety of purposes. Despite this great variety of areas and purposes, general trends can be observed that allow for categorization of the use of simulations. The categories used here are motivated by the seminal book chapter by Tuncer Ören in 2009 for a textbook intended to be used for the first M&S engineering curriculum: using models to communicate and preserve knowledge and use this to teach and train and to gain new insights and ultimately become a knowledge repository. As Ören (2009a) points out, there are two aspects when simulation is applied: *experience* and *experimentation.*

Experience is something we know now and have gained and want to provide to others. We gain experience by being exposed to or involved in something. Simulation can create the necessary emergent environments to expose students to and involve them in things or events that are not as dangerous or costly as the real experience or event they simulate. We can use such simulations for teaching, training, or even entertaining, but in any case, we use M&S to provide experience.

We conduct experimentation to gain insight. We want to understand some phenomenon better, or we want to find a solution of a given problem, or we simply want to conduct some what-if analysis to better understand our alternatives.

Finally, Tuncer Ören was among the first to recommend using M&S to not only capture knowledge but also to generate new knowledge. Knowledge repositories use M&S to store knowledge, process knowledge, and actually derive new knowledge as cybernetic tools.

5.3.1 Using M&S to Provide Experience

One of the most mature application domains of M&S is defense (Tolk 2012). Combat modeling and distributed simulation was supported by the military–industrial complex with significant funding allowing for lots of research and innovation. In addition, soldiers have used simulation for centuries. Soldiers used wooden weapons to gain experience in fighting without having to fear being seriously wounded. Officers used wargames, like the *Kriegsspiel* developed by Baron von Reisswitz, the War Counselor in Prussia in 1811, which was a tabletop covered in model terrain representing a miniature version of the battlefield in which wooden blocks showed the placement of units, rules, and charts describing actions and outcomes (Loper and Turnitsa 2012). Using simulation to make this experience more realistic was a logical step. Weapon simulators allowed for individual training and reduced risks as well as the cost of life ammunition significantly. Tank crews were trained in tank simulators so well that in the Operation Desert Storm in February 1991, the U.S. Army defeated the Iraqi Republican Guards although they were outnumbered and outgunned. The reason for this success in the *Battle of 73 Easting* was that the American troops trained intensively before the engagement using distributed simulators in a virtual recreation of the expected battlefield (Banks 2009). Using simulation for training has become so predominant in the armed forces that other domains are often not even perceived to be important.

Another M&S customer, also interested in a realistic simulation of the environment and the effects of actions taken by the user in a safe manner, is the entertainment industry! Theme parks like Universal Studios and Disney World use simulations to provide their guests a range of experiences that they cannot get anywhere else.

Gaming utilizes simulation more and more as well. The worldwide video game marketplace, which includes video game console hardware and software, online, mobile, and PC games, will reach $93 billion in 2013, up from $79 billion in 2012, according to online publications of Gartner, Inc. on their website.

Another domain of increasing interest is health care (Levine et al. 2013). Standardized patients, mannequin-based simulators, haptic simulators, and virtual environments are integrated into the education of medical professionals to provide them with experiences without endangering the student or the patient. Okuda et al. (2009) report on two studies that have shown the direct improvement of medical outcomes from the use of simulation: Residents trained on laparoscopic surgery simulations showed improvement in procedural performance in the operating room, and residents trained on simulators were more likely to follow advanced cardiac life support protocols when treating cardiac-arrest patients. They also show benefits in many other areas; however, additional systematic research on the utility of simulation is still needed.

In summary, simulation has been successfully applied to train and educate students in various domains. They gained motoric skills, procedural skills, decision-making skills, and even high-level operation skills. Systematic studies on the effectiveness and efficiency are being published.

Connecting this use of M&S to our earlier sections, we observe that for all these examples, we have an established and accepted theory that we use to build our models on and that we can use to validate our simulations against. We know how bullets fly, so we can build a valid gun simulator. Knowing how certain medicine influences blood flow and heart rate allows us to build a mannequin that exposes this behavior to train nurses, doctors, and other medical professionals. As a rule, we observe that whenever we know a solution, we can use this solution to provide the desired experiences to the user in the form of the simulation.

5.3.2 Using M&S to Gain Insight

In the previous section, simulations and simulators were shown to represent systems or components that are well understood for the purpose to train students to act and behave in the best way. The approach described in this section is different. Here, we want to gain insight into something, looking for solutions or simply broader understanding. In this case, we use simulations to analyze problems, to evaluate possible engineered solutions, or to find good control parameters. In all these cases, our solution space is not fully defined, and we still need to acquire an understanding of the phenomenon of interest. The premise is that we do not have a solution; instead, we use simulation to find one. We may have requirements constraining the solution space, but we have no specification for a point solution. In other words, we use M&S for theory building, as foretold by Tuncer Ören in his comprehensive and integrative view (2009b).

In his dissertation, Padilla (2010) used an agent-based simulation approach to represent three components of understanding—knowledge, worldview, and problem. This construct was used to successfully explain existing theories of understanding, as published in the literature. Both types of understanding could be explained and modeled by this construct. When Padilla defined the guiding rules on how knowledge, worldview, and problem are interconnected and ran several simulations to find out if the two theories could be reconstructed by these simple construction rules, the simulation actually produced three types of understanding; two types that are known from literature and a third type that so far was not described but made perfect sense when evaluated by subject matter experts. In other words, by using existing theory components, the simulation not only supported the guiding theory but also generated new constructs that are valid under the given constraints and that were accepted as new theoretical contributions.

Simulation for experimentation to gain insight is also a powerful tool. As Tuncer Ören (2005) described it, "One of the superiorities of simulation over real-system experimentation is that simulation is possible in all cases when experimentation on real system cannot be performed. Furthermore, in simulation experimental conditions can include cases that cannot and should not be performed on real systems." We can use powerful heuristic tools, like genetic algorithms, to calibrate our simulation parameters to create new trajectories, behaviors, or even new entities and fit the simulation to empirical data. Theory building using simulation is possible, and we are starting to use this capability.

5.3.3 Using M&S as a Knowledge Repository

The third usage category for M&S is one that has not been referred to in textbooks, although it is connected with using M&S for training and teaching—using M&S as a knowledge repository. From the first section of this chapter, we already learned that Rosen (1998) understands modeling as the essence of science and the habitat of all epistemology. In this chapter, we introduce the viewpoint that simulations can be understood as executable theories (or hypothesis if they have not been verified with empirical observations).

In their introduction to their book *Information and the Nature of Reality*, Davies and Gregersen (2010) describe a long tradition of using the pinnacle of current technology as a metaphor for the universe on the search for universal truth. They observe,

> In ancient Greece, surveying equipment and musical instruments were the technological wonders of the age, and the Greeks regarded the cosmos as a manifestation of geometric relationships and musical harmony. In the 17th century, clockwork was the most impressive technology, and Newton described a deterministic clockwork universe, with time as an infinitely precise parameter that gauged all cosmic change. In the 19th century the steam engine replaced clockwork as the technological icon of the age and, sure enough, Clausius, von Helmholtz, Boltzman, and Maxwell described the universe as a gigantic, entropy-generating heat engine, sliding inexorably to a cosmic heat death. Today, the quantum computer serves the corresponding role. Each metaphor has brought its own valuable insights; those deriving from the quantum computation model of the universe are only just being explored. (Davies and Gregersen 2010, p. 3–4).

Using M&S, we now can define a concept—whether it has a real-world reference or not—by its axioms and rules as an executable simulation and "bring it to life" using animation and visualization and use emergent virtual environments to make the user part of this creation. This is a powerful approach to understand things that are, that could be, or that could not be. Quantum computation will be a powerful technology that will support the next generation of even more powerful M&S applications, but it will be the application that will drive our imagination and understanding of the universe. These M&S applications will become the pinnacle of technology of our epoch. They will become the knowledge repository that cannot just be understood by study; it can be experienced and actively integrate the user to discover new theories and close existing gaps in an interactive interplay.

At the end of this section, we want to come back to the notion of simulation as the third pillar in science. In this section, we have shown that simulation can support theory building as well as theory testing, hence it is a useful method in support of theory building and experimentation, but to claim it to be a new pillar of science may be an exaggeration. Nevertheless, simulation changes and will continue to change our view on science. It is therefore even more important to clearly understand what simulation can do and also what simulation cannot do. As simulation systems are computer programs, they are ruled by computability and computational complexity constraints. The last section will "poor some water into the wine" by looking at some limits and constraints we have to be aware of.

5.4 Limits and Constraints

> One of the best things to come out of the home computer revolution could be the general and widespread understanding of how severely limited logic really is. (Frank Herbert, Krieger 2002, p. 102)

The famous artificial intelligence (AI) researcher Huber L. Dreyfus is well known for his two books on the limits and constraints of computers in his work: *What Computers Can't Do: The Limits of Artificial Intelligence* (1972) and *What Computers Still Can't Do: A Critique of Artificial Reason* (1992). While many of his colleagues were overly enthusiastic, Dreyfus pointed to known limits that are founded in the nature of computers. While some researchers openly dreamed about computers with real intelligence or even with a soul, Dreyfus pointed to the works of Turing, Church, Gödel, and other pioneers of computer science that were often overlooked by his colleagues. As a result of too enthusiastic and unbounded promises, AI could not keep up with the expectations. Many valuable methods and heuristics survived and are successfully applied today, but overall, AI became a disappointment. A nice summary of important AI debates has been compiled by Graubard (1988).

Did we make—or are we making—a similar mistake with M&S? Are we setting the expectations too high? Do we make promises that we cannot fulfill? Are there significant limits and constraints that we have to know when dealing with M&S? To answer these questions, we will look into computability and decidability, computational complexity, and algorithmic information theory from a simulationist's point of view: what do they imply for M&S? While the first sections of this chapter focused on the role of models and modeling, our focus in the following section will be on simulation, in particular computer simulation as the executable expression or version of models.

5.4.1 Computability and Decidability

A simple interpretation of the term *computability* is the ability of a function to be executed on a computer. Computer science established several models of computations that are used to give a more precise definition of these terms; the best known are the Lambda calculus, the Turing machine, and recursive function theory (Davis et al. 1994).

From a practical sense, it boils down to functions that have a limited and discrete range and domain. If range and domain are not limited, they cannot be handled in the finite space of the computer. If they are not discrete, they cannot be mapped to the digital space. We often treat noncomputable functions as computable by simply using an approximation; we artificially limit range and domain to fit (or we receive an overflow error), and we discretize the arguments. For many practical purposes, this is sufficient, but we always have in mind that we are just working with approximations. Deterministic chaotic functions, e.g., will follow completely different trajectories if the initial starting points are different. Even if the starting points are arbitrarily close to each other, after several iterations the results will differ significantly. No matter how dense our digitalization mesh is, there are always spots that are not part of it, which means we cannot realistically capture chaotic functions with computable functions. This is just one example of many. As computer simulation is limited to using computable functions, all of these limitations apply.

Another aspect is that not every problem can be decided by algorithms that can be executed on computers. In 1931, Kurt Gödel's incompleteness theorem shocked the academic world. Up until his proof, mathematical logic was considered to be the key for unambiguous, consistent, and complete descriptions of knowledge. Gödel showed that a logical system that is powerful enough to allow for mathematical reasoning will necessarily comprise axioms that *are* true but that *cannot be proven to be true* within the system. Another interpretation is that complex and powerful logical systems can be either complete or consistent but not both. If the system is complete, it comprises statements that make the system fail. If we exclude these statements to reach consistency, the system is no longer complete.

Turing applied a similar idea to show that problems do exist that cannot be decided by a computer. He used the *halting problem* to demonstrate this: If we have a computer program and the input data, can we write a general program that decides if the program will halt with this input, or will it go into an infinite loop? Turing argued as follows: He assumed a program that solves the halting problem would exist. If so, it would return the Boolean value "true" if the program holds with the input data and "false" if this is not the case. He then applied this program to itself, meaning he used the data describing the program as input data to the program. Next, he constructed a new program by extending this halting decision program working on its own data and going into a loop if the returned value is "true" and halting the program if the returned value is "false." If we now apply the halting decision program to this new program, we run into a paradox. If we apply this new program to itself and it goes into an infinite loop, it does so because the halting decision program returns "true," which means that it does not go into an infinite loop. Similarly, if it stops and returns "true," it does this because the halting decision problem decided that it would loop. As the only assumption Turing made was that he can write the halting decision problem and this assumption leads to the paradox, the assumption must be wrong. Ergo, we cannot write a program that generally decides the halting problem! The halting problem is undecidable, which really means that it cannot be decided. There is no existence of an algorithm that can help, so the best we can do is to look for a good heuristic. This is a matter of mathematics, not of computing power. No matter how fast and powerful our computers will become, we will never be able to use them to decide the halting problem. Computers can't do this!

There are many more examples of undecidable problems, such as questions like "Will the system terminate?", "Are two modeled actions order independent or do I have to orchestrate them?", "Is the specification complete?", "Is the specification minimal?", or "Are two specifications functionally equivalent, in other words, do they deliver the same functionality?" As simulation systems are computer programs, they also cannot decide undecidable problems. If we know from computer science that a problem is undecidable, we do not have to develop a simulation to solve it, as we cannot succeed. However, simulation systems may help with coming up with better heuristics or engineering approaches that do not solve a problem generally but provide good enough results for all practical purposes.

The simulationist must therefore know his theory as well as possess an engineering mindset to understand what he/she can make work and what can just be approximated. Computability and decidability must be well understood.

5.4.2 Computational Complexity

The next hurdle for a simulationist is computational complexity. Only because a problem is computable and decidable, it does not mean it can be solved in a practically reasonable time. Generally, computational complexity studies the use of resources, namely, computer memory and computing time. Computational complexity is interested in the order of magnitude of functions to describe their general behavior. This order is often defined using the letter "O," and the resulting notation is known as the *big O* notation. The seminal paper of Stephen Cook (1971) introduced complexity classes of *P*, *NP*, *NP-complete*, and *NP-hard*. He distinguished between whether a problem is verifiable in polynomial time, which means I have an answer and want to know if it is a correct one, and whether a problem can be solved in polynomial time. From a practical standpoint, problems that are known to be NP hard or NP complete are not generally solvable with computers. Interestingly, the fundamental *P* versus *NP* problem is still a major unsolved problem in computer science.

For simulationists, the question of reuse is very interesting. Intuitively, component-oriented design offers a reduction in the complexity of system construction by enabling the designer to reuse appropriate components without having to reinvent them. However, Page and Opper (1999) showed that this intuition is wrong. Although determining if a collection of components satisfies a set of requirements becomes feasible under certain assumptions, we still have to solve a potentially computationally intensive problem. Selecting the right component to fit into a bigger solution is a nontrivial NP class task that cannot be generally solved or left to technology. Again, the simulationist has to know his theory or he is in danger of trying to solve the insolvable with the computer simulation.

5.4.3 Chaotic Dynamical Systems

Chaos theory describes the behavior of chaotic dynamical systems. They received a great deal of attention two decades ago, including from the simulation community

(Dewar et al. 1991) as the behavior of chaotic systems cannot be predicted over longer periods, which generally challenges the analytic usability of computer-based analytics including simulation.

The predominantly used definition of chaos was suggested by Devaney (1989), who proposed that for a metric space X, a continuous map $f: X \rightarrow X$ is said to be chaotic in X when

1. f is topologically transitive
2. The periodic points of f are dense in X
3. f has sensitive dependence on initial conditions

In laymen's terms, chaotic systems are bounded and show some degree of order, but they are not predictable over time as even points that are close to each other will follow different trajectories over time. The system can also not be decomposed due to the topological transitivity: whenever we look only at components or subspaces, we are losing interactions that characterize the system.

Whenever a system is truly chaotic, the initial points can be arbitrarily close to each other, and still they will be all over the map after several iterations. We cannot apply our intuitive idea of interpolation: that two points close at the beginning will continue to stay close over time.

The implications for digital computer simulations are significant: no matter how precise our computer will be, it still is only able to represent a discrete subset of possible initial conditions. No matter how precisely we measure our initial value, there is always an element of error and imprecision connected with this. No matter how hard we try, we cannot reliably predict the long-term behavior of any truly chaotic dynamical system! This does not exclude using simulation to gain better understanding of the behavior of chaotic systems, but it limits the use of simulation to predict the outcome of operations in the long term. Palmore (1996) gives several examples from the defense domain, but this problem of unpredictability exists in all application domains as all of them are affected by these limits and constraints.

5.4.4 Algorithmic Information Theory and Model Theory

The last part of this section on limits and constraints of simulations goes a little bit deeper into the mathematical foundations and their implications for the epistemology of simulation.

Algorithmic information theory (Chaitin 1987) is a branch of mathematics that proposes that a theorem deduced from an axiomatic system cannot contain more information than the axioms themselves. To get to this insight, the classical theory of information (Shannon 1948) had to be extended from pure information to algorithms that can produce this information. While Shannon focused on encoding of symbols and syntactical expressions, Chaitin extended these ideas to complete programs. However, the programs are not seen as syntactical expressions but as formal languages that produce sentences. If two theorems produce the same sentences, they are equivalent. A program that produces a series of sentences bears therefore the same information as an enumeration of all the sentences it produces. Algorithmic information theory seeks the shortest and most compact form to produce information as this is the most efficient form to communicate capabilities.

For the epistemologically interested simulationist, these findings mean that all a simulation system can produce is already in the simulation system and the input data. Computer programs can only transform, not create. As such, we cannot create new knowledge with simulation. However, we may be able to discover new insights that were not as obvious before the transformation by the simulation. We may know the details, but simulation systems may help us to experience the big picture that emerges from this detailed knowledge we already have. Simulation cannot help us to understand what we do not know, but it can help us to discover what we already know.

Another branch of mathematics of particular interest to simulationists is *model theory*. Model theory is a subset of mathematics that focuses on the study of formal languages and their interpretations and is recognized as its own branch since around 1950. It applies logic to the evaluation of truth represented using mathematical structures. In other words, the way we model truth using mathematical structures can lead to different interpretations: what is evaluated to be true in one representation can be false in another one. Ultimately, model theory deals with answering questions regarding consistency of formal language interpretations, i.e., result in the same truth statements for the same logical statements. Simulation systems are written in programming languages, which are a subset of formal languages. The findings of model theory have direct implications for simulation systems in general. Two results of model theory that are directly applicable in support of interoperability challenges are Robinson's joint consistency theorem and Łoś's theorem.

Robinson's joint consistency theorem simply states that the union of two formal language expression sets is satisfiable under a common interpretation of truth if and only if their intersections are consistent. In other words, there is only one interpretation of truth valid in both models. It is possible that two formal languages are using different symbols and structures and the resulting sentences are not comparable. To cope with these cases, Łoś's theorem generalizes the idea of consistency by expanding the expression sets through the Cartesian product and defines filters that allow the comparison in a common equivalent representation. It allows us to find out if a mediation function exists, and if so, Robinson's joint consistency theorem can be applied to the results. This small subset of model theory insights shows that the mathematical foundations of interoperability and composability are already laid and have far-reaching implications, such as the introduction of a "common information exchange model" can never solve the underlying systemic problem of interoperability: the need for a consistent representation of truth in all participating components.

These examples of limits and constraints have been chosen to show the necessity for simulationists to understand their mathematical roots. M&S is a very powerful discipline, but without a strong mathematical foundation, we are in danger to oversell the potential and give bad advice, which would be unethical. In the long term, it is also bad practice from a business perspective to promote exaggerated M&S capabilities, like promising to solve undecidable problems, as we cannot deliver such promised solutions. In summary, the mathematical foundations can become the anchor point for the diverse body of knowledge of M&S that Tuncer Ören so eagerly contributes to.

5.5 The Future of Modeling and Simulation

> Prediction is very difficult, especially if it's about the future. (attributed to Niels Bohr)

Despite all these constraints, the future of M&S has always been bright in Tuncer Ören's mindset (Ören 2002). In current days, simulation is dealt with more and more in the media as well. The Nobel prize mentioned earlier is one example, but another story caught also many people's attention. Satell (2013) contributed an article to the technical section of the Forbes magazine on "Why the Future of Innovation is Simulation." In this article, he describes the following anecdote:

> In business life, Mitt Romney was known for his acumen, strong work ethic and keen eye for talent. He carried over these practices to his political career and his campaign team was similarly bright and indefatigable. They analyzed past trends, developed a theory of the case and executed their strategy efficiently. They had only one chance to get it right.
>
> His opponent, the incumbent President Barack Obama had a different approach. He created an entire division of young, unkempt, over-caffeinated data junkies with little experience in business or politics. They had no set theory of the case, but instead ran 62,000 simulations per night and continuously updated their approach.
>
> The result is now clear to just about everyone on the planet. The smartest guys in the room were no match for terabytes of data and smart algorithms. There is no one "theory of the case" anymore, but thousands of them, being run constantly. The point isn't to be right, but to become less wrong over time.

In general, Satell observes that the idea of "one correct common picture" or "one common theory" is no longer the real objective. It is much more about finding a set of less wrong models than it is about finding the one right model. Weather simulation is a good example: instead of trying to find one good forecast, new approaches are looking at a number of forecast models and creating a set of most likely predictions. Instead of trying to find the optimal, we are starting to look more at not being completely off.

This trend was observed in my own philosophical reflections as well (Tolk 2013b), and in Tolk et al. (2013), we pointed to the ideas of Tuncer Ören and colleagues on multimodeling as exactly the best approach (Yilmaz et al. 2007) when it is rooted in the mathematical foundations of model theory. Weather modeling is an application of these ideas well known to everyone who followed the forecasting of hurricanes: several models are executed in parallel and visualized on a common map showing the possible paths of the hurricane as predicted by the various models. Based on these possible paths, homeowners can make their decision on how to prepare for the impact and if they want to leave their house or not. Nothing speaks against using this method for other prediction analyses as well.

If we now include the idea of Big Data and marry it with multimodeling, the future of M&S remains indeed bright, and many ideas of Tuncer Ören will continue to make it from his visionary concept to the tool sets of engineers, scientists, scholars, and decision-makers. While Big Data exposes correlations we were not aware of, running simulation on these data sets will add the component of dynamic developments and possible trends that cannot be derived from snapshots alone.

5.6 Concluding Remarks

> "So this, then, was the kernel of the brute!" Faust Part 1: Studierzimmer (1808)
> – Johann Wolfgang von Goethe

Within this chapter, we looked at the question of how to learn something from using M&S. We observed a short history of science as a series of models that were used to capture knowledge. In doing so, the need for a clear conceptualization framework becomes clear as well as how simulationists can use ontological and epistemological means to capture this. As we use models, we use them to provide experience or to gain insight and ultimately to capture our knowledge. As we use computer simulations to accomplish these goals, we have to be aware of the limits and constraints.

Philosophy of simulation answers *WHAT* we are doing. It completes simulation as a science and engineering discipline answering *HOW* we do something and the code of ethics for simulationists answering *WHY* we are doing our task. Each simulationist must be aware of these three pillars of our profession. Tuncer Ören contributed as a titan in this field to all three interrogatives with questions, ideas, and answers (Ören et al. 2002; Ören and Yilmaz 2013). His students and protégés are carrying this flame on in order to mature M&S as a discipline carried by engineering, ethics, and philosophy.

Acknowledgments I have to thank several colleagues and friends for their support over the years that helped to shape the ideas described in this chapter, in particular Dr. Saikou Diallo, Dr. Jose Padilla, Dr. Robert King, Dr. Mamadou Seck, and Dr. Charles Turnitsa, who accompanied my journeys into the philosophy of science. I also thank Dr. Randall Garrett, Mr. Jay Gendron, and Mrs. Emily Edwards for their help to summarize these ideas in a coherent and comprehensible form.

Finally, my utmost thanks go to Dr. Tuncer Ören himself. He was my mentor and tutor from the day we met and, more important than this, a friend who made me aware of the imperative of keeping engineering, philosophy, and ethics in balance. I truly hope that one day, my students and colleagues will look with similar admiration at my legacy as I look at his still ongoing work today: He is truly a titan in our domain!

References

Alexander HG (1998) The Leibniz-Clarke correspondence: together with extracts from Newton\'s Principia and Optics. Manchester University Press, Manchester

Banks CM (2009) What is modeling and simulation? In: Sokolowski JA, Banks CM (eds) Principles of modeling and simulation: a multidisciplinary approach. Wiley, Hoboken, pp 3–24

Box GEP, Draper NR (1987) Empirical model building and response surfaces. Wiley, New York

Chaitin GJ (1987) Algorithmic information theory. Cambridge University Press, Cambridge

Cook SA (1971) The complexity of theorem-proving procedures. In: Proceedings of the third annual ACM symposium on theory of computing. ACM Press, New York, pp 151–158

Davies P, Gregersen NH (2010) Information and the nature of reality: from physics to metaphysics. Cambridge University Press, Cambridge

Davis M, Sigal R, Wevuker EJ (1994) Computability, complexity, and languages: fundamentals of theoretical computer science. Academic, Boston

Devaney RL (1989) An introduction to chaotic dynamical systems, 2nd edn. Addison-Wesley Publishing Company, Reading

Dewar JA, Gillogly J, Juncosa ML (1991) Non-monotonicity, chaos, and combat models. Rand Corporation, Santa Barbara, R-3995-RC

Dreyfus HL (1972) What computers can't do: the limits of artificial intelligence. Harper & Row, New York

Dreyfus HL (1992) What computers still can't do: a critique of artificial reason. MIT Press, Boston

Elzas MS, Zeigler BP, Ören TI (1989) Modelling and simulation methodology: knowledge systems paradigms. Elsevier Science Inc., Amsterdam

Frank U, Troitzsch KG (2005) Epistemological perspectives on simulation. J Artif Soc Soc Simul 8(4). http://jasss.soc.surrey.uk/8/4/7.html

Goldman SL (2006) Science wars: what scientists know and how they know it. Lehigh University, The Teaching Company, Chantilly

Graubard SR (1988) The artificial intelligence debate. MIT Press, Cambridge, MA

Gruber TR (1995) Toward principles for the design of ontologies used for knowledge sharing. Int J Hum Comput Stud 43(4–5):907–928

Grune-Yanoff T, Weirich P (2010) The philosophy and epistemology of simulation: a review. Simul Gaming 41(1):20–50

Harman PM, Harman PM (1982) Energy, force and matter: the conceptual development of nineteenth-century physics. Cambridge University Press, Cambridge

Hofmann M, Palii J, Mihelcic G (2011) Epistemic and normative aspects of ontologies in modelling and simulation. J Simul 5:135–146

Kant I, Jaki SL (1981) Universal natural history and theory of the heavens. Scottish Academic Press, Edinburgh

Krieger RA (2002) Civilization's quotations: life's ideal. Algora Publishing, Chicago

Kuhn TS (1962) The structure of scientific revolutions. University of Chicago Press, Chicago

Levine AI, DeMaria S Jr, Schartz AD, Sim AJ (2013) The comprehensive textbook of healthcare simulation. Springer, New York

Loper ML, Turnitsa CD (2012) History of combat modeling and distributed simulation. In: Tolk A (ed) Engineering principles of combat modeling and distributed simulation. Wiley, Hoboken, pp 331–355

National Science Foundation (NSF) Blue Ribbon Panel (2006) Report on simulation-based engineering science: revolutionizing engineering science through simulation. NSF Press, Washington, DC

Newton I, Motte A, Cajori F, Thompson SP (1955) Mathematical principles of natural philosophy, vol 34. Encyclopaedia Britannica, Chicago

Okuda Y, Bryson EA, DeMaria S Jr, Jacobson L, Quinones J, Shen B, Levine AI (2009) The utility of simulation in medical education: what is the evidence? Mt Sinai J Med 76(4):330–343

Ören TI (1977) Simulation – as it has been, is and should be. Simulation 29(5):182–183

Ören TI (1984) Forward to the book: multifaceted modelling and discrete event simulation, by B.P. Zeigler. Academic, London
Ören TI (2000) Responsibility, ethics, and simulation. Trans Soc Comput Simul 17(4):165–170
Ören TI (2002) Future of modelling and simulation: some development areas. In: Proceedings of the 2002 summer computer simulation conference, San Diego, pp 3–8
Ören TI (2005) Toward the body of knowledge of modeling and simulation (M&SBOK). In: Proceedings of the interservice/industry training, simulation, and education conference (I/ITSEC). Paper 2025, Orlando
Ören TI (2009a) Uses of simulation. In: Sokolowski JA, Banks CM (eds) Principles of modeling and simulation: a multidisciplinary approach. Wiley, Hoboken, pp 153–179
Ören TI (2009b) Modeling and simulation: a comprehensive and integrative view. In: Yilmaz L, Ören TI (eds) Agent-directed simulation and systems engineering. Wiley-VCH, Weilheim, pp 3–136
Ören TI, Yilmaz L (2013) "Philosophical aspects of modeling and simulation". Ontology, epistemology, and teleology for modeling and simulation. Springer, Berlin/Heidelberg, pp 157–172
Ören TI, Zeigler BP (1979) Concepts for advanced simulation methodologies. Simulation 32 (3):69–82
Ören TI, Elzas MS, Smit I, Birta LG (2002) Code of professional ethics for simulationists. In: Proceedings of the summer computer simulation conference, Society for Modeling and Simulation, San Diego, pp 434–435
Padilla JJ (2010) Towards a theory of understanding within problem situations. Doctoral thesis at Old Dominion University, Frank Batten College of Engineering and Technology, Norfolk
Page EH, Opper JM (1999) Observations on the complexity of composable simulation. In: Farrington PA, Nembhard HB, Sturrock DT, Evans GW (eds) Proceedings of the winter simulation conference, vol 1. ACM Press, New York, pp 553–560
Palmore JI (1996) Dynamical instability in combat models: computer arithmetic and mathematical models of attrition and reinforcement. Mil Oper Res 2(1):45–52
Popper KR (1935) Logik der Forschung [The Logic of Scientific Discovery]. Springer, Berlin
Rosen R (1998) Essays on life itself. Columbia University Press, New York
Satell G (2013) Why the future of innovation is simulation. Forbes Magazine, 15 July
Shannon CE (1948) A mathematical theory of communication. Bell Syst Tech J 27:623–656
Silver GA, Miller JA, Hybinette M, Baramidze G, York WS (2011) DeMO: an ontology for discrete-event modeling and simulation. Simulation 87:747–773
Squazzoni F (2009) Epistemological aspects of computer simulation in the social sciences. Springer, Berlin/Heidelberg
Teo YM, Szabo C (2008) CODES: an integrated approach to composable modeling and simulation. In: Proceedings of the 41st annual simulation symposium. IEEE CS Press, pp 103–110
Tolk A (2012) Engineering principles of combat modeling and distributed simulation. Wiley, Hoboken
Tolk A (2013a) Interoperability, composability, and their implications for distributed simulation: towards mathematical foundations of simulation interoperability. In: Proceedings of the ACM/IEEE DS-RT 2013 conference, Delft, pp 3–9
Tolk A (2013b) "Truth, trust, and turing – implications for modeling and simulation". Ontology, epistemology, and teleology for modeling and simulation. Springer, Berlin/Heidelberg, pp 1–26
Tolk A, Diallo SY, Padilla JJ, Herencia-Zapana H (2013) Reference modelling in support of M&S—foundations and applications. J Simul 7(2):69–82
von Goethe JW (1808) Faust. Eine Tragödie. In: Goethe's Werke, vol 8. JG Cotta Books, Tübingen
Winsberg E (2010) Science in the age of computer simulation. The University of Chicago Press, Chicago
Yilmaz L, Ören TI, Lim A, Bowen S (2007) Requirements and design principles for multisimulation with multiresolution, multistage multimodels. In: Proceedings of the winter simulation conference, Washington, DC, pp 823–832

Chapter 6
Managing Hybrid Model Composition Complexity: Human–Environment Simulation Models

Hessam S. Sarjoughian, Gary R. Mayer, Isaac I. Ullah, and C. Michael Barton

6.1 Introduction

It is becoming commonplace to use multiple types of models together for simulating multifaceted systems across many scientific disciplines. Indeed, in recent years, some approaches (referred to as multimodeling or multiformalism modeling) have been developed for representing a complex system as a set of subsystem models. Among these, there has been an interest in developing hybrid methods where structures and behaviors of models are explicitly accounted for. Furthermore, theories and approaches are proposed to define the *interactions* among *heterogeneous* model types. However, modeling a system this way brings about composition complexity that must also be managed. The complexities of hybrid modeling resulting from the interactions of the composed models can be reduced using interaction models, an approach referred to as *polyformalism* modeling (Sarjoughian 2006). Independently developing and utilizing such interaction models provides additional flexibility in system model design, modification, and execution for both the subsystem models and the resultant hybrid system model. This paper discusses the use of the polyformalism model composition approach for

H.S. Sarjoughian (✉)
Department of Computer Science and Engineering, Arizona State University, Tempe, AZ, USA
e-mail: Sarjoughian@asu.edu

G.R. Mayer
Department of Computer Science, Southern Illinois University, Edwardsville, IL, USA

I.I. Ullah
School of Human Evolution and Social Change, Arizona State University, Tempe, AZ, USA

C.M. Barton
Center for Social Dynamics and Complexity, Arizona State University, Tempe, AZ, USA

L. Yilmaz (ed.), *Concepts and Methodologies for Modeling and Simulation*, Simulation Foundations, Methods and Applications,
DOI 10.1007/978-3-319-15096-3_6

researching human–environment dynamics with direct support for managing the complexity which results from subsystem model interactions within this domain.

The dynamics of a system (and therefore its complexity) arises from its individual subsystems and their interactions with one another. A basic concept when modeling systems that are built from subsystems and exhibit compound dynamics is to model each subsystem separately and then combine them to represent system models (see Wymore 1993; Fishwick 1995; Zeigler et al. 2000; Ptolemaeus 2014). However, decomposition of a system model into subsystem models or inversely composing the system model from the subsystem models is difficult (Davis and Anderson 2004). Models can be conceptualized and defined to be connected to one another in many ways. Tuncer Ören has developed a list of coupling types and associated terms (Ören 2014). It details structural input/output relationships with types for coupled models with considerations such as dynamic changes in input/output couplings and model parts that can be defined using the system-theoretic approach (Wymore 1993; Ören and Zeigler 2012).

Engineering researchers have been developing useful and practical concepts, techniques, and tools to support hybrid modeling for embedded devices (e.g., Karsai et al. 2004) and manufacturing and planning enterprises (e.g., Huang et al. 2009). In natural sciences, research in hybrid modeling (e.g., social–ecological systems) is relatively new. For these areas of science as well as others such as system biology (e.g., Kirschner et al. 2014), it is important to manage the complexity of the individual and composed models.

To define a system model that consists of distinct subsystem model types (e.g., differential and difference equation system specification (Zeigler et al. 2000)), it is necessary to devise appropriate model abstractions from the system and its subsystems based on the kinds of questions expected to be formulated and answered by the simulation model. The model descriptions for the subsystems can range from simple to complex. For instance, a subsystem may be represented by a set of discrete-time components having hierarchical relationships. Alternatively, its dynamics may be abstracted to a single difference equation. The choice of modeling formalism for each subsystem model has a direct relationship to the complexity of the system model dynamics. This is because two disparate modeling formalisms can differ significantly in the kinds of structure and behavior that they represent and simulate (e.g., agent (Epstein and Axtell 1996) and cellular automata (Anderson 2006)).

Another main consideration in developing hybrid models is the intrinsic characteristics of the system to be modeled. For example, the structures and behaviors of socioecological systems—as compared, for example, with discrete manufacturing systems—are not as well understood and may be difficult to accurately model at desirable levels of resolution. It can also be argued that hybrid modeling of a socioecological system poses unique challenges given the disparity of its subsystems' dynamics. The subsystem models can be quite expressive, and combining them together results in complex model structure and behavior beyond those that are contained in each of the subsystem models. Contributors to model complexity include the type of data and how it is processed in the individual (subsystem) model components and the kinds of interactions sanctioned among those components. For example, consider modeling a socioecological system with two subsystems: a village with some households that can be described as a rule-based, discrete-event agent model and a landscape providing food for a village that can be described as a discrete-time cellular automata model (Mayer et al. 2006; Mayer 2009; Barton et al 2012; Ullah et al. 2008; Ullah 2012). Such a composite socioecological system model may best answer the research questions if the agent and cellular automata use different spatial resolutions for the landscape. Similarly, since the time-based dynamics of the agent and cellular automata models can be different, their dynamics and interactions must be synchronized (Mayer and Sarjoughian 2008, 2009). The differences in spatial data of the subsystem models should be modeled separately to allow a spatial representation that best suits each model's abstraction. Furthermore, the interactions between agent and cellular automata models should be flexible and configurable to allow for modification in research design and data collection. The ability to formulate the interactions of such subsystem models in a stand-alone model is a central factor in managing the complexity of this hybrid system model.

The complexity of a human–environment model can be explained in terms of how its subsystems are modeled and composed. Since each subsystem model acts as both producer and consumer, there is bidirectional interaction in the hybrid system model. This paper considers the use of a third model, referred to as an interaction model (IM). Such a model explicitly describes the subsystems' interactions and is formalized using polyformalism modeling (Sarjoughian 2006). A structural and behavioral independence between all three models allows a suitable modeling formalism to be chosen for each model. Figure 6.1 shows a hybrid model concept of a discrete-event, rule-based agent model composed with a discrete-time, cellular automaton landscape model (Mayer et al. 2006; Mayer 2009; Ullah 2012). The arrows represent communication between the models. Note that there is no direct communication between the two composed models. The third model, which composes the two subsystem models and facilitates their communication, is called an Interaction model (IM). An important benefit is the ability to partition and map domain knowledge to the models (e.g., use an agent model to represent the humans, a cellular automata (CA) model to represent the environment, and an agent–CA interaction model to represent synchronized control and data exchanges between the agent and cellular automata models). The agent–CA interaction model may use a known formalism that is as specific or general as needed (e.g., discrete-event) to properly represent the interaction between the composed subsystem models. By maintaining separation of the models, researchers and practitioners are better able to study the subsystem models separately as well as examine their interactions methodically. The use of an interaction model in this fashion is referred to as polyformalism composition (Sarjoughian 2006).

6.2 Hybrid Human–Environment Modeling Categories

Various approaches (including systems and agent-based modeling) have grown from within scientific communities for building human–environment simulation models. A variety of established and popular toolkits have been examined during

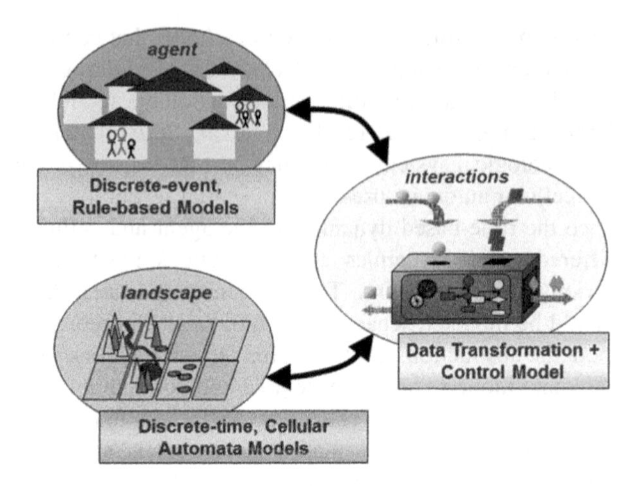

Fig. 6.1 An Human-Interaction-Landscape model concept

the course of this research in terms of their capabilities for hybrid agent and landscape process models (Mayer 2009; Ullah 2012). The most applicable can be divided into agent-centric, landscape-centric, and theory-centric categories. The agent-centric and landscape-centric development environments are primarily built by casting domain-specific knowledge and modeling concepts to programming languages, whereas theory-centric environments emphasize the use of general-purpose modeling formalisms with proven execution protocols. Next, some major approaches that belong to these categories are discussed from the perspective of hybrid human–environment modeling.

6.2.1 Agent-Centric

Various toolkits specifically for agent modeling have been under development since the 1990s and, in recent years, adopted by researchers and practitioners in social and environmental sciences (Ören and Yilmaz 2009). For agent–landscape domains, Mason (Luke et al. 2005), NetLogo (Tisue and Wilensky 2004), Repast (North et al. 2005), and Swarm (Minar et al. 1996) are generally used. Each offers a specialized development environment with foci on certain types of agents that are combined in specialized ways with spatial, particularly 2D/3D, models. Generally, agents are implemented as basic software objects where each agent generates output after receiving input. These software-centric toolkits are developed using domain concepts. They also vary in their features such as levels of support for data visualization, experimentation configuration, and graphical user interfaces. These toolkits essentially require using templates for developing models which may be modified for intended application domains as in social sciences.

The agent interactions are managed through interfaces that are defined in terms of software concepts and techniques (Parker et al. 2003). Toolkits are developed based on object-oriented programming languages such as Java. While agent modeling toolkits can be interfaced with external landscape model tools such as GIS (see next section), the resulting integrated tools suffer from two possible problems. The first is that these tools may become more difficult for modelers to use unless they have expertise in software development and in particular reverse software engineering due to lack of access or availability to software specifications (e.g., see Lytinen and Railsback 2012). That is, expertise in software design, implementation, and testing in programming and software engineering is necessary when agent and landscape models need to be changed and especially when relationships between agent and landscape models are to be modified or created.

The second challenge is in determining correctness of the system model. This is because software-centric toolkits do not have a formal modeling and simulation underpinning. Instead, they are grounded in a combination of domain-level knowledge, agent modeling concepts, and software engineering techniques. A consequence is the difficulty in having confidence in correctness of the interactions between agent and nonagent model components which in turn weakens simulation validation. It is important to note that some studies (e.g., Lytinen and Railsback 2012) have compared computational efficiency of some agent-based modeling toolkits suggesting they may be used interchangeably (i.e., simulation results from these tools are indistinguishable). Another study, however, shows that such agent-centric toolkits may produce substantially different dynamics (Bajracharya and Duboz 2013). Correctness refers to the relationships between models to have correct structural and behavioral syntax and semantics. Object and object-oriented models must represent time and sanction acceptable interactions (e.g., how to resolve confluence of multiple inputs and outputs arriving at instance of time). To provide formal grounding to such models, it is suggested to add formal modeling methods as a layer atop such toolkits (Cicirelli et al. 2013). However, placing constraints on such simulation infrastructures is not straightforward and is likely to result in untamed complexity (Sarjoughian and Zeigler 2000).

6.2.2 *Landscape-Centric*

Geographic information systems (GIS) are a popular environment for modeling landscape processes. The Geographic Resources Analysis Support System (GRASS) (GRASS 2014) is a well-known, open-source GIS (Neteler and Mitasova 2004) that includes a georeferenced data management system that supports examination and modification of large data sets. It is widely used throughout government, research, and educational communities that deal with large amounts of geospatial data. Landscape-centric development environments, including GRASS, have the capability to support efficient map algebra calculations on very large data sets. However, they are not well suited for modeling rule-based agents or, more generally, component-based models. The main weakness of GRASS, from a hybrid human–environment model perspective, is its inability to represent detailed models of human agents and their integration with one another. While mathematical representations such as differential equations can be used with the GRASS map algebra functions, the concept and constructs of agents are difficult to conceptualize, create, and execute within the GRASS development environment and the scripting languages typically employed (Ullah 2012). Furthermore, the map algebra theory upon which GRASS is developed cannot directly support the representation of structured systems such as cellular automata models (Mayer 2009).

6.2.3 Theory-Centric

Theory-based approaches have mathematical formalisms within which every model has known structural and behavioral properties. A proof of correctness can be applied to models specified with the formalism and to the models interfaced within. However, theory-based approaches are general in order to remain applicable to a broad array of domain applications. For example, the Discrete Event System Specification (DEVS) (Zeigler et al. 2000) is a modeling framework for describing discrete-event systems. The formalism defines two types of models—atomic and coupled—with an abstract simulator responsible for executing the model specification with toolkits such as DEVS-Suite (Kim et al. 2009; ACIMS 2009). An atomic model is the basic abstraction that exhibits autonomous and reactive behavior. A coupled model contains one or more atomic or coupled models. Its behavior is determined by the models it contains and how they are hierarchically composed. The relationships between models are established via the use of couplings that employ message passing between models' input and output ports.

Cellular DEVS, an extension of DEVS, uses regular tessellations to model cell structures as a collection of individual automata (e.g., Wainer 2006). DEVS could be used to represent agents, while Cellular DEVS can represent the landscape processes. These two formalisms are closely related, and together they can support modeling human–environment systems. The strength of this approach is its formalism with built-in interface for composing agent and landscape models together. The use of this approach weakens as large-scale models must be developed. The landscape model may encompass a very large data set—on the order of millions of cells. However, while the hierarchical structure of DEVS fits well for an agent model, the cost of managing each cell as an individual object can grow exponentially (Kincaid et al. 2006). While it could be argued that parallel implementation can increase execution efficiency for large-scale cellular models (e.g., Vasconcelos and Zeigler 1993), an environment such as GRASS is still more efficient for the type of landscape process dynamics considered in this work. In the late 1990s, GRASS was integrated with DEVS-C++ to support simulation of complex adaptive systems containing georeferenced movement of satellites, aircrafts, and ground assets (Hall 2005). This approach uses high-level architecture (HLA) where different simulations are integrated (IEEE Std 1516-2010 2010). Integration-based approaches such as HLA primarily support simulation interoperability (the coordination of different simulation executors) instead of model composability (e.g., Sarjoughian and Zeigler 2000; Davis and Anderson 2004).

6.3 Human–Environment Model Composability

The study of many interesting systems requires developing multimodels with distinct abstractions. One method, as proposed here, is to define the structure and behavior of every model and its composition with other models according to modeling formalisms. A modeling formalism can be used to describe the external and internal parts of a model using proper syntax and semantics independent of its software design and implementation. Some researchers, however, develop hybrid human–environmental models that do not employ formal modeling and simulation methods. These models can generate dynamics that are difficult to explain. The modeling categories presented in the previous section are extended from the conceptual and software implementation aspects, for example, to assign agent models to spatial models (Howe and Diggory 2003; Luke et al. 2005). However, methodical model composition must use more rigorous definitions centered on the formalisms that are used. It is useful to note that although modeling in such environments can be sanctioned to comply with monolithic formal modeling and simulation methods, it is impractical to achieve grounded heterogeneous model composition as described below.

Building on monoformalisms, the superformalism, metaformalism, and polyformalism multimodeling approaches (Sarjoughian 2006) have been considered for heterogeneous model composition (see Fig. 6.2). An attempt at a best-fit of the categories discussed in the previous section into these multiformalism modeling approaches can be made. It is useful to say "best-fit" for a model whose behavior and structure are only limited by programming language concepts and, perhaps, object-oriented design, which are typically difficult to compartmentalize within the more strict modeling formalism composition approaches. In the superformalism modeling approach, different subsystem models may be described in closely related modeling formalisms and mapped to a single, higher-level specification. For example, continuous and discrete models may be composed using a mixed continuous/discrete modeling formalism (e.g., Praehofer 1991). This includes the DEVS and Cellular DEVS model described in the previous subsection. Using metaformalism modeling approaches such as (Karsai et al. 2004) and (Mosterman and Vangheluwe 2004), subsystem models that are described using different formalisms may be transformed into a third, common modeling language such as Discrete Event System Specification (DEVS) and Unified Modeling Language (UML) models. Domain-specific needs drive the specific data that incur the transformation. For both superformalism and metamodeling approaches, any add-ons must be capable of transforming to the core constructs. In the polyformalism modeling approach, an interaction model is introduced to describe data and control exchanges at varying levels of detail between two models that are specified using disparate modeling

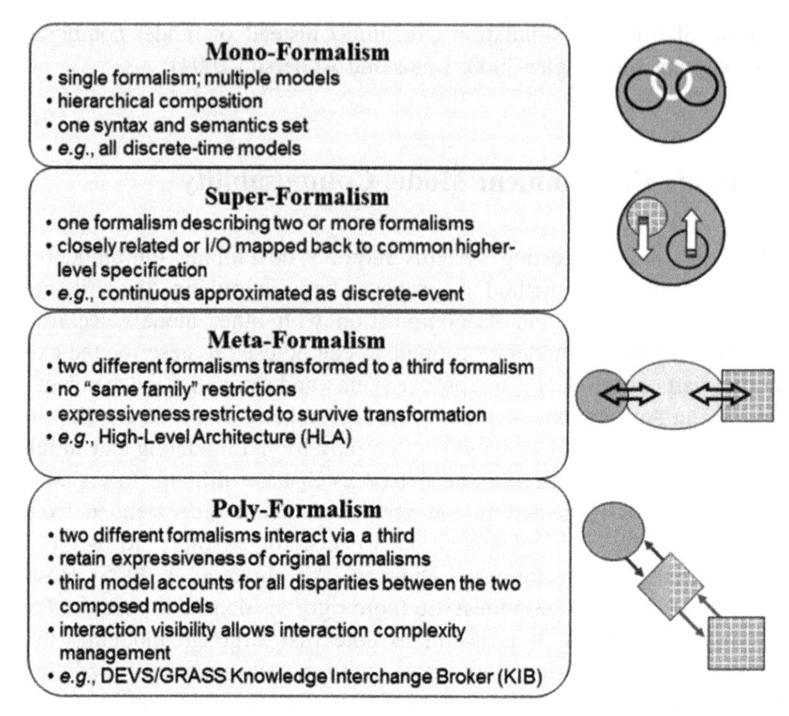

Fig. 6.2 A classification of modeling formalisms

formalisms. The interaction model is consistent with the composed models in that the composed models retain their formal properties without the need to incorporate specific knowledge of the structure and dynamics of the other composed model.

6.3.1 Composition Complexity

Many systems exhibit traits that researchers define as complex. As decomposition of a system is a common approach to managing complexity, multimodeling techniques prove useful. All three model composition approaches discussed above can capture the complex behaviors of multimodel systems and thus simulate their dynamics. However, a way to simplify the often difficult task of describing the interactions between the composed models is an important consideration in choosing which multimodeling composition approach to use. Importantly, system and subsystem complexities must be accounted for within these descriptions.

Modeling the interactions between disparate model types can be viewed in terms of complexities that are associated with (i) software toolkits and (ii) modeling approaches. When describing agent or landscape dynamics through the use of software components, the complexity of agent-centric and landscape-centric approaches can be attributed to their software toolkits. The reason is that software complexity contributors (i.e., discreteness of model components, model component design, management of model development process, and the flexibility of model components to evolve) underlie the complexity of developing hybrid models (Booch and Young 2006; Grimm et al. 2006). Consequently, the complexity of toolkit approaches such as Mason, Netlogo, and Repast can be mainly described in terms of software complexity (Rand et al. 2005). Therefore, development of agent and landscape models must be understood in terms of software components instead of model abstractions that are based on modeling formalisms.

As stated above, the complexity of hybrid model development can also be considered in terms of the modeling approach that one chooses. Algorithmic, deterministic, and aggregate complexities have been proposed in the context of social and natural sciences (Manson 2001). Algorithmic complexity refers to the difficulty in solving a problem mathematically as well as simplifying that solution (a domain-dependent endeavor). Deterministic complexity involves the ability to determine the outcome of the system given changes in initial conditions and input. These two kinds of complexity comprise a holistic view of a model. In contrast, aggregate complexity emphasizes the importance of synergy arising from the interactions among the subsystems of a system. The concept of aggregate model complexity (i.e., composition complexity) is especially relevant for multimodeling approaches that have been under development for engineered systems and, to a lesser extent, for socioecological systems. In some cases (e.g., agent/cellular DEVS (Ntaimo et al. 2004)), the complexity of hybrid model development may be explained in terms of both a modeling approach and the software toolkit. With an emphasis on the composition of models, this paper's focus is aggregate model complexity and choosing a model composition approach that helps to manage it. To this end, the degree of effort in hybrid modeling is also included. The modeling effort is influenced by three factors: (i) expressiveness of the modeling formalisms and their composition, (ii) simplicity of model development, and (iii) availability of robust and efficient simulation tools. All of these factors are in turn influenced by domain knowledge (e.g., human–environment modeling).

Of the three model composability approaches, polyformalism composition best suits human–environment modeling when the system model is prone to modification (e.g., testing different model subsystems or subsystem configurations). Social science is often beset by partial data sets and contested meanings of the data at hand. It is also often not fully understood how the system works. Examining system dynamics theories may be the crux of the research. In these cases, it is useful for the researchers to able to build and modify the system model to test system dynamic hypotheses and/or modify the system as more data becomes available. This may sometimes mean changing the abstractions of the models or the subsystem models. Polyformalism composition enables each subsystem model to be developed using its own formalism. Thus, the models can be as expressive as the modeler and domain require. Polyformalism composition also provides the domain experts (e.g., anthropologists and geologists) with the capability to work on their subsystem models independently of the other modelers. This makes model development easier. Furthermore, this modeling approach directly exposes the interaction between the subsystem models, allowing the modelers to have explicit and exact control over the interaction and visibility into the resultant behaviors of both the subsystem models and the hybrid system model itself. This in turn simplifies the management of aggregate complexity.

6.3.2 Polyformalism Model Composition

The polyformalism approach introduces the *knowledge interchange broker* (KIB) concept to compose multiple modeling formalisms (Sarjoughian 2006). To model interactions between models that are described in disparate modeling formalisms, the KIB can be used to specify input/output mappings (including transformations) and synchronization. The purpose of the KIB is to succinctly define model data and control interactions. The polyformalism composition approach does not predetermine which formalism is used to describe either the human model or the environmental model. Furthermore, it does not require that either model have any structural or behavioral knowledge of the other subsystem model. What it specifies is a third model (i.e., an interaction model, an instance of the KIB) being created that explicitly models the interaction between the two subsystem models. In this way, the two subsystem models remain independent of each other. This reduces the complexity of managing model development as each may be handled separately with minimal concern for the other subsystem beyond domain application. Furthermore, explicit visibility and management of the composition is provided. This enables more control over the complexity generated by the interaction, including direct input and output for experimentation (Mayer and Sarjoughian 2008; Godding et al. 2007; Sarjoughian and Huang 2005; Huang 2008).

The reason for explicitly modeling interactions between modeling formalisms has important implications. For example, it is important to know what it means to inject data and control from an external source that may not have the same approach to model specification and execution. One model may have an innate concept of time, while the other may not. Consider, also, what it means for a rule-based agent model to inject data into a CA model not at predefined discrete time steps (Mayer and Sarjoughian 2008, 2009). The modeler must ensure that an external input does not arbitrarily modify the state of the CA model. All state changes must be in accordance with the rules of that model's formalism; otherwise, correctness and validity of that model is suspect. To achieve model composability, the main goal centers on having appropriate concepts and methods to compose different model types such that both the disparate models and their interactions can be described using model abstractions with proper syntax and semantics. The composed models must be specified correctly—i.e., (i) correctness is ensured according to the domain-neutral modeling formalism of each subsystem model and (ii) validation is ensured according to the domain-specific model descriptions (Mayer et al. 2006;

Huang et al. 2009; Godding et al. 2007). Once the subsystems are formally defined, an interaction model can be developed to explicitly model the interaction.

The interaction model composes the two subsystem models by accounting for all aspects of the two models being composed. These are grouped into formalism and realization aspects (see Fig. 6.3). The formalism refers to model specification and execution (i.e., mathematical descriptions of the system and the machinery to execute the model). The specification and execution layers are not specific to any one model instantiation. Rather, they describe any model that conforms to the formalism. The realization aspect encompasses software architecture design and implementation. The architecture layer refers to software design (e.g., described in the Unified Modeling Language (UML) (Booch and Young 2006)) that can be forward engineered to specific programming languages (e.g., Java, C, Lisp). Design considerations can be as simple as converting an integer from one model into a double for another or as complicated as handling synchronous versus asynchronous input and output exchanges. The implementation layer specifies software library and programming language choices, for example. This layer refers to the implementation details that comply with the software architecture and are driven by the system's domain knowledge. For a socioecological model, this could involve how the scale and resolution of a model relates to other models, for example, an environmental model scaled to employ millions of data elements, each at a 10,000 m^2 resolution, interacting with an agent model using only a few hundred agents working at a 100 m^2 resolution. The modeler must consider the significance to an agent's movement possibly existing entirely within one cell of the environment model.

While the implementation-specific layer may have the most explicit representation of the domain, all other layers also are influenced by domain knowledge. It is the domain that should dictate what formalisms are suitable to develop the model of a system, how the model is executed, and how the model experimentations are to be conducted. By considering the domain across all the layers of the model, the model can better represent the domain under study. The domain experts will be capable of developing the system model provided that a suite of subsystem model parts for a given domain are already (or can be) developed. Furthermore, the complexities of the domain dynamics must be systematically represented within the model and, just as importantly, managed and simulated.

It is through the context of the domain and the design of experiments that an understanding of how the two subsystem models can interact is gained. From this understanding, the modeler can choose an appropriate formalism for the interaction model. The IM may be developed using either of the modeling formalisms used for the subsystem models or may employ a distinct formalism. The realization of the IM is derived from its chosen formalism. The dynamics of the IM are dependent upon the realizations of the two composed subsystem models. It is within the realization of the IM that the interactions between the subsystem models occur through data transformation, synchronization, concurrency, and timing of the KIB (Sarjoughian and Huang 2005). This approach affords the ability to segregate the two subsystem model formalisms and realizations. Furthermore, it allows the

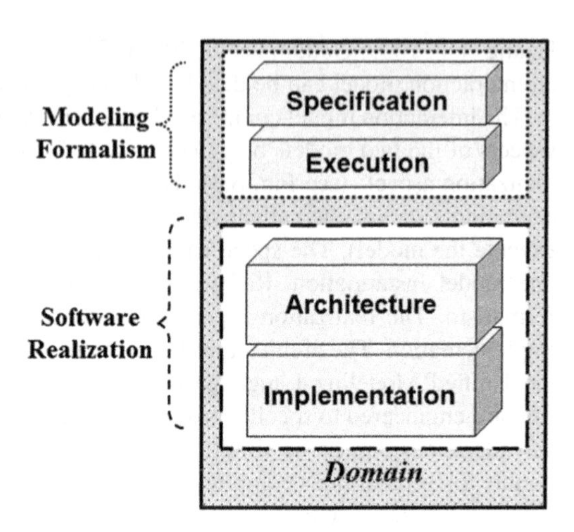

Fig. 6.3 A conceptual depiction of separation of a model's formalism from its realization

domain-specific details for each subsystem and their interactions to exist within their respective models.

6.4 Hybrid Agent–Landscape Model

The exemplar human–environment model is a simulation of human farmers living in the Penaguila Valley, Spain, during the early Bronze Age (Mayer et al. 2006; Mayer 2009; Ullah 2012; Arrowsmith et al. 2006; Soto et al. 2007). A conceptual illustration of this system is shown in Fig. 6.1. The agent model represents human farmers who employ cropping cycles in a rain-fed subsistence agropastoral system. The landscape model is a process-based, cellular automata simulation of surface runoff flow, sediment detachment (erosion), sediment transport, and sediment deposition. The agent model is dependent upon some characteristics of the landscape model to make decisions and provide for its survival. The actions that an agent takes in order to survive may then impact the landscape model. These impacts change the dynamics of the landscape model and, quite frequently, the same characteristics upon which the agent depends. A clearer delineation between complexity emerging from within a subsystem model (i.e., algorithmic and deterministic) and that which results from interaction with the other composed model (i.e., aggregate) is gained by directly expressing these models' interactions within a separate model.

Separating these internal and external complexities (with respect to a subsystem model) has other benefits. The algorithmic and deterministic complexities are more the result of modeler abstractions and implementation choices. These choices are dependent upon the experimentation setup and the domain. Thus, one may argue that there is some degree of modeler control over these complexities. On the other hand, aggregate complexity typically materializes as emergent behavior. As this behavior is not explicitly defined by the modeler and its results are often nondeterministic, there is much less modeler control. Further, as the emergent behavior must be analyzed at the hybrid system level, there needs to be visibility and flexibility in the hybrid model design in order to test and modify the subsystem models such that the resultant emergent behavior, assuming that it is desired, produces valid results.

6.4.1 Agent Model

The agent model is a combination of discrete-event, rule-based agents and models that define an agent's relation to another (Mayer et al. 2006; Mayer 2009; Mayer and Sarjoughian 2009). The discrete-event timing nature of the agents signifies that the agent conducts its actions in a discrete fashion but its current action may be interrupted by external factors. The result of such an interruption is dependent upon the agent's current state at the time.

The agent is a representation of a human household in which all members have common goals and share resources managed by the household. The model maintains population and allows for growth (positive and negative). The goal of each household is simply survival. A household is given a caloric requirement based on a per capita value. Additionally, the household is able to provide labor based upon all or a percentage of its population. To support its population, a household must feed its members by farming wheat and barley. The wheat is consumed directly by the household, while the barley is used to feed livestock that produce milk and meat, which add to a household's caloric intake. In each cycle, the household calculates how much of each crop is required and how much land is required to produce it. Assuming that the household has enough labor to cultivate for the projected crop, it attempts to do so. If not, it attempts to cultivate as much as its current laboring populace will allow. To farm, households conduct a survey of their surrounding landscape. They assign a value to land based upon the attributes of soil depth, land cover, and distance. Once the land is assigned a value, a household makes a plan for that cycle which meets its needs (or comes as close as possible) using the most valuable land. A probabilistic birth rate and death rate modified by the ratio of yield to need is used to determine a household's yearly change in population (if any) (Cowgill 1975; Wood et al. 1998).

Land cover is a discrete-time abstraction of both the plant growth on the soil surface and the health of the soil overall. Healthier soil supports more plant growth. The households desire a moderate level of land cover which is indicative of soil that is healthy enough to support a large crop without having to put too much effort into clearing the land of trees and shrubs to cultivate it. If the land cover is above the desired amount, the households will reduce it to a desired level for farming, indicative of clearing the land.

6.4.2 Landscape Model

The landscape model (Ullah 2012) is based upon well-known and broadly applied algorithms that calculate sediment flux at points on a landscape given the supply of sediment at that point and the topographic characteristics of the areas up- and downslope from that point (Braun et al. 2001; Dietrich et al. 2003; Hancock 2004; Mitasova and Mitas 1993; Willgoose 2005; Barton et al. 2004). These flux calculations also rely on the results of flow accumulation. The model simulates discrete flow of water and its accumulation in each cell. The sediment flux at each point on the landscape is then transformed into an elevation change through the Unit Stream Power Erosion Deposition (USPED) equation (Mitasova and Mitas 1993), a three-dimensional modification of the Revised Universal Soil Loss Equation (RUSLE) (Warren et al. 2005). In the landscape model, the net erosion/deposition of the sediment is estimated from the sum of the change in the sediment transport capacity in the x and y directions. The sediment transport capacities are modified by a transport factor Tr derived from a modification of the standard RUSLE. It is the product of the rainfall intensity and three erosion resistance factors. These erosion resistance factors are a soil erosion resistance factor based upon soil composition, a vegetation erosion protection factor based upon vegetation's ability to hinder rainfall and surface flow, and an erosion prevention practices factor (e.g., terraces and check dams). The rainfall intensity factor is computed by an equation that combines monthly precipitation amounts and is expressed as a map of rainfall intensity for each area in the landscape. The value is calculated from retrodicited precipitation values for the middle of the early Neolithic (7,000 BP (Before Present)) for southern Spain (Bryson and McEnaney-DeWall 2007). All the erosion resistance factors are scaled from 0 to 1, with 0 being nonerodible and 1 being unprotected from rainfall intensity.

The landscape model also includes a simple vegetation model (Ullah 2012). Vegetation is coded on a 50-year timescale. Any degraded land cover, if left unmodified by the agents, is regenerated in stepwise fashion through a simplified succession sequence. Bare ground will grow back through grass to scrub and eventually to forest cover in a 50 year period if no human impact occurs during the regeneration cycle (Pardo and Gil 2005). The land cover values are then translated to vegetation erosion protection factor values based upon linear regression of the known relationship between "classic" land cover types and vegetation erosion protection factor values in Mediterranean environments.

6.4.3 Interaction Model

The IM encompasses all the interactions occurring between the agent and the landscape models as sketched in Fig. 6.1. Figure 6.4 shows the interactions between the agent and landscape models as a separate part having its own specification,

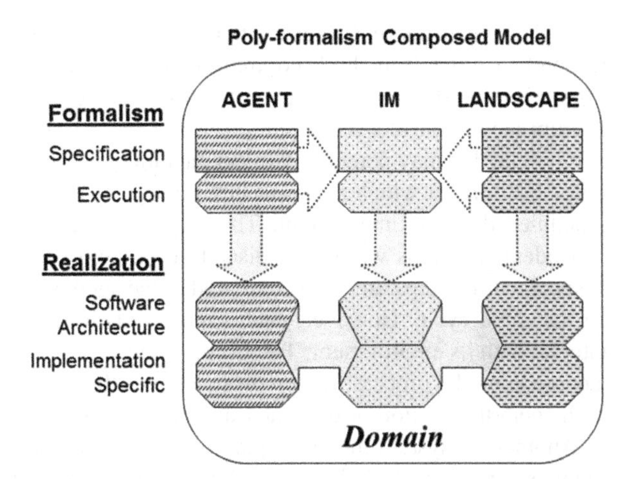

Fig. 6.4 A schematic view of the hybrid DEVS-IM-GRASS environment

execution, software, and implementation (Mayer 2009). The IM is developed in part in the DEVS-Suite simulator (ACIMS 2009) and conforms to the abstract communication regime defined for parallel DEVS. The IM's data transformations and control follow the KIB method since the interactions between the IM and GRASS cannot be defined in the DEVS formalism.

Household farming impacts the environment by reducing the land cover. When coupled with the landscape evolution model, this has the effect of increasing soil erosion in that area. As soil erodes, less crop yield is produced by the same land. However, each year that land is allowed to fallow, the land cover increases. This represents some health returning to the soil. As plant matter grows back, it also has the effect of decreasing soil erosion, thus possibly increasing soil depth. An emergent result of this logic is that the households may exhibit the behavior of cycling their planting sites—farming in one area the in the first cycle and then farming in another area in a later cycle and allowing the previous site to fallow. Additionally, the village footprint also impacts erosion. While no crops are grown within the village itself, villages were often established near water sources. The erosion change caused by the village often has effects on the environment downstream.

In the current hybrid model, there are four interactions in each simulation cycle: the agent view of the landscape, the agent impacts on the landscape, the agent's harvesting of the crops, and the landscape processes affected by agent actions. It is these four interactions which define the aggregate complexity for this hybrid model. It is through the explicit modeling of these interactions, the usage of aggregation/disaggregation, timing control, and other mechanisms (described below) by which this complexity is managed.

It should be noted that the IM does not initiate interaction with a subsystem model on its own. It is a model of the interaction between the subsystem models, and as such, it sends data or a control message to a subsystem model as a result of input from the other. However, the IM must be cognizant of the differences in timing mechanisms and data content for each subsystem model. The IM must respond to input from both composed models and, in the case of the discrete-event agent model, may not know the exact timing of such inputs. Therefore, it is appropriate to consider a discrete-event modeling formalism for the IM. With this in mind, the IM may potentially inject data into one of the subsystem models at any time. The time at which an event input is injected may not align with a regularly scheduled discrete-time instant. This poses a potential problem since the time-dependent functions within the discrete-time landscape model must be executed at specific time steps. In the current model, the agent and landscape models operate on the same cycle. However, each agent sends independent control messages to interact with its environment. The IM then aggregates all agent input to a particular landscape model process. It then disaggregates the output of the process to provide each agent the portion of the landscape model result with which it is concerned.

Another subsystem model disparity that the exemplar IM accounts for is environmental resolution (Ullah et al. 2008). The agent manages cells with a 25 m^2 resolution (5 m $\times$ 5 m). The landscape models use a coarser granularity of 100 m^2 (10 m $\times$ 10 m). Thus, each agent action impacts four landscape model cells. This has important implication during each of the view, farm, and harvest interactions. As stated above, the agent uses soil depth, land cover, and a distance cost from the village to determine the land value. During viewing, the IM provides the same landscape data to the one or more agents that occupy a specific landscape cell. The distance cost calculation is special. It is not just the length between two points; it is weighted by the type of terrain over which the agent must travel to get to that point. The distance cost is therefore a value created by an association between an agent model attribute (village location) and a landscape model attribute (terrain) and, as such, is an attribute of the IM itself.

When an agent specifies a farming action, the IM will aggregate the impact from the agent's four "farmed" cells to the representative cell in the landscape model. When an agent decides to harvest a crop, it receives back from that action an amount of crop harvested as yield. This yield is dependent upon the environmental factors in which it was grown (soil depth, rainfall, etc.). The crop itself is neither part of the landscape model nor of the agent model. However, for this attribute of the hybrid model to be useful, it requires data from both subsystem models and is therefore well suited to be formulated within the IM. Thus, it is the IM that intercepts an agent's harvest request, queries the landscape model for environmental values which impact crop yield, calculates the actual yield, and provides this value to the agent. In turn, the IM collects agent impact data (farming and village site data) and provides this to the landscape model in order to facilitate the erosion model dynamics. Agent impacts are another model characteristic which, while generated as a result of agent decisions, makes little sense on their own in either the context of agent decision making or in pure landscape dynamics. So again, this is another attribute of the hybrid model that is best suited within the interaction model.

As stated previously, the IM provides visibility into the hybrid model interactions and behaviors. Consider that the agent model uses a deterministic, rule-based approach to drive its basic behaviors. However, it also uses stochastic methods to resolve conflicts between agent farming locations. This leads to nondeterministic results when it comes to individual household land holdings and, ultimately, survival. The landscape model also uses deterministic processes to model the environmental behavior. However, the exact process for each subsystem is unknown. Researchers strive to include more details into the subsystem models in order to test their understanding of these real systems. Thus, the subsystem model's algorithmic complexity has grown due to the addition of a large number of variables.

While the range of input variables and initial conditions can be a priori formulated when each subsystem model is considered independently, it becomes less so when the two models interact. The impact that each variable has on the hybrid system behavior can become very difficult to determine. This is exacerbated by the fact that the agents working with landscape data exhibit emergent behaviors in farming practices such as farming location and land management (i.e., farmed or fallowed). This in turn leads to different erosion patterns as a result of the agent's actions. By using an interaction model, the researchers are able to model and manage the data passing between the subsystem models by focusing on things such as boundary values and input/output relation (i.e., sensitivity analysis) for a range of values. Furthermore, having an IM reassures the modeler that only specific data is being passed between the subsystem models. This helps to ensure that as the internal complexity of each model rises in the form of algorithmic or deterministic complexity, the aggregate complexity remains controlled. Thus, if the emergent behavior is invalid, a much smaller subset of processes and variables need be examined and modified. Finally, another important factor is the scalability of the IM since the amount of data exchanges and the frequency of subsystem models' interactions can be separately handled.

6.4.4 Software Realization

The software architecture for interaction modeling and simulation must account for disparate software languages and constructs (Mayer et al. 2006). The agent model and the IM are developed in DEVS-Suite simulator (ACIMS 2009), a Java language implementation of the parallel DEVS modeling formalism. The landscape model is developed in GRASS (GRASS 2014). GRASS modules are written in C. Scripts (and functions) may be written in any scripting language such as Bash or Python depending on the system functionality that the modeler requires (e.g., file management, use of regular expressions, etc.). To run a GRASS script, a DEVS model component uses the Java `Runtime.exec()` command to execute it.

Each GRASS module is independent and therefore has its own interface, but the modules do not continuously run. They accept input, return output, and terminate in a noninterruptible fashion. The output from the modules is only provided to the standard output stream (and sometimes, standard error stream). Thus, to get return data, the IM must capture and parse the data from the standard output buffer and then insert that data into a DEVS message object. This approach is complicated by the fact that GRASS, being an open-source project initially developed during the early 1970s during the time of command-line interfaces, has output that is typically preformatted for ASCII viewing and each module has its own unique output format. The GRASS community is working on standardizing such variations. It is managerial modeling tasks such as buffer parsing, which are not directly related to the simulation with which the researchers are concerned, that can tightly couple two composed models and restrict the approach used in subsystem model development. In light of the fact that the GRASS implementation may have to change, the IM implementation may also require modification. The use of the interaction model frees both subsystem models from the burden of data transformation and mapping, allowing their developers to focus on the best architecture for each.

6.5 Simulation

The simulation models presented in this paper are designed to investigate the impacts of farming in the early Neolithic (1–17,000 BP) of southern Spain (MedLand 2014). The environmental conditions of the model were set according to what is believed to be the conditions: moderate Mediterranean rainfall regime, Mediterranean terra rossa soils, and an initial land cover of Mediterranean oak woodland. Because archeological evidence is rather limited, data from ethnoarcheological investigations of similar modern farmers has been used to devise the agent model (Mayer et al. 2006). With the simulation described in this paper, the MedLand research group sought to better understand the workings of the hybrid human–landscape model and how it represented the underlying villagers and environment subsystems. Questions that were focused on included

1. Can the environment sustain a moderately sized village population using a known archeological site?
2. How much do certain factors contribute to population sustainment?
   - Soil depth (and associated erosion)
   - Labor force versus the household size and caloric requirements
   - The number of households
3. What impacts do the farmers have on the landscape?
4. What are the impacts of different data exchanges between agent and landscape (e.g., selected data elements and data size/resolution) and control frequency of data exchanges?

6.5.1 Hybrid Model Configuration

In one scenario, one village site was chosen to allow examination of how topography influences the outcome of the model (Ullah 2012). The village site is located in an area that was actually a village in the Neolithic period (as best we can tell from data collected during a field survey). We reconstructed the paleotopography (by filling in the deep barrancos) and used the early Neolithic climate data. The details of some of the major initial conditions are given below. Many of these values may be visually modified at the start of the simulation (Mayer 2009). The intent is to highlight the number of variables in the hybrid model, many of which may have profound impacts on the simulation results and increase the difficulty of managing the model's complexity.

In the simulation, the village consists of 10 households, each with an initial population of 6 people. The initial birth rate is set at a 3 % probability, and the initial death rate is set at 2 % (Cowgill 1975; Wood et al. 1998). In the event of a particularly good or bad year (when the yield-versus-need ratio is high or low, respectively), the birth rate changes at a rate of 1 % per year, while the death rate would inversely change at 5 % per year. The maximum birth and death rates were capped at 5 and 100 % probability, respectively. The minimum birth and death rates were 0 and 2 % probability, respectively.

Maximum yield of wheat and barley (in a soil depth greater than or equal to 100 cm) was set to 460 kg/Ha. Minimum yield (in a soil depth less than or equal to 13 cm) was 168 kg/Ha (Thomson et al. 1985). Yields for soils of depths between 13 and 100 cm were scaled by the power regression formula: $Y = 43.951 \times D^{0.522071}$, where Y is the yield and D is the soil depth in centimeters (Sadras and Calvino 2001). Households were set to initially expect a yield of 450 kg/Ha in the first year and from then on to expect the same yields as were actually produced in the prior year. As extensive grazing has not yet been added to the human model, the impact of herd animals (goats and sheep) is only modeled by the amount of barley (grains, straw, and chaff) needed for supplemental herd feeding over the fall/winter months.

Each member of the household populace required 1,000,000 kcal of wheat per year and 250,000 kcal of milk and meat per year. Wheat was considered to be directly consumed in the form of bread and porridges, while barley was considered only as fodder for herd animals that produced meat and milk products to be consumed by the households. Therefore, after harvest loss, seed reserve loss, and processing loss, wheat provided agents with 3,500 kcal/Ha, while barley only provided agents with 213.87 kcal/Ha (Thomson et al. 1985). In the simulation, only 50 % of the population was able to produce labor, and each productive person was considered able to produce 300 man-days per year. Wheat and barley farming required 50 man-days per Ha per year (Simms and Russell 1997).

In order to speed land patch querying and village-level land negotiations (currently the most time-costly operation in the agent model), the village in the current simulation was given a "viewable" territory based on a 10 min anisotropic walking-time cost radius from the village center, which translates into an area of about 160 *Ha*. Village area was calculated using the population density formula, $A = V_p/D$, at the start of the simulation and every time the village population changed. A is the area occupied by the village in square meters, V_p is the current village population, and D is the population density coefficient of people per square meter. The current simulation used a D value of 0.0159. Additional details and in particular details of the archeological data, landscape modeling, and evaluation of the hybrid human–landscape dynamics highlight the scale and complexity of modeling and the necessity of hybrid model correctness for simulation validation (Ullah 2012).

Figure 6.5 shows the simulated agropastoral landuse at three stages run for 200 cycles (at a temporal resolution of one cycle per year). The extent of the households' impacts on their environment was assessed by comparison with a control landscape model. Figure 6.5-(a, b) at year 5, (c, d) at year 100, and (e, f) at year 200 show the impact of households on the landscape and the resultant landcover. We can observe an emergent pattern of concentric "rings" of landuse around the village that results in corresponding "rings" of landcover in various stages of regrowth. This pattern emerges as agents encounter a series of thresholds in their land allocation logic, whereby they use portions of land until fertility (or grazability) is reduced to the point that they release it and move outward in search of new plots (note the difference between year 5 and year 100). Eventually, the discarded plots regain fertility and vegetation, and since they are closer, agents abandon the plots located further away in favor of those that are closer. This creates the "rings," and as the agents' population increases (and they need more and more land to satisfy their food requirements), a series of superimposed rings of fallowed and actively cultivated/grazed land develops (see the landcover and impact patterns for year 200).

6.5.2 Results

The simulation study provides some clear answers to the questions posed in the previous subsection. Realistic values derived from ethnographic data, archeological data, and other research studies were used. The simulation results revealed hybrid model dynamics that could be directly attributed to complex interactions between the individual household and landscape models. The domain experts are able to understand and explain model composition complexities that previously could not be examined at such a level of detail and scale for the agent and landscape models.

1. The environment *can* sustain a moderately sized village population using known archeological sites. However, the extent to which the population is successful at surviving is dependent upon the output of both the conflict resolution and population growth (i.e., birth and death) stochastic processes.
2. How much do certain factors contribute to population sustainment?

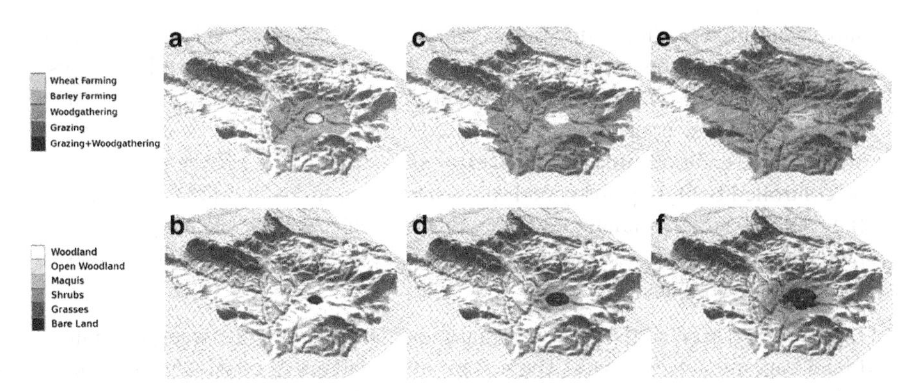

Fig. 6.5 Penaguila Valley with village locations and erosion impacts

- Soil depth (and associated erosion) only plays a significant role if the depth falls within the minimum and maximum values for yield. If soil depth is beyond the maximum for yield, the household does not notice any changes in soil depth with relation to its yield.
- In examining labor force versus the household size and caloric requirements, careful consideration must be given to the initial values of the variables. The wrong values either provide too easy a scenario for the households or one in which it is impossible to survive.
- An impact of changing the number of households is the change in the distance from the household to the farmland. This is because as the number of households increases, some households have to travel further to find viable land for farming. Additionally, due to the stochastic nature of how households were assigned management rights to a land area, some unlucky households were left with farmland that produced lower yields than average, and therefore, their populations did not do as well. With these impacts, the rules used to model how households determine where to farm (e.g., soil quality and distance) could be more thoroughly tested and compared to anthropological data.

3. Farming is shown to increase landscape erosion. However, a village site, which is considered to be compacted soil, decreases erosion. The extent of the impact is dependent upon the environmental conditions surrounding the area. For example, if an area would naturally incur deposition of soil, then farming may just cause the soil depth to be near static. On the other hand, an area already prone to erosion would decline rapidly, forcing the farmers to soon give up their farmland areas and seek better ones.
4. The different data exchanges between the human and landscape subsystem models show the importance of maintaining their separation via an interaction model. An important idea is that each model need not expose all its data to its external environment. It provides the data it wants to the IM, and the IM may then manipulate that data as needed to conform to the input of the other

subsystem model. This reduces the complexity within each subsystem model. Furthermore, with respect to frequency of data exchanges, each model may use its own timescale. The IM may then manage data transformation and exchange frequency at the appropriate times. Reducing data exchanges can improve simulation execution time (performance) since interaction between models can be controlled both in quantity and frequency. For our development environment (i.e., DEVS-Suite simulator), aggregating multiple, similar exchanges (households desiring to examine the same land, for example) proved an efficient way to improve performance.

In answering each of these, examining the data exchanged between the human and landscape models was key. There were many factors to consider in determining, for example, why a population decreased. By examining the interaction between the two subsystem models, it might be seen that yield values were plentiful for supporting the household's population. This then leads the researcher to examine the internal processes of the human model, such as the amount of land that could be farmed and the stochastic population growth. For example, even when the average birth rate is higher than the average death rate, it is possible that random number generation will produce more deaths than births. Alternatively, it might be found that yield values would not support the population even after the household tried farming many different lands. This then leads the researcher to examine the landscape soil depths to determine why they became so low. In short, it is the visibility offered by the interaction model which leads the researcher along a logical path to understanding why the hybrid model is behaving the way it is. In another simulation setting, one village located at high elevation in a relatively steep valley flank was added. In this kind of scenario, larger-scale agent models and thus a higher degree of interactions make the role of the IM more significant.

6.6 Conclusions

Multimodeling has become more commonplace across many scientific research areas. This approach allows modelers to represent larger, more complex systems with smaller, less complex ones that work together. Also, while developing these subsystem models, the modeler may find that each is best represented by a different formalism. A difficulty then arises in composing these disparate subsystem models into a system model. Furthermore, some level of complexity invariably comes about from the interaction between the subsystem models, which is reflected in the overall behavior of the system. Thus, it is incumbent on the modeler to decide upon a multimodeling approach that methodically and correctly composes the subsystem models into a system model.

Polyformalism model composability is a multimodeling approach with several advantages. First, it ensures that each subsystem model is loosely coupled with its composed model to provide flexibility in design and to make changes with minimal impact on the other model. Also, each subsystem retains its formalism expressiveness to provide the best description of the subsystem model. This can also reduce the algorithmic and deterministic complexities by allowing the modelers more flexibility in the design of the subsystem models. Second, polyformalism composition explicitly models the interaction between the two composed models using a third model. This provides visibility into and strict management of the interaction and, therefore, the aggregate complexity of the system model. Third, the IM provides a point at which to implement simulation tools common to both models without burdening the architecture of either composed model. The applicability of the polyformalism concept for three or more very different modeling approaches along with multiple interaction models is an ongoing research (Sarjoughian et al. 2013). Composing disparate modeling formalisms can benefit from coupling model concepts (Ören 2014) and metamodeling concepts that are aimed as automating generation of simulation code (Sarjoughian and Markid 2012).

These advantages are most beneficial to scientific communities whose purpose for modeling involves complex systems in which the dynamics of the system itself are being explored. In this paper, we detailed a study of a complex human–environment system. Both human behavior and environmental dynamics involve many factors. Some of these factors are well understood, while others are the subject of ongoing research across different application domains. The resultant hybrid model is sensitive to the complexities within each subsystem model, and their interactions add more of their own. Furthermore, there is a need to compare simulation results between models with different initializations, different input regimes, different behavior, and different structures to best understand the significance of the system data that is available.

In addition, the interaction model's relation to the two subsystem models facilitates stand-alone data observation using suitable visualization tools. When dealing with simulation output data that represents over one million dynamic landscape cells and the actions of some 100 agents in various locations throughout the landscape, visualization tools provide a means for the modeler to quickly assess the outcomes of the simulation. The IM, being a model of the interaction, offers clear and concise points to capture data of the subsystem models' interaction. This data may then be sent to visualization tools as necessary. The current hybrid modeling environment (i.e., DEVS-Suite/GRASS) offers rudimentary support for simulation initialization, data collection, and observation (Mayer 2009). However, advanced visualization capabilities are planned.

In order to implement a hybrid model using polyformalism composition, a modeler (or group of modelers) must be familiar with both subsystem models—from both a formalism and realization perspective—and the domain. This person or group has the responsibility of developing the interaction model. This includes selecting the formalism to which the IM is developed and working with both subsystem modelers to determine how to define the interaction including how those aspects of the hybrid system fit within the IM. Also, these interaction modelers must then design, develop, and maintain the IM. As with any good software architecture, the latter requires some preplanning in the design phase based upon the anticipated changes to both the subsystem models and the interactions themselves.

6.6.1 Other Observations

Another key feature of using polyformalism composition for human–environment modeling is usability. Developing distinct, detailed subsystem models entails the need to apply domain expertise from two very different domains. The benefit of this is that it opens the door to diverse multidisciplinary endeavors. Domain experts may normally work within very different communities. A potential downside is that domain languages, tools, and methodologies may vary widely between the two subsystem model domains. As a result, the IM offers an opportunity to provide a common means to describe data access and simulation control that either community may feel comfortable using to experiment with the entire hybrid model.

For example, the MedLand project models are being developed by groups of researchers from different scientific disciplines. The agent and IM models are being developed by computer scientists and the landscape model by social scientists. Each has had experience building their respective models previously and within that specific development environment. The instance simulation model of the human–landscape IM is developed jointly. The use of the interaction model eliminates the need for the modelers to incorporate specific details of the other model into their design and implementation. It also eliminates the need for one or both of the modelers to explicitly incorporate data and control mechanisms into the models in order to facilitate interaction with the other model. To enable all modelers to execute the simulation and perform experimentation, a centralized graphical user interface (GUI) has been developed that parameterizes the initial values for each of the models.

Another benefit of an interaction model is a level of generality derived from composing two formalisms (see Fig. 6.4). Due to the uniqueness of the semantic behind different formalisms and the domain dependency of the subsystem models' interaction, a generalized concept for the creation of an interaction model needs to be specialized and realized. A unique IM is built for each multiformalism pair and domain. For instance, there are different approaches required to manage the interaction between a discrete-event model and a Model Predictive Control (MPC) optimization model than between a discrete-event model and a cellular automaton. Furthermore, if an IM composes a discrete-event model that represents human agents that farm and a cellular automaton that represents farmland, it will likely require a different IM to manage a discrete-event model representing robotic aircraft flying over a cellular automaton representing enemy land, "different" in this case being in terms of the specific data transformation and control mechanisms used. Both interaction models (for the farmer and aircraft) are likely to employ the same approach to composing the discrete-event model with the cellular automaton from a formalism interaction perspective. Once it is semantically defined how two formalisms interact at the application domain level, then this level of interaction, like the formalisms themselves, is domain independent.

Acknowledgments This research is supported by National Science Foundation grant #BCS-0140269 and #DEB-1313727. We would like to thank the entire MedLand team for their help and partnership.

References

ACIMS (2009) DEVS-Suite simulator. Retrieved from http://devs-suitesim.sourceforge.net/

Anderson JA (2006) Automata theory with modern applications. Cambridge University Press, New York

Arrowsmith J, DiMaggio E, Barton C, Sarjoughian H, Fall P, Falconer S, Ullah I (2006) Geomorphic mapping and paleoterrain generation for use in modeling holocene (8,000 1,500 yr) agropastoral landuse and landscape interactions in Southeast Spain. AGU Fall Meet Abstr 1:0453

Bajracharya K, Duboz R (2013) Comparison of three agent-based platforms on the basis of a simple epidemiological model. In: Proceedings of the symposium on theory of modeling & simulation – DEVS Integrative M&S symposium, Society for Computer Simulation International, San Diego

Barton CM, Bernabeu J, Aura JE, Garcia O, Schmich S, Molina L (2004) Long-term socioecology andcontingent landscapes. J Archaeol Method Theory 11(3):253–295

Barton CM, Ullah II, Bergin S, Mitasova H, Sarjoughian HS (2012) Looking for the future in the past: Long-term change in socioecological systems. Ecol Model 241:42–53

Booch G, Young B (2006) Object oriented analysis & design with application. Pearson Education India

Booch G, Maksimchuk R, Engle M, Young B, Conallen J, Houston K (2007) Object-oriented analysis and design with applications, 3rd edn. Addison-Wesley Professional, Reading

Braun J, Heimsath AM, Chappell J (2001) Sediment transport mechanisms on soil-mantled hillslopes. Geology 29(8):683–686

Bryson, R.A. and K. McEnaney-DeWall (eds), (2007). A Paleoclimatology Workbook: High Resolution, Site-Specific, Macrophysical Climate Modeling & Template CD. Mammoth Site of Hot Springs, SD, Hot Springs, SD. CCR #930

Cicirelli F, Furfaro A, Nigro L, Pupo F (2013) Agent methodological layers in repast simphony. Proceedings of the 27th European Conference on Modelling and Simulation, ECMS 2013, Ålesund, Norway

Cowgill GL (1975) On causes and consequences of ancient and modern population changes. Am Anthropol 77(3):505–525

Davis PK, Anderson RH (2004) Improving the composability of DoD models and simulations. J Def Model Simul Appl Methodol Technol 1(1):5–17

Dietrich WE, Bellugi DG, Sklar LS, Stock JD, Heimsath AM, Roering JJ (2003) Geomorphic transport laws for predicting landscape form and dynamics. Geophys Monogr 135:103–132

Epstein JM, Axtell RL (1996) Growing artificial societies – social science from the bottom up (Vol. Washington: Brookings Institution Press). MIT Press, Cambridge

Fishwick PA (1995) Simulation model design and execution: building digital worlds. Prentice Hall PTR, Upper Saddle River

Godding G, Sarjoughian HS, Kempf KG (2007) Application of combined discrete-event simulation and optimization models in semiconductor enterprise manufacturing systems. In: Proceedings of the 39th conference on Winter simulation conference, pp 1729–1736

GRASS (2014) Geographic resources analysis support system. Retrieved from http://grass.itc.it/
Grimm V, Berger U, Bastiansen F, Eliassen S, Ginot V, Giske J, Rossmanith E (2006) A standard protocol for describing individual-based and agent-based models. Ecol Model 2(1):115–126
Hall S (2005) Learning in a complex adaptive system for ISR resource management. In: Spring simulation multi-conference. Society of Computer Simulation International, pp 5–12, San Diego, CA.
Hancock GR (2004) Modelling soil erosion on the catchment and landscape scale using landscape evolution models a probabilistic approach using digital elevation model error. In: 3rd Australian New Zealand soils conference, University of Sydney, Australia
Howe T, Diggory M (2003) A topological approach toward agent relation. In: Proceedings of the agent 2003 conference on challenges in social simulations, Chicago
Huang D (2008) Composable modeling and distributed simulation framework for discrete supply-Chain systems with predictive control. Arizona State University, Tempe
Huang D, Sarjoughian H, Wang W, Godding G, Rivera D, Kempf K, Mittelmann H (2009) Simulation of semiconductor manufacturing supply-Chain systems with DEVS, MPC, and KIB. IEEE Trans Semicond Manuf 22(1):164–174
IEEE Std 1516-2010 (2010) IEEE standard for modeling and simulation's high level architecture (HLA)-framework and rules. IEEE Press
Karsai G, Maroti M, Ledeczi A, Gray J, Sztipanovits J (2004) Composition and cloning in modeling and meta-modeling. IEEE Trans Control Syst Technol 12(2):263–278
Kim S, Sarjoughian HS, Elamvazhuthi V (2009) DEVS-suite: a simulator supporting visual experimentation design and behavior monitoring. In: Proceedings of the 2009 Spring simulation multiconference, high performance computing & simulation symposium, San Diego
Kincaid CA, Mohanty SP, Mikler AR, Kougianos E, Parker B (2006) A high performance ASIC for cellular automata (CA) applications. International conference on information technology, Los Alamitos
Kirschner DE, Hunt CA, Marino S, Fallahi-Sichani M, Linderman JJ (2014) Tuneable resolution as a systems biology approach for multi-scale, multi-compartment computational models. Wiley Interdiscip Rev Syst Biol Med 6(4):225–245
Luke S, Cioffi-Revilla C, Panait L, Sullivan K, Balan G (2005) Mason: a multi-agent simulation environment. Simul Trans Soc Model Simul Int 82(7):517–527
Lytinen SL, Railsback SF (2012) The evolution of agent-based simulation platforms: a review of NetLogo 5.0 and ReLogo. In: Proceedings of the fourth international symposium on agent-based modeling and simulation
Manson S (2001) Simplifying complexity: a review of complexity theory. Geoforum 32:405–414
Mayer GR (2009) Composing hybrid discrete event system and cellular automata models. Arizona State University, Tempe
Mayer GR, Sarjoughian HS (2008) A composable discrete-time cellular automaton formalism. First workshop on social computing, behavioral modeling, and prediction. Phoenix, pp 187–196
Mayer GR, Sarjoughian HS (2009) Composable cellular automata. Trans Soc Model Simul Int 11–12:735–749
Mayer GR, Sarjoughian HS, Allen EK, Falconer S, Barton M (2006) Simulation modeling for human community and agricultural landuse. Spring simulation multi-conference, Huntsville, AL
MedLand (2014) Landuse and landscape socioecology in the mediterranean basin: a natural laboratory for the study. Tempe: https://shesc.asu.edu/medland/. Retrieved June 2014
Minar M, Burkhart R, Langton C, Askenazy M (1996) The Swarm simulation system: a toolkit for building multi-agent simulations. Santa Fe Institute, Santa Fe
Mitasova H, Mitas L (1993) Interpolation by regularized splines with tension: I. theory and implementation. Math Geol 25:641–655
Mosterman PJ, Vangheluwe H (2004) Computer automated multi-paradigm modeling: an introduction. Simulation 80(9):433–450

Neteler M, Mitasova H (2004) Open source GIS: a GRASS GIS approach. 2nd edn, The Kluwer international series in Engineering and Computer Scienc: Volume 773. Kluwer Academic Publishers/Springer, Boston, MA
North M, Howe T, Collier N, Vos J (2005) The repast symphony development environment. In: Proceedings of the agent 2005 conference on generative social processes, Argonne National Laboratory, Chicago, pp 159–166
Ntaimo L, Zeigler B, Vasconcelos M, Khargharia B (2004) Forest fire spread and suppression in DEVS. Trans Soc Model Simul Int 80(10):479–500
Ören TI (2014) Coupling concepts for simulation: a systematic and comprehensive view and advantages with declarative models. Int J Model Simul Sci Comput 5(2):1–17
Ören TI, Yilmaz L (2009) Agent-directed simulation and systems engineering and management. (Ören TI, Yilmaz L eds) Wiley series in systems engineering, Wiley-VCH, Berlin, Germany
Ören TI, Zeigler BP (2012) System theoretic foundations of modeling and simulation: a historic perspective and the legacy of. A Wayne Wymore 88(9):1033–1046
Pardo F, Gil L (2005) The impact of traditional land use on woodlands: a case study in the spanish central systems. Hist Geogr 31(3):390–408
Parker D, Manson S, Janssen M, Hoffmann M, Deadman P (2003) Multi-agent systems for the simulation of land-use and land-cover change: a review. Ann Assoc Am Geogr 93(2):314–337
Praehofer H (1991) System theoretic formalisms for combined discrete-continuous system simulation. Int J Gen Syst 19(3):226–240
Ptolemaeus C (2014) System design, modeling, and simulation using Ptolemy II. www.ptolemy.org
Rand W, Brown D, Riolo R, Robinson D (2005) Toward a graphical ABM toolkit with GIS integration. In: Proceedings of the agent 2005 conference on generative social processes, models, and mechanisms, Chicago, Illinois
Sadras V, Calvino P (2001) Quantification of grain yield response to soil depth in soybean, maize, sunflower, and wheat. Agronomy 93:577–583
Sarjoughian HS (2006) Model composability. In: Proceedings of the 38th conference on Winter simulation conference, pp 149–158, Monterey, California
Sarjoughian HS, Huang D (2005) A multi-formalism modeling composability framework: agent and discrete-event models. Ninth IEEE international symposium on distributed simulation and real-time applications, Washington, DC
Sarjoughian HS, Markid AM (2012) EMF-DEVS modeling. Symposium on theory of modeling and simulation – DEVS integrative M&S symposium, Orlando
Sarjoughian HS, Zeigler BP (2000) DEVS and HLA: complementary paradigms for modeling and simulation? Trans Soc Model Simul Int 17(4):187–197
Sarjoughian HS, Smith J, Godding G, Muqsith M (2013) Model composability and execution across simulation, optimization, and forecast models. In: 3rd international workshop on model-driven approaches for simulation engineering, SpringSim multi-conference, San Diego
Simms SR, Russell KW (1997) Bedouin hand harvesting of wheat and barley: implications for early cultivation in southwestern Asia. Curr Anthropol 38(4):696–702
Soto M, Fall P, Barton CM, Falconer S, Sarjoughian HS, Arrowsmith R (2007) Land cover change in the southern levant: 1973 to 2003. ASPRS Southwest technical conference, Arizona State University, Tempe, AZ
Thomson EF, Bahhady F, Termanini A, Mokrebel M (1985) Availability of home-produced wheat, milk products and meat to sheep-owning families at the cultivated margin of the NW Syrian steppe. Ecol Food Nutr 19:113–121
Tisue S, Wilensky U (2004) Netlogo: design and implementation of a multi-agent modeling environment. In: Proceedings of the agent 2004, Chicago, Illinois
Ullah II (2012) The consequences of human landuse strategies during the PPNB-LN transition: a simulation modeling approach. Arizona State University, Tempe
Ullah II, Mayer GR, Barton CM, Sarjoughian HS, DiMaggio E (2008) Ancient mediterranean landscape dynamics: hybrid agent and goespatial models as a new approach to socioecological simulation. Society for American archaeology symposium, Vancouver, Canada

Vasconcelos MD, Zeigler BP (1993) Modeling multi-scale spatial ecological processes under the discrete event systems paradigm. Landsc Ecol 8(4):273–286
Wainer G (2006) Applying Cell-DEVS methodology for modeling the environment. Trans Soc Model Simul Int 82(10):635–660
Warren SD, Mitasova H, Hohmann MG, Landsberger S, Skander FY, Ruzycki TS, Senseman G (2005) Validation of a 3-d enhancement of the universal soil loss equation for preediction of soil erosion and sediment deposition. Catena 64:281–296
Willgoose G (2005) Mathematical modeling of whole landscape evolution. Annu Rev Earth Planet Sci 33(1):443–459
Wood J, Cowgill GL, Dewar RE, Howell N, Konigsberg LW, Littleton JH, Swedlund AC (1998) A theory of preindustrial population dynamics: demography, economy, and well-being in malthusian systems [and comments and reply]. Curr Anthropol 39(1):99–135
Wymore AW (1993) Model-based systems engineering: an introduction to the mathematical theory of discrete systems and to the tricotyledon theory of system design, vol 3. CRC press
Zeigler BP, Praehofer H, Kim TG (2000) Theory of modeling and simulation, 2nd ed. Academic Press, Inc., Orlando, FL, USA.

Chapter 7
Transformation of Conceptual Models to Executable High-Level Architecture Federation Models

Gürkan Özhan and Halit Oguztüzün

7.1 Introduction

An early insight into the importance of a model-based approach for the field of modeling and simulation was offered by Tuncer Ören and his colleagues. As early as 1979, Ören and Zeigler set forth their concepts for the design and implementation of advanced simulation methodologies. In Ören and Zeigler (1979), they contend, "A successful methodology will allow the modeler to think in the modeling terms most familiar to him and suitable to the relevant characteristics of the real system." This is exactly what we call domain-specific modeling today. They go on to argue, "The computer's role here will be to translate the description of the model, whatever its format, into a standardized representation suitable for undergoing the manipulations of the model file." The format or representation of the model in the present setting is defined by a metamodel, and the translations they refer to corresponds to our model-to-model transformations. Finally, in anticipation of automated model-to-text transformations, they add, "Subsequently, a compiler must be available to convert the model description into a procedural code for simulation (the target language may well be a conventional simulation language)."

Ören advocated "model-based simulation" initially in Ören (1984), later elaborated in Ören (2009). He identifies three types of model-based activities: model building, model-based management, and model processing. Model transformation and to a lesser extent behavior generation are two of the model processing activities that are the subjects of the present chapter. In Ören (2002), he called for exploratory research for the interoperability of next-generation HLA federates from the following perspectives: "(1) Domain-specific and graphic specification environments would be useful in specifying simulation studies. (2) A translator, to be developed

G. Özhan (✉) • H. Oguztüzün
Department of Computer Engineering, Middle East Technical University, Ankara, Turkey
e-mail: gurkan.ozhan@nato.ncia.nato.int

L. Yilmaz (ed.), *Concepts and Methodologies for Modeling and Simulation*, Simulation Foundations, Methods and Applications,
DOI 10.1007/978-3-319-15096-3_7

only once, can transform graphic problem specification expressed in a high level specification language. (3) The specifications and not the computer codes should be integrated and corresponding code should be generated by a program generator." The present work can be viewed as an attempt to answer his call.

In this chapter, we present a formal, declarative, and visual model transformation methodology from a domain conceptual model (CM) to a distributed simulation architecture model (DSAM). The produced simulation model is then fed into a code generator to generate source code that can be executed on a distributed simulation runtime infrastructure. The presented mechanism is generic in the sense that the proposed abstract CM template can be extended and specialized into a domain-specific CM and transformed following the necessary tweaking on the domain-specific parts of the transformation rules.

Specifically, this chapter introduces a two-phase automatic transformation of a field artillery conceptual model (ACM) into a high-level architecture (HLA) federation architecture model (FAM) into executable distributed simulation code. The approach followed in the course of this work adheres to the principles of model-driven engineering (MDE). The ACM and the FAM conform to their metamodels, which are separately built with the generic modeling environment (GME) tool. These two metamodels are composed of data and behavior parts, where the behavior representation in both is based on live sequence charts (LSC). The ACM-to-FAM transformation is carried out with the graph rewriting and transformation (GReAT) tool and partly hand coded. Code generation from FAM is accomplished by employing a model interpreter that produces source code for each member federate. After weaving the necessary computational aspects and compiling the code for a federate application, it is ready for execution on an HLA runtime infrastructure (RTI). Generalizing the ACM-to-FAM transformation case study, we propose a set of key design principles and an implementation framework as a step forward in achieving generic conceptual model (CM) transformations for publish/subscribe (P/S)-based distributed simulation.

The model-driven engineering (MDE) approach (Schmidt 2006) is becoming prominent in software and systems engineering, bringing a model-centric approach to the development cycle in contrast to today's mostly code-centric practices. A well-known MDE initiative is the model-driven architecture (MDA) of the object management group (OMG). Model transformations are considered to be the heart of MDA, where the platform-independent model (PIM) of a system to be constructed is transformed into a platform-specific model (PSM), which can be translated to executable code (Sendall and Kozaczynski 2003). An earlier manifestation of MDE, Model Integrated Computing (MIC) (Ledeczi et al. 2001a), relies on metamodeling to define domain-specific modeling languages and model integrity constraints. The metamodel (also referred to as a paradigm) is then used to automatically compose a domain-specific model-building environment to create, analyze, and evolve the system through modeling and generation. In this approach, a crucial point is generation, in which (domain-specific) models are transformed into lower-level executable and/or analysis models. Model transformation techniques and tools are essential in realizing the generation process.

7.1.1 Motivation and Scope

In the last decade, there has been a considerable proliferation of literature on model transformations and a rapid dissemination of the MDE approach (Sendall and Kozaczynski 2003; Agrawal et al. 2006; Bézivin 2009). As such, a recent interest has been shown by the modeling and simulation (M&S) community (Ören 2009; Kewley and Tolk 2009; Özhan and Oguztüzün 2013; Garcia and Tolk 2013; Bocciarelli et al. 2012). However, accounts of the nuts and bolts of the practical MDE application are missing. Usually, transformations are applied between narrow domains, avoiding realistic concerns and practical difficulties (Maoz and Harel 2006), and they are single-step source models to target model transformations (Garcia and Tolk 2013). Also, some transformations are achieved either within the same domain (mostly operating on a single model) or between two highly similar, tightly coupled domains. The term *domain* is used in the sense of an area of interest such as field artillery or HLA-based distributed simulation. Last, although descriptions of model transformations of either the data or behavior (Maoz and Harel 2006; Bontemps et al. 2005) are abundant, reported works placing both aspects on the same footing in an integrated fashion are rare.

The methodology is applied in a case study where the CM is the field artillery conceptual model (ACM) (Özhan et al. 2008) and the DSAM is the federation architecture model (FAM) (Topçu et al. 2008), which is a domain model of high-level architecture (HLA)-based distributed simulation. The model transformation paradigm is graph based, where the transformation rules are defined over the source and target metamodels. The transformation process is continued with a model interpretation of the produced FAM to generate Java/AspectJ code that can be executed on an HLA runtime infrastructure (RTI) (IEEE 2000a).

From an implementation perspective, our work can be considered as a sequence of applications of the MIC approach. It is intended as an MDE-based end-to-end systems development endeavor from the conceptual model to executable simulation code, promoting a rigorous use of model transformations.

7.1.2 The Context of the Transformations

In order to clarify the purpose and provide a referential overview of the process, this section outlines the two-phased transformations within the context of the case study. The first phase is a model-to-model transformation, and the second is a model-to-code transformation. The ACM is a PIM of the part of reality (i.e., field artillery domain) with which the simulation is concerned. The FAM is a PSM, where the platform is the HLA-RTI in our case. The model transformer produces a FAM from an ACM by executing the ACM2FAM transformation rules, and the Aspect/Java-based code generator produces executable code from that FAM.

ACMM and FAMM are the metamodels of ACMs and FAMs, respectively. The two metamodels (and consequently the models) consist of data and behavior parts. The metamodel of live sequence charts (LSCs) and message sequence charts (MSCs) (Damm and Harel 2001; ITU-T 2004; Topçu et al. 2008) are used for behavioral representation in both metamodels. The two data models are integrated with the behavioral model in such a way that the top-level data model elements are extended from a set of designated LSC and MSC model elements. The transformations are defined in such a way that first, the data model transformation is conducted, followed by the behavioral model transformation. In the second phase, the federate application code generator produces executable simulation code and supporting artifacts such as the federation object model (FOM) document data (FDD) from the FAM. This chapter is more focused on the first phase transformations.

It may seem plausible to directly produce HLA federate codes from the conceptual model instead of passing through two phases of model transformations. Our approach is more appealing in several ways. First, the ACM rests at a higher conceptual level, while federation code is at a lower, more detailed level. FAM on the other hand is at an intermediary level, serving as a bridge between the two models. It has a clearer mapping from ACM and to the federation code. This makes the transformations more modular and maintainable. Second, the components of a FAM, consisting of the HLA object model template (OMT) model and intra-federation behavioral model, are useful artifacts in their own right as they are used by the RTI and HLA federation development tools. Furthermore, once a FAM, which is machine processible, is made available, it can be used as an input for further activities such as optimization, debugging, verification, and validation.

The transformations are defined over the metamodels of the source and target domains expressed in a Unified Modeling Language (UML)-based notation. The relationships between the models and the metamodels are summarized in Fig. 7.1.

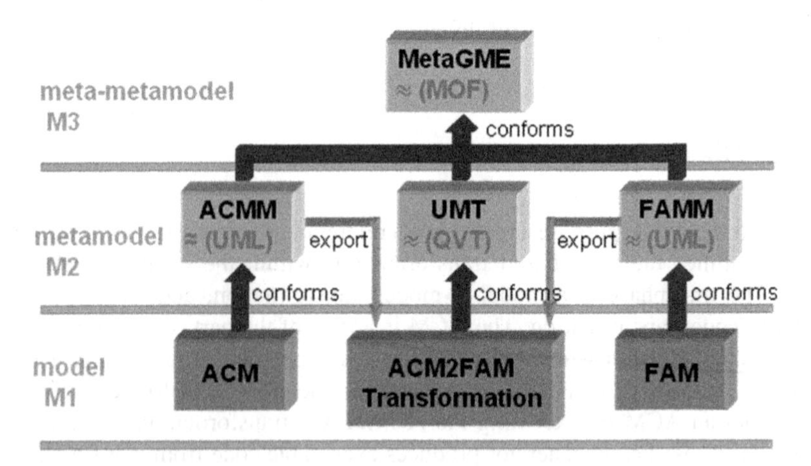

Fig. 7.1 The three layers of modeling used in this work in relation to MDA

The modeling and model transformation activities and products in this work are formally defined due to model conformance relationships. The figure also associates the different levels of models used in this work to modeling layers of the OMG and the MDA standards (OMG 2006, 2007, 2011).

7.1.3 Organization

The remainder of this chapter is organized as follows. Section 7.2 provides background information to the modeling and the model transformation tools that were used as well as the source and target domain models. Section 7.3 outlines a conceptual framework independent of the details and jargon of the specific domains, tools, and technologies. Section 7.4 presents the highlights of the ACM2FAM transformation case study. Section 7.5 contains the discussion and the assessment of this work. Section 7.6 provides an overview of the related works in the literature. Finally, Sect. 7.7 concludes and points to the directions of future research.

7.2 Background

This section provides brief background information on GME, the modeling tool used in building both the source and target metamodels and the models, and GReAT, the model transformation tool. The section also presents short narrations on the source and target domains and introduces the corresponding metamodels.

7.2.1 Generic Modeling Environment

Generic Modeling Environment (GME) (Ledeczi et al. 2001a) is a configurable toolkit for creating domain-specific modeling and program synthesis environments. The configuration is achieved through metamodels specifying the modeling language (i.e., "paradigm" in the GME vernacular) of the application domain. The paradigm defines the family of models that can be created using the resulting modeling environment.

The metamodel for each domain-specific modeling language is defined using the UML-based metamodeling language named MetaGME, which has the same role Meta-Object Facility (MOF) (OMG 2006) plays in UML2 (OMG 2007). When a metamodel is registered in GME, GME provides a domain-specific model-building environment. The generated environment is then used to build and manipulate domain models.

GME further provides a generic application programming interface (API) called BON2 to access the models by paradigm-specific interpreters. Using the API, developers are able to programmatically traverse and manipulate a GME model with the same set of capabilities provided by the visual GME environment. The federate code generator of the second phase transformation is implemented in such a way.

7.2.2 Graph Rewriting and Transformation

Graph rewriting and transformation (GReAT) (Agrawal et al. 2006) is a graph-based transformation language for model-to-model transformations with precise formal and executable semantics. GReAT has a high-level control flow language built on top of the graph transformation language with sequencing, nondeterminism, hierarchy, recursion, and branching constructs. GReAT uses UML (OMG 2007) class diagrams and Object Constraint Language (OCL) to represent the source and target domains of the transformations and integrity constraints over those domains. The GReAT metamodel, the UML Model Transformer (UMT) paradigm, comes bundled with the GME installation. By creating models conforming to this paradigm in GME, it is possible to define model transformations.

GReAT defines a production (i.e., rule in UMT terms) as the basic transformation entity. A production contains a pattern graph that consists of pattern vertices and edges. The pattern graph consists of elements from the source and target metamodels and elements that are newly introduced inside the transformation model (such as cross-links or globals). Each pattern object has a *bind*, *delete,* or *new* designation that specifies the role it plays in the transformation. Bind is used to match objects in the graph. Delete is also used to match objects in the graph, but afterward, they are deleted from the graph. New is used to create objects after the pattern is matched. The execution of a rule involves matching every pattern object marked either bind or delete. If the pattern matcher is successful in finding matches for the pattern, then for each match, the pattern objects marked delete are deleted from the match, and objects marked new are created.

Sometimes, the patterns themselves are not sufficient to specify the exact graph parts to match. Then, other nonstructural constraints on the pattern are needed. These constraints or preconditions are expressed in a guard and are described using OCL. Attribute-mapping elements provide values to the attributes of newly created objects or modify the attributes of existing objects. Attribute mapping is applied to each match after the structural changes are completed.

The model transformation language is supported through the GReAT execution engine. The engine basically inputs the transformation definition (i.e., rules and sequencing) and a source model to automatically produce a corresponding target model. A high-level overview of the model transformation architecture of this work is illustrated in Fig. 7.2. The shaded components are parts of GReAT, and the rest are artifacts developed within the scope of this work. The engine provides access to a generic programming API called the Universal Data Model (UDM) (Bakay and Magyari 2004) that can be used to further manipulate and fine-tune the generation.

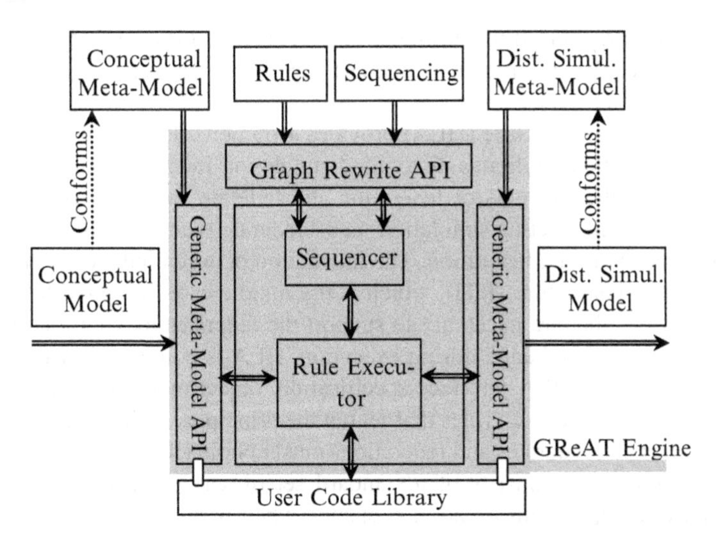

Fig. 7.2 Overview of the architecture of the model transformer

This mechanism works by invoking user code library methods from within transformation rules.

7.2.3 Field Artillery and High-Level Architecture Domains

A brief account of the field artillery domain provided here is gleaned from US Army field manuals (FM 6–30 1991; FM 6–40 1996). The general mission of field artillery is to destroy, neutralize, or suppress the enemy by cannon, rocket, and missile fire and to help integrate all fire support assets into combined arms operations. Observed fire is carried out by the coordinated efforts of the field artillery team, which is composed of the forward observer, the Fire Direction Center (FDC), and several firing sections of the firing unit. The basic duty of the forward observer, who is considered to be the eyes of the team, is to detect and locate suitable indirect fire targets within his zone of observation. In order to start an attack on a target, the forward observer issues a call for fire (CFF) request to the FDC. It contains the information needed by the FDC to determine the method of attack.

As it is likely to miss the target in the first round of fire, the common practice is first to conduct adjustment on the target. Usually, the central gun is selected as the adjusting weapon. After each shot is fired, the observer provides correction information to the FDC based on his spotting of the detonation. Once a target hit is achieved, the observer initiates the fire for effect (FFE) phase by noting this in his correction message. FFE is carried out by all the weapons of a firing unit firing together with the same fire parameters as the last adjustment shot. After the designated number of rounds is fired, the observer sends a final correction including surveillance information. If the desired effect on the target is achieved, the mission ends. Otherwise, the observer may request repetitions or restart the adjustment phase if deemed necessary.

The high-level architecture (HLA) provides a framework based on the publish/subscribe pattern that facilitates distributed simulation interoperability and reuse. HLA introduces the concepts of federation and federate, where a federation is a composable set of interacting simulations and a federate is an individual application that participates within a federation. The interaction between federates is managed by a runtime infrastructure (RTI), which is the middleware that provides a set of software services that are necessary to support the federates by coordinating their operations and data exchange during execution. HLA is comprised of three components, namely, the HLA interface specification, object model template (OMT), and HLA rules. There are ten rules that set out the principles of the HLA in terms of responsibilities that federates and federations must uphold (IEEE 2000a). The HLA interface specification consists of a standard set of services and interfaces that federates use to support information exchange when participating in a federation execution (IEEE 2000c). It is defined in terms of a set of functions specified through an API. Practically speaking, an RTI implements the HLA interface specification but is not itself part of the specification. The OMT provides a means of documenting key information about the federates and the federation (IEEE 2000b).

7.2.4 Field Artillery and Federation Architecture Models

ACMM, the metamodel of ACM, serves as a conceptual model for the field artillery observed fire domain. In other terms, it can also be called a field artillery mission space model, and it is developed using the GME tool. Registering ACMM as a paradigm in GME yields a domain-specific language for the formal definition of an observed fire mission (i.e., an ACM) such as *fire for effect*. An ACM is the input source model for a transformation. ACMM consists of a behavior component and a data component. The data model addresses certain aspects of tactical rather than technical fire direction. The entire top-level domain entities in the data model are specialized from NATO's Joint C3 Information Exchange Data Model (JC3IEDM 2007). Please refer to Özhan et al. (2008) for the details of ACMM and the model of a fire for effect mission in visual LSC notation.

FAMM, the metamodel of FAM, describes the architecture of an HLA-compliant federation. Like ACMM, it is formulated in GME and as such provides a domain-specific language for the formal representation of an HLA federation when registered in GME. FAMM is also comprised of behavior and data components. The data portion covers the HLA interface specification (IEEE 2000c) and the OMT (IEEE 2000b). Please refer to Topçu et al. (2008) for a thorough presentation of FAMM.

The behavioral parts of both ACMM and FAMM are essentially the LSC metamodel (Topçu et al. 2008). As LSC is extended from MSC, the metamodel

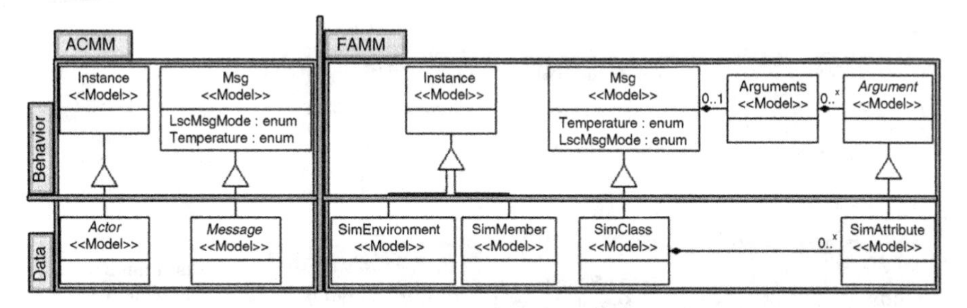

Fig. 7.3 Overview of data and behavior integration in ACMM and FAMM

covers the MSC metamodel in its core. MSC is a visual language that specifies the behavior of a concurrent system focusing on the communication aspect. The metamodel covers all the standard MSC features (ITU-T 2004) and the proposed LSC extensions (Damm and Harel 2001), most notably the distinction between optional and mandatory behavior as a coherent whole.

Figure 7.3 shows a high-level overview of the integration of data and behavior submodels of the ACMM and FAMM model components, which have been separately developed and seamlessly integrated (see Sect. 7.3.1 for model details). Effectively, the behavioral model relies on the data models for the definition of domain-specific data types. The observable behaviors in a field artillery mission (HLA federation in a FAM) are represented by means of specialized LSCs. This is achieved by extending the relevant data model elements from the behavioral model elements in the sense of the UML inheritance (OMG 2007).

7.3 The Conceptual Framework for Model Transformation

This section provides a conceptual framework for this work including the metamodeling and model transformation. The content is abstracted as much as possible to facilitate comprehensibility and appeal to a broader range of readers and potential adopters. We present a formal multistage model transformation endeavor from a domain conceptual model (CM) to a distributed simulation architecture model (DSAM) and from that to executable simulation code and supporting artifacts. The end-to-end transformation process, which is elaborated in the subsequent sections, is depicted in Fig. 7.4. CMs and DSAMs are formally built in compliance with the metamodels conceptual metamodel (CMM) and distributed simulation architecture metamodel (DSAMM), respectively. The transformation is defined over these metamodels. The CM encompasses the (static) domain entities and the (dynamic) behavior of the source domain. The same holds for the DSAM in relation with the target domain.

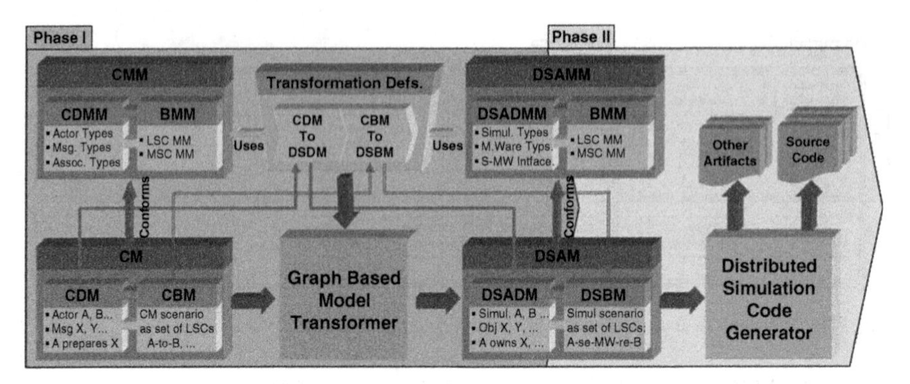

Fig. 7.4 The overall model transformation process

7.3.1 The Conceptual and Distributed Simulation Architecture Models

This section introduces the data portions of the conceptual and distributed simulation architecture metamodels and the behavioral (i.e., LSC) metamodel.

7.3.1.1 The Conceptual Data Model

The data portion of the CMM, the conceptual data metamodel (CDMM), consists of a set of domain entities called actors that are able to perform computations and send/receive messages (to/from other actors and the environment) on a one-to-one or multicast basis. The communicated messages are collections of domain information extracted from authoritative sources and composed in different granularities. The messages are categorized as being *durable* or *nondurable*. This durability distinction facilitates the transformation definitions for the target models of the distributed simulation domain such as HLA. A durable message represents information that is intended to be kept and maintained by the receiver. A nondurable message represents information that is meant to be used immediately and is then discarded by the receiver (barring, of course, logging).

The upper-level model elements of the CDMM and their associations are shown in the UML diagram given in Fig. 7.5. In the figure, the elements labeled with stereotype Model are the primary building blocks of the communicated data and can be organized recursively to represent composite structures. The data model is built of Messages, Actors, and DurableDataStore folders. The messages, durable or nondurable, are stored in the Messages folder. The durable data messages are further specialized into *instantiation*, *update*, and *delete* types. Since the objects corresponding to durable data messages need to be retained, they are kept in the DurableDataStore folder. An *instantiation* type of durable message contains the original copy of the durable data that was placed in the store the first time.

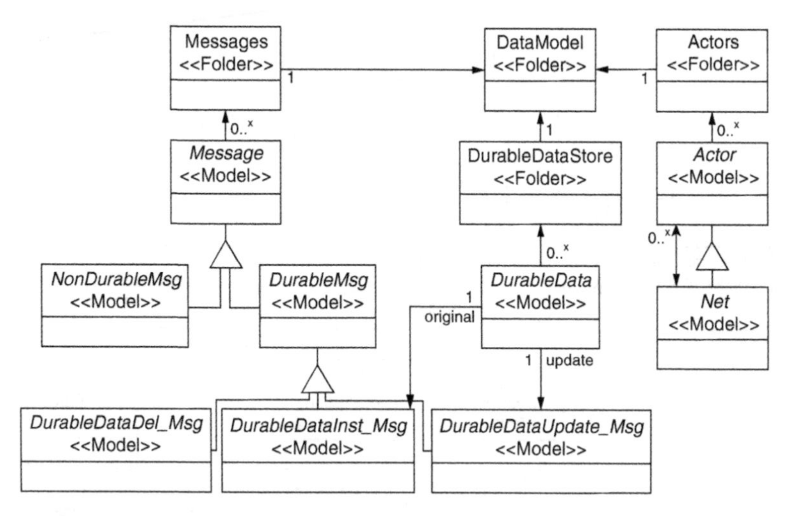

Fig. 7.5 The upper level CDMM elements

Subsequent *update* messages contain template objects that are used to update the effective copy residing in the store. The *delete* message indicates the corresponding durable data to be deleted from the store. The Actors folder contains the domain elements of type Actor and Net. Net is a special kind of Actor that represents a collection of actors where any message coming into a net is assumed to be delivered to all the actors included in the net.

7.3.1.2 The Distributed Simulation Architecture Data Model

The DSAMM data model is the distributed simulation architecture data metamodel (DSADMM) that consists of elements that collectively define the static view of a set of autonomous and loosely coupled interoperating simulations. The interactions are mediated via the simulation infrastructure or middleware. The middleware functions as the overarching manager that knows about the identities and data exchange interests of the participating simulations and orchestrates all the communication traffic. The DSADMM defines the structure and organization of the communicated data as classes of simulation objects categorized by having lifetimes of single interactions or the whole simulation.

An overview of the DSADMM is depicted in Fig. 7.6. The Connection stereotype is used to define association classes between model elements. The simulation data model consists of the simulation environment, a number of member applications (SimMember), and a simulation data exchange model element. The data exchange model contains instances of simulation classes, which represent the data structures communicated within the simulation environment. The simulation class is specialized into simulation objects and simulation interaction types, of which the former is intended to model durable information and the latter to

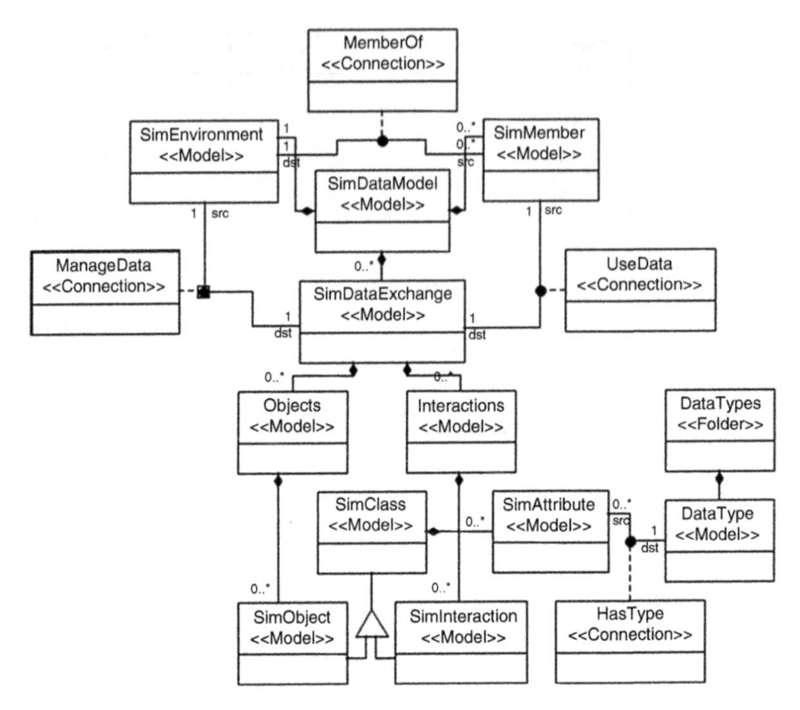

Fig. 7.6 Prominent DSADMM elements

model instantaneous events. The simulation classes contain attributes having data types defined in the specific distributed simulation domain. For instance, HLA has a default set of simple, enumeration, array, and record data types. The simulation environment *manages* the overall communication taking place among the member applications. The member applications *use* a set of simulation class instances that they produce or consume for exchanging data with each other.

7.3.1.3 The Behavioral Model

The behavioral metamodel (BMM) is used in both the source and target models. BMM is a representation of the LSC/MSC formalism, which shares many constructs with the UML sequence diagrams (OMG 2007). The behavioral metamodel is capable of representing the discrete communication behavior of many practical systems consisting of components exchanging messages independently of the domain. This communication aspect of the system behavior is particularly emphasized when taking the LSC modeling perspective. A simplified view of the upper-level elements of the BMM and their associations are provided in Fig. 7.7.

According to BMM, the behavioral specification of a system is captured in a single MSC document, which consists of a document head and one or two document bodies. The document head includes declaration lists for the instances,

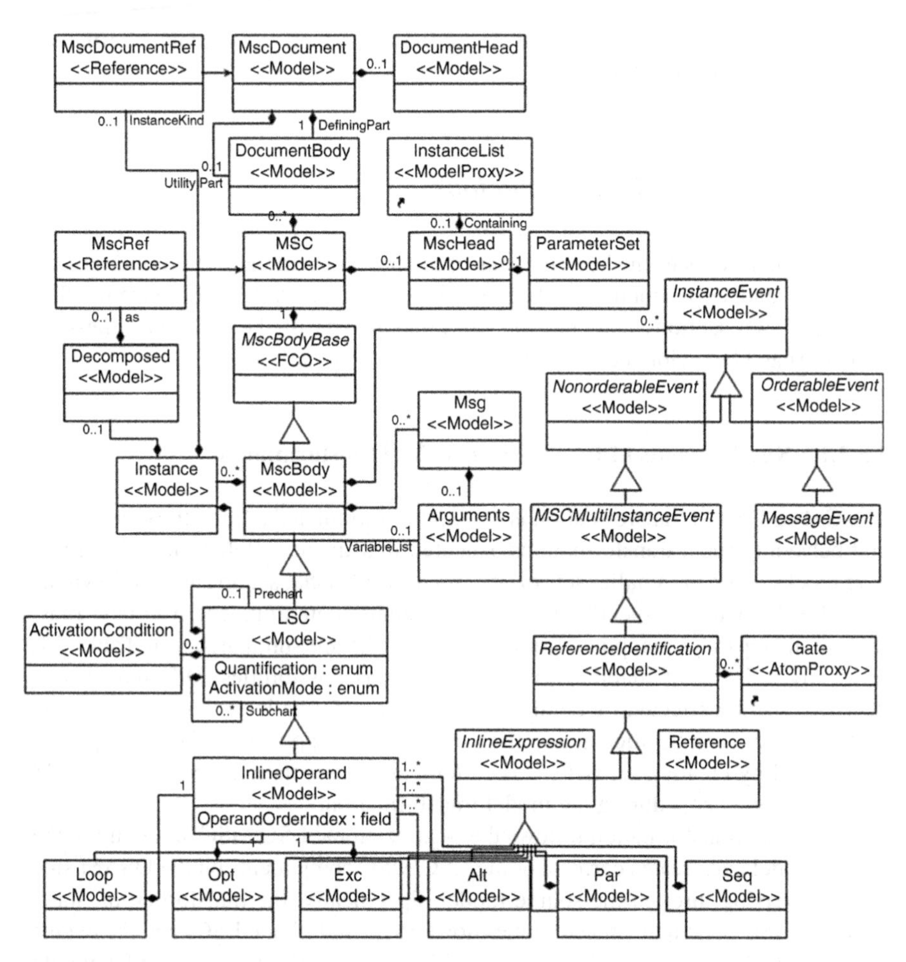

Fig. 7.7 Simplified view of upper-level BMM elements

messages, and timers used in the document and optionally a reference to another document from which it is "inherited" (not shown in the figure). The body part of the document is modularized into a set of MSCs, each of these, similar to the MSC document but only pertaining to its own scope, has a head and a body. LSC is the most commonly used MSC body type and is the primary means for representing the behavioral specification of the system being modeled. The LSC contains, besides other items, a set of references to the instances that interact with each other using a variety of instance events. An important event group from a model transformation perspective is the message event, which provides the mechanism to exchange data between the instances in the form of LSC messages. LSC is recursively defined and is allowed to refer to other MSCs in order to support the modularization of large behavioral descriptions. The inline operand is the main building block of the nonorderable multi-instance type of events called inline expressions. Inline expressions include constructs for defining loop, optional, exceptional, alternative, parallel, and sequential flows.

7.3.2 *The Model Transformations*

The section introduces the key elements of the transformations and summarizes the mappings and correspondences between the CM and DSAM components. The following two sections elaborate on the crucial parts of the data (i.e., message structure) and behavior (i.e., communication) transformations.

7.3.2.1 Key Elements of the CM-to-DSAM Transformation

The CM-to-DSAM transformation is essentially formulated around the core of data and behavior model transformations executed in sequence. Before and after these core blocks come the smaller sets of pre and post rules that set up and tear down the stage for the platform-specific distributed simulation environment. The key transformation steps and mappings from CM to DSAM are summarized in Table 7.1. Evidently, this set of mappings is one of various possibilities. Different design decisions can be effected by defining different transformation rules. For example, multiple actors can be allocated to the same member application. These mappings are marked by creating *cross-links* (i.e., temporary associations) between the CM and DSAM elements during the model transformation.

The model transformation follows the organizational LSC/MSC hierarchy of the source model and creates corresponding LSC/MSC components on the DSAM side while traversing through the source model. Specifically, the transformation starts from the MSC document and continues down to the individual LSCs and the events inside the LSCs (see Fig. 7.7 for the LSC/MSC structure). Since the top-level data model elements are extended from the LSC elements, the LSC transformation also implicitly covers the data model elements. Due to this LSC-centric approach, the CM-to-DSAM transformation is essentially an LSC transformation.

7.3.2.2 Transforming Message Structures

The transformation of CM message structures to their corresponding simulation classes is performed during the data model transformation step of the CM2DSAM transformation. In a nutshell, all nondurable (i.e., with a life span of only a message transmission period) messages are transformed to interaction classes, and durable messages (i.e., with a life span of the whole simulation execution unless deliberately deleted) are transformed to object classes.

The user code library facilitates data model transformations in several ways. A CM message can be deeply structured, possibly with optional child objects

Table 7.1 A partial view of the mappings from Conceptual Model to Distributed Simulation Architecture Model

CM component	DSAM component
Actor/Net	Member application
Non-durable message	Simulation interaction
Durable message	Simulation object
<NA>	Simulation environment is brought in as a collection of communicating member applications
Actor-actor non-durable comm.	Member application to member application communication via the simulation environment (running the middleware), using a pair of send/receive interaction messages
Actor-actor durable comm. (instantiation type)	Member application to member application communication via the simulation environment, using three pairs of register/discover object, request/provide attribute update and update/reflect attributes messages
Actor-actor durable comm. (update type)	Member application to member application communication via the simulation environment, using a pair of update/reflect attributes messages
Actor-actor durable comm. (deletion type)	Member application to member application communication via the simulation environment, using a pair of delete/remove object messages
<NA>	Default distributed simulation data types are brought in to be used by the simulation class attributes
<NA>	Simulation environment initialization is introduced in a preliminary LSC by creating the environment, having the member applications join the environment, declaring member application data exchange interests and other sorts of simulation-specific in itializations
<NA>	Simulation environment shut down is brought in to the final LSC by resigning the registered member applications from the simulation middleware and destroying the simulation environment
Other CM LSC components	Similar DSAM LSC components

(i.e., having containment cardinality *0..m*). Handling such a message transformation solely through graph matching requires many (exponential in the number of descendants in the worst case) transformation rules to address all the possible pattern combinations. On the other hand, with the user code library, only one rule is needed no matter how many children exist with whatever cardinality a message structure may have. The rule matches the top CM message element, creates an empty DSAM class, and invokes the code library, which programmatically builds

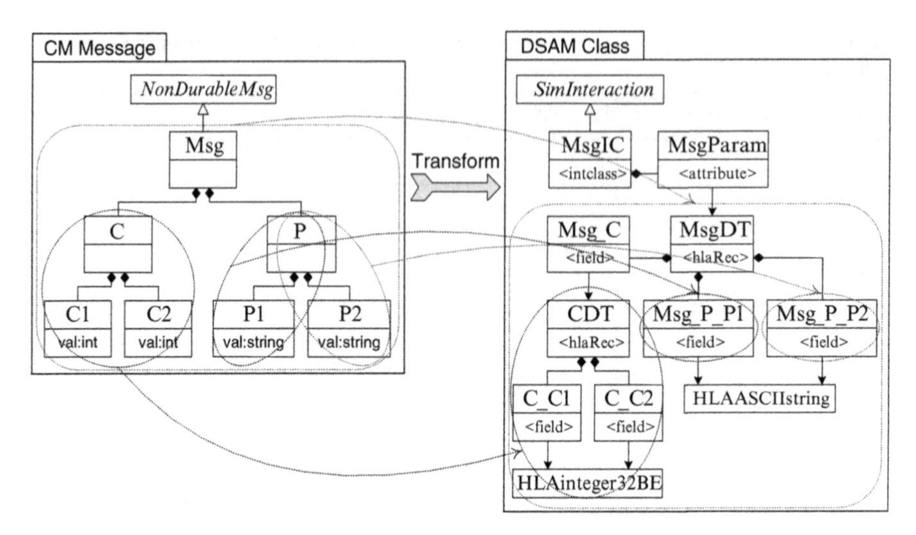

Fig. 7.8 An abstract view of a CM message to DSAM class (in HLA-OMT) transformation

the DSAM class contents as it traverses the CM message structure. The other reason for using the code library is to obtain a considerable performance gain by directly executing the C++ code instead of running many transformation rules with minor variations. (Be reminded that the subgraph isomorphism problem, which is involved in every pattern-matching step, is NP complete.)

Transforming a CM message into a DSAM class (more specifically, object model template (OMT) class in HLA terms) and its attributes is conceptually presented in Fig. 7.8. The figure only demonstrates a nondurable CM message transformation. Durable message transformation is undertaken in a similar way. On the left side is a CM message structure named Msg having two components named C and P, both of which have a couple of child elements. Each leaf child has one attribute named val of the type shown. It is assumed that C is a common component, which is possibly reused by other CM messages, and P is a noncommon component specific to the message in question.

The transformation rule creates the MsgIC interaction class on the right-hand side through pattern matching. Then, the user code library creates an attribute and its data type for the class. The CM message content is transformed into an HLA (fixed) record data type. Each common message component is transformed into a field of the main record type, which in turn has a record data type mimicking the common content. This common child record type is made reusable for subsequent rules. All the other noncommon parts of the message structure are flattened as direct fields of the main record type having appropriate primitive/simple types, with the field name reflecting the message structure hierarchy. Conventionally, the field name of a record type consists of a string of concatenated message structure element names separated by "_" from the leaf to the top CM element corresponding to the record. Alternative message transformation approaches are discussed in Özhan et al. (2010).

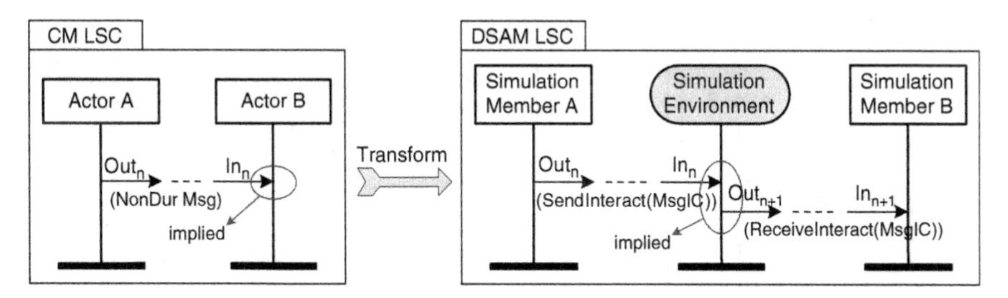

Fig. 7.9 An abstract view of a non-durable CM message comm. to DSAM transformation

7.3.2.3 Transforming Message Communications

The crux of the model transformation work is the transformation of a typical one-to-one communication between the actors of a CM. The nondurable and the three kinds of durable message types are used in message communications. Each kind is transformed using a specific transformation rule set.

A simplified and abstract view of the CM LSC-to-DSAM LSC transformation involving a nondurable message event transmission is illustrated in Fig. 7.9. The LSCs in the figure are represented in graphical LSC notation (Damm and Harel 2001). The figure also demonstrates the mappings of the CM actors and message types to their DSAM counterparts.

For every CM actor, a corresponding member application (federate in HLA terms) with the same name is defined. The simulation environment, which is specific to the DSAM domain, corresponds to the runtime infrastructure (RTI) of HLA that mediates and monitors the communication between the federates during the simulation execution (IEEE 2000a). These DSAM LSC instances are actually created as part of the data model transformation before the behavior model transformation step. They are later referenced from within the LSCs. The CM messages and DSAM classes are included as parameters of the in/out events.

According to the LSC specification, for every out-event sent by a sender, a corresponding in-event is received by the receiver (provided that the out-event is not lost in the meantime, meaning that it is "hot") (Damm and Harel 2001). In our behavioral metamodel, the definitions of out-event and in-event include both the source and target elements of the event. Since both parties of the out (in) event are known, we discard the declaration of the other corresponding in (out) event and thus reduce the total number of events by half but still do not sacrifice any communication semantics. It is assumed that the corresponding event is implicitly there.

The communication architecture of DSAMM requires that an actor A to B out-event transmission of a CM should be represented as member application A sending an out-event to the simulation environment first and the simulation environment sending another out-event to member application B (see the right part of Fig. 7.9). Since the single out-event communication in a CM is transformed into a pair of out-event and in-event communication in the DSAM, if the out-event has execution

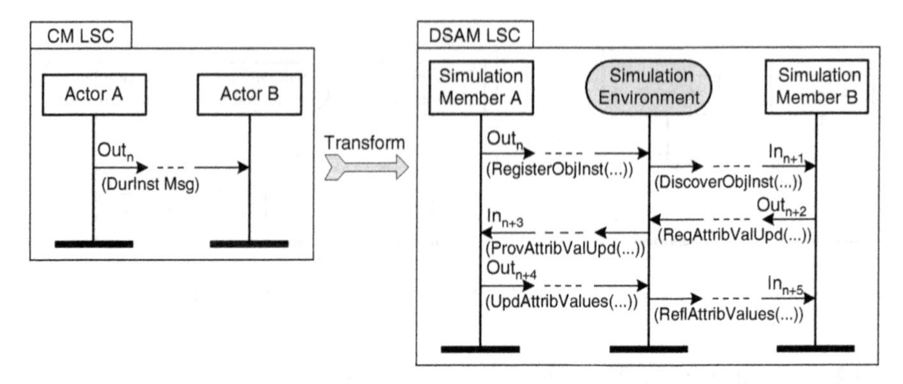

Fig. 7.10 An abstract view of a durable-instantiation CM message communication to DSAM transformation

order index n, then the following in-event is given a higher execution order index, say $n+1$, reflecting this discrete communication behavior.

Similarly, Fig. 7.10 summarizes how a durable *instantiation* type of CM message communication is transformed into a series of HLA messages between the corresponding member applications and the simulation environment. In accordance with the HLA specification, the owner of the object class first registers the object with the simulation environment, which in turn informs the interested parties of its availability. Then, the owner is requested to provide the values for the object. Finally, the owner sends value updates to the simulation environment, which in turn delivers the requested information to the subscribed members.

All these six messages are time ordered and assigned increasing execution order indices along the vertical axis as seen in the figure. The other two durable message types, namely, the *update* and *delete* types, are similarly transformed, each having a pair of out- and in-events with the parameters listed in Table 7.1.

7.4 Case Study: ACM to FAM Transformation

This section demonstrates the CM2DSAM transformation conceptually described in Sect. 7.3 in a real-life case study. The source model is the artillery conceptual model (ACM) of the field artillery observed fire domain (FM 6–30 1991; FM 6–40 1996). The target model is the federation architecture model (FAM) of the HLA-based distributed simulation domain (IEEE 2000a, b, c). Both the source and target models adhere to their metamodels ACMM (Özhan et al. 2008) and FAMM (Topçu et al. 2008). This section presents excerpts of the transformation rules developed in the Generic Modeling Environment (GME) (Ledeczi et al. 2001b) tool using the graph rewrite and transformation (GReAT) language notation (Agrawal et al. 2006). Figure 7.11 illustrates the overall modeling and model transformation context in a nutshell. The data part of ACMM defines the field artillery domain

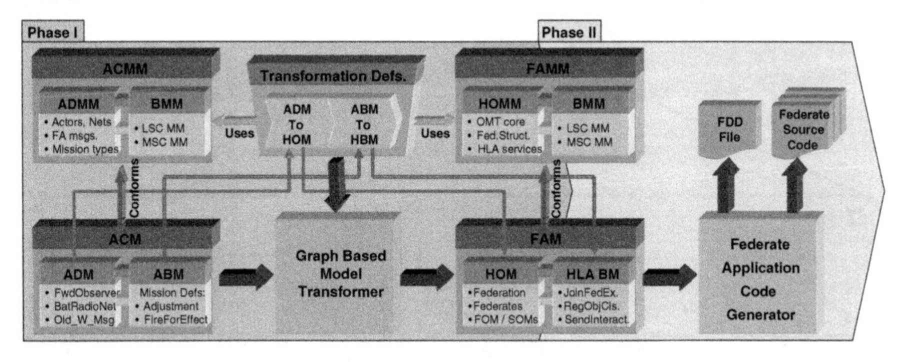

Fig. 7.11 Overview of ACM to FAM to executable code transformation

Table 7.2 Metrics for the ACM2FAM transformation

Transformation expression	Count	Transformation expression	Count
Block	64	Test	13
ForBlock	4	Case	55
Rule	187	Expression Reference	21

entities, and its behavior part defines the observed fire missions in LSC form. Likewise, the data part of FAMM defines the field artillery entities as federates, the federation, and HLA messages, and its behavior part defines the fire missions as intercommunicating federates via the RTI, again in LSC form.

The transformation definition is comprised of a set of major transformation blocks, which contain other blocks, transformation rules, cases, or expression references. Table 7.2 summarizes the metrics for the overall ACM2FAM transformation indicating 64 blocks, 4 for-blocks, 187 rules, 13 tests (with 55 cases), and 21 references to other rules in total. The definitions of these GReAT language constructs can be found in Agrawal et al. (2006).

7.4.1 *Data Model Transformation*

Data model transformation corresponds to the structural part of the ACM2FAM transformation. From a FAM perspective, it aims to construct the federation object, the federate objects, and the FOM for the federation. The main DataModelTr block is shown in Fig. 7.12. It is composed of two inner blocks named ObjectModelTr and FederationStructureTr that are executed sequentially in that order.

7.4.1.1 Object Model Transformation

Object model transformation basically transforms the set of messages communicated among the domain actors in field artillery missions into HLA classes. As a

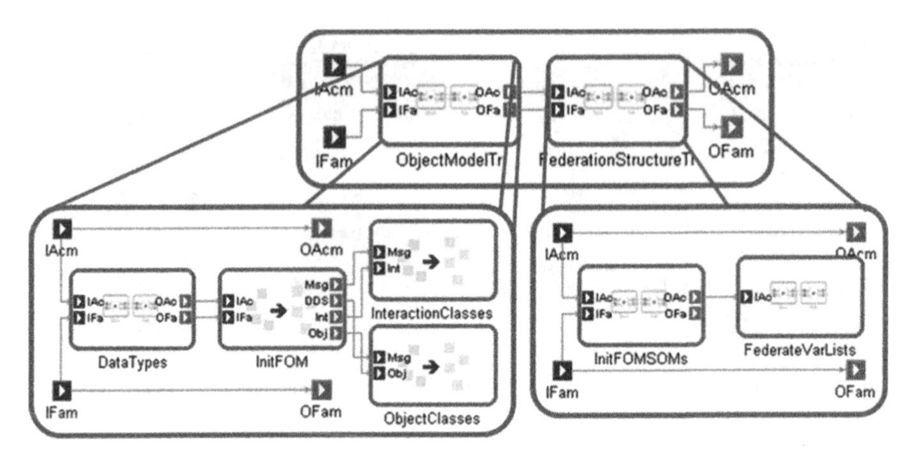

Fig. 7.12 The main DataModelTr block

preliminary step in the field artillery message–to–OMT class transformation, the DataTypes block creates all the standard HLA-OMT data types (IEEE 2000b). The InitFOM rule creates containers for the interaction classes and object classes and an empty FOM element, which is later filled with interaction and object classes. These two kinds of OMT classes are the key elements in FAM data model in that they are used frequently throughout the rest of the rules. Then, the transformation flow splits into two parallel branches where the interaction and object classes are concurrently created. At this point, the user code library is invoked to manipulate the bound objects using the generic UDM API (Bakay and Magyari 2004). The user code library executes the actual field artillery message to OMT class transformation programmatically.

The InteractionClasses rule is provided in Fig. 7.13, where black model elements indicate a pattern to match and blue colored elements designate the new elements to be created. The code snippet inside an AttributeMapping element is executed after the rule pattern is matched and the structural modifications on the matched model elements are made. It invokes the user code library's message transformation method.

7.4.1.2 Federation Structure Transformation

The federation structure transformation constitutes the second part of the data model transformation shown in Fig. 7.12. It instantiates the singleton federation object together with a reference to the FOM that was previously created. It also transforms every field artillery actor and net to a corresponding HLA federate along with a reference to an associated SOM. In this work, the SOMs per federate are left as stubs and are not developed any further. The FOM is sufficient to capture all the OMT objects participating in the federation execution. Indeed, the FOM is what an RTI needs to run a federation (IEEE 2000a). Finally, cross-domain associations

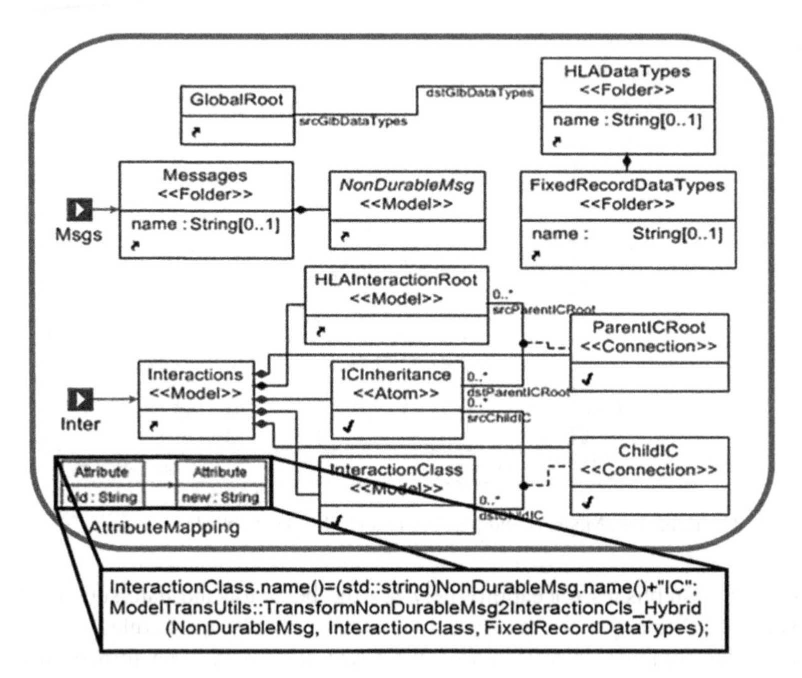

Fig. 7.13 The InteractionClasses rule

(has-correspInst) establish references from each actor/net of the ACM to its corresponding federate of the FAM. These temporary links later function as a key enabler during the transformation of the message communications in the behavioral model transformation step.

7.4.2 Behavioral Model Transformation

The behavioral model transformation is the larger and more challenging part of the overall ACM2FAM transformation. It uses the resulting objects of the data model transformation as the instances and message parameters in the LSCs that are being produced. Since the behavioral model transformation is so large, consisting of more than 60 transformation blocks and over 180 rules, only the prominent parts are presented here.

The cascaded main behavioral model transformation blocks are given in Fig. 7.14. The dashed projector lines indicate the existence of other nonshown transformation blocks in between the connected blocks. The AscGlobalHlaMeths block fetches the method library of FAM that contains predefined HLA methods for the management of *federation*, *declaration*, *object*, *ownership*, and *time*. The AscInstanceOfAcm block basically creates is-InstanceOf associations between the instances that stand for the same actor element in ACM. This chain of

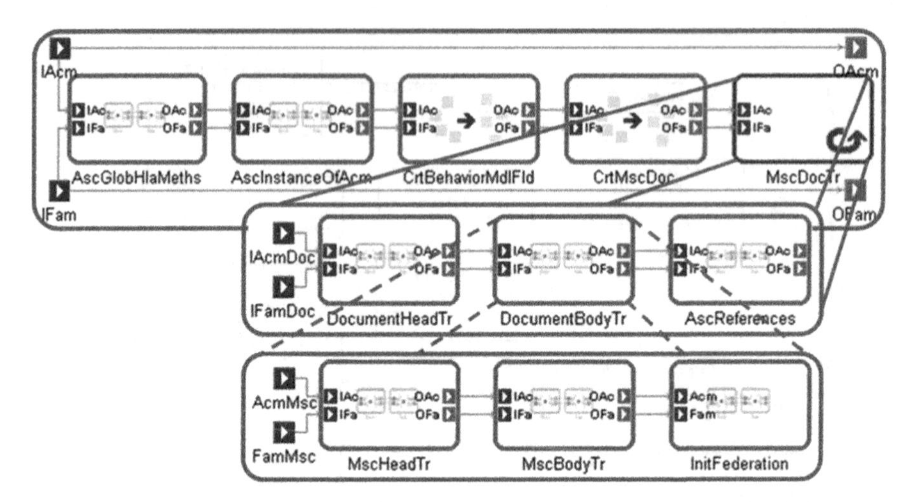

Fig. 7.14 The BehavioralModelTr and MscDocTr and MSCTrans blocks

associations establishes traceability between the behavior and data submodels of ACM and facilitates pattern matching in a number of subsequent rules.

A similar approach is followed on the FAM side as the transformation rules progressively construct the target model. The CrtBehaviorMdlFld and CrtMscDoc rules create a FAM behavioral model folder and an MSC document underneath it provided that their corresponding counterparts are matched in the ACM. A hascorrespMscDoc association is established between the ACM and FAM MSC documents since there can be more than one MSC document in a source model and in such a case this association is necessary to keep track of MSC references in different documents and during *instance decomposition*. See Sect. 7.4.2.2 for a discussion of instance decomposition.

7.4.2.1 MSC Document Transformation

The transformation blocks and rules within MSCDocTr match and traverse the structure delineated by the MSC metamodel to create a FAM MSC document from an ACM MSC document. The DocumentHeadTr block handles the data definition, message declaration, instance declaration, and timer declaration parts of the document head of the FAM being constructed. The instance declaration part creates federate objects and a federation object *derived* from the corresponding counterparts found in the federation structure portion of the FAM data model.

A *derived* object (Bakay and Magyari 2004), which is a deep copy of another compound object, is created inside the attribute-mapping code by invoking a UDM API method. This at the same time creates a kind of inheritance association where the attribute values of the derived object are maintained in synchronization with the values of the corresponding attributes in the *archetype* object (Ledeczi et al. 2001b) (i.e., the topmost object in the hierarchy) as long as they are only modified through their archetype. Once an attribute's value is modified directly on the derived object, the attribute becomes desynched from the archetype. This approach enables the behavior model being contentwise backed up by the data model. Any attribute update to federate objects in the data model will be automatically propagated down to the derived objects in the MSC document and from them to the further derived objects in the MSCs of the behavioral model.

DocumentBodyTr transforms the utility and defining parts of an MSC document. Note that it is necessary to handle the utility part first because the MSCs of the utility part are referenced from within the defining part. The MSC document body transformation essentially boils down to an MSC transformation. The attribute-mapping code in the rule copies the chart order index of the ACM MSC to the FAM MSC. The chart order index, although not an integral part of the MSC metamodel, is a crucial *annotation* that facilitates the construction of model interpreters, particularly the code generator, by providing the processing order of the MSCs at runtime. The rule finally delivers both MSCs to the MSCTrans block for further construction of the FAM MSC.

7.4.2.2 MSC Transformation

The MSC transformation block MSCTrans consists of three consecutive steps that handle the MSC head and body transformations and initialize the federation after the completion of the former two. In this work, we only use LSC as the MSC body and only allow LSCs and inline operands within the LSCs. The MSC body transformation eventually hands over the flow of execution to the LSC transformation after the creation of a stub LSC element.

MSC Head Transformation

The functionality of MSCHeadTr is to prepare the instances used in the FAM MSC by looking at the instances found in the corresponding ACM MSC. Derived FAM MSC instances are created from the corresponding FAM document head instances and are associated with the ACM MSC instances (through has-correspInst) and with the FAM document instance archetypes (through is-instanceOf). Thus, structural and one-to-one correspondences are established between and inside ACM and FAM MSCs. This principle is followed throughout the MSC document, LSC, data element, and event transformations.

The MSC head transformation partly handles *instance decomposition*. The MSC specification (ITU-T 2004) states that an instance can be viewed as an abstraction of a whole MSC document (representing a system component) that is participating in a higher-level system; hence, it is the mechanism for hierarchical decomposition. The outcome of instance decomposition is the introduction of a separate MSC document per decomposed instance and a new MSC for every MSC in the higher-level document that the decomposed instance participates in, which describes the MSC from the perspective of that instance. This noteworthy part of the ACM2FAM transformation is thoroughly described in Özhan and Oguztüzün (2013).

Federation Initialization

Before moving to MSC/LSC transformation, this section fast-forwards to explain the federation initialization on the FAM side. The federation initialization is undertaken in the InitFederation block after an MSC document is transformed head- and bodywise (see Fig. 7.14). It is a postprocessing step following the full transformation of all the LSCs in the document. This is a part of the behavioral model transformation pertaining only to the FAM domain; that is, there are no associations to ACM in the transformation rules apart from the identification of the instances involved. The block handles the four preliminary federation execution activities of creating a federation execution, joining federates to the federation execution, and initializing time and declaration management.

7.4.2.3 LSC Transformation

The LSC transformation is the place where the nuts and bolts of the transformation of field artillery interentity communications to federate interactions via the HLA RTI are defined. The LSC transformation process is carried out in the LSCTrans block as shown in Fig. 7.15, which is the largest component of the ACM2FAM transformation. Each pass of the block inputs an ACM LSC and a stub FAM LSC and step by step constructs the FAM LSC as the transformation proceeds through the internal blocks. Please refer to ITU-T (2004) for a comprehensive description of the MSC/LSC concepts and elements mentioned here.

The execution order of the sub-blocks does not matter except for the second and last blocks. The InstanceRefTr creates the necessary references to federate

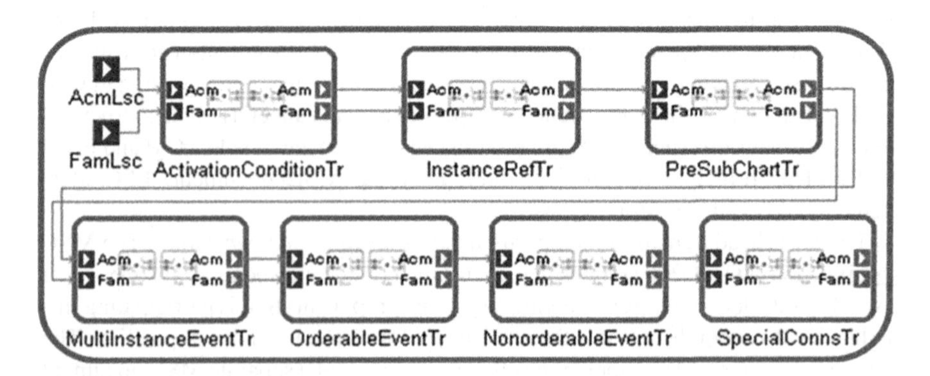

Fig. 7.15 The LSCTrans block

instances in the FAM LSC by inspecting the ones found in the corresponding ACM LSC. Since these instances are referred to in most of the subsequent rules, InstanceRefTr must be executed before them. The last block, SpecialConnsTr, creates associations between a pair of *instance events* within the LSC and thus needs to be executed after ensuring that all such events have been created. The activation condition is a boolean expression that indicates the start predicate for a chart, and its transformation is performed in ActivationConditionTr. The definitions of the LSC transformation blocks are generally based on the instance event types of the elements inside the LSC. These blocks are briefly explained in the remaining part of this section.

Precharts and *subcharts* are child LSCs that have special role names on the containment associations with their parents. The PreSubChartTr block handles the transformation of precharts and subcharts in an LSC. The block ends with a recursive call to the LSCTrans block in order to continue the transformation for the child element, which is yet another LSC.

The transformations of *multi-instance events*, which constitute a set of frequently used elements including condition, otherwise, inline expression, and reference (to an MSC), are handled in the sub-blocks and rules of the block MultiInstanceEventTr. Multi-instance event transformation rules, although numerous, are quite straightforward and intuitive in that they basically create a similar event in the FAM that corresponds to the matched ACM event. Since the inline operand specializes from LSC (Topçu et al. 2008), the paired ACM and FAM inline operands are recursively fed into the LSCTrans rule for further processing as LSCs.

The *nonorderable events* constitute the set of instance events that do not have an explicit ordering of execution. A relative execution order among the events of an instance is implicit along the axis line of an instance; that is, the events attached higher up along the axis execute before the ones that are attached lower along the axis. However, no claim can be made about the execution order of two disjoint events on separate instance axes without using explicit ordering. Inside the NonorderableEventTr block, the input elements are matched and dispatched to one of the handler rules according to the type of the ACM nonorderable event. Each handler rule performs the actual transformation of a specific non.orderable event type.

Orderable events are generally the most frequently used set of events found in the behavioral model of a field artillery scenario. One of the orderable event types, the *message event*, plays a key role for the communication among the LSC instances (i.e., actors in an ACM and federates/federation in a FAM). The top-level OrderableEventTr block is shown in Fig. 7.16. The block starts by matching and dispatching an ACM orderable event to the appropriate rule or block to create its FAM counterpart.

The kinds of orderable events handled are action, create, timer, method, and message. After these events are transformed, any general ordering relations are applied in the GeneralOrderTr block.

Message events implement communication flows in behavior models. These rules are driven by conditions that take the type and structure of the input ACM

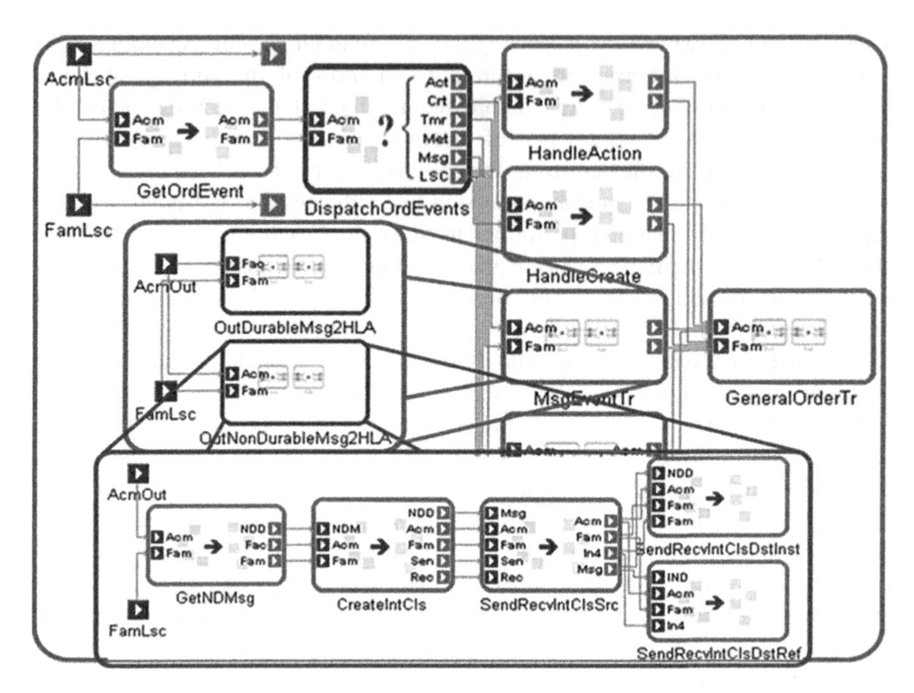

Fig. 7.16 The OrderableEventTr block

element and are the primary parts where platform (i.e., HLA)-specific content is created. Consequently, message event rules are usually more involved than other types of rules. MsgEventTr performs the transformation of *durable* and *nondurable* ACM messages in two parallel blocks. Nondurable message transformation is relatively simpler because a nondurable message transmission in ACM maps to two HLA message transmissions in FAM, whereas a durable message transmission can map up to six.

Out- and in-events are the two kinds of message events and are the conjugates of each other in that every out-event from A to B implies a corresponding in-event sourced in B and targeted on A. In our implementation, if an ACM actor A sends an out-event message to actor B, then federate A sends an out-event message to the federation, and one discrete step later, the federate B receives an in-event message from the federation. Since the federation element is already included in these two federate events, the in-event to and out-event from the federation are not explicitly created for the sake of simplicity.

The SendRecvIntClsSrc rule shown in Fig. 7.17 actually defines the federate-to-federate HLA method transmissions via the federation.

It first creates a message out-event and associates it with the source instance (i.e., federate) using an *ordered* connection. Then, it associates the out-event to the send interaction method using a *special* connection. Finally, it associates the send interaction method to the federation instance using an *address* connection. A similar set of steps is defined for the receive interaction method from the federation

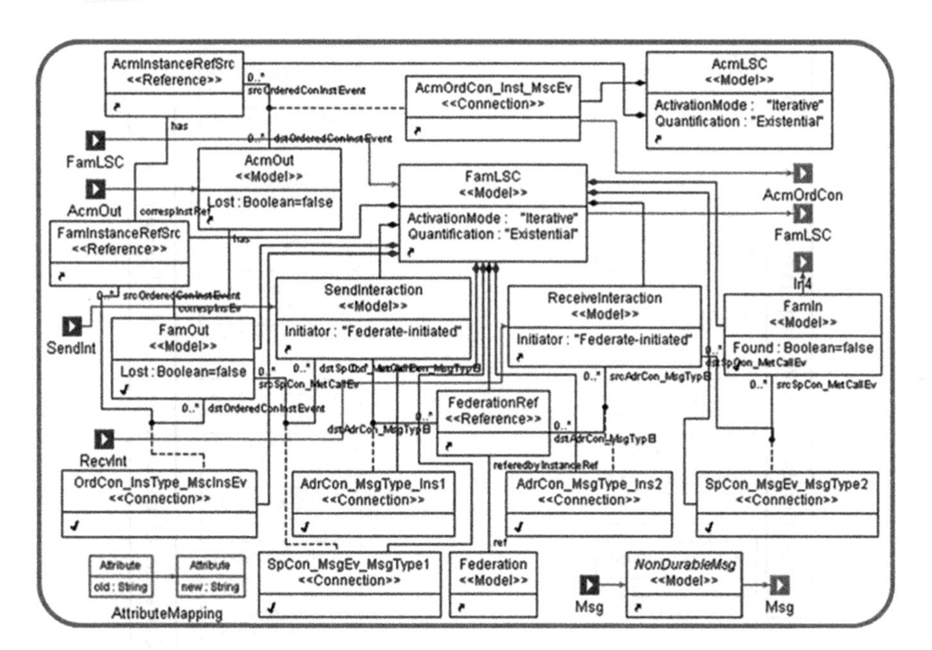

Fig. 7.17 The SendRecvIntClsSrcrule

to the target federate. First, the receive interaction method is associated with the federation instance using an address connection. Then, an in-event message is created and associated with the receive interaction method using a special connection. The last part of the out-event transformation is carried out by one of the two parallel rules SendRecvIntClsDstInst and SendRecvIntClsDstRef (see Fig. 7.16). They associate the new FAM in-event either to a target instance or to an MSC reference.

The outcome of the nondurable message transformation is illustrated in Fig. 7.18, showing the partial view of an ACM LSC and its generated FAM LSC (in abstract syntax). The name of a model element is shown below the element, and its type is shown as its stereotype. In the source LSC, it is seen that an Oid_W_Msg message out-event is sent from FwdObserver to BatteryFDC. In the produced LSC, this corresponds to two HLA message event transmissions.

The figure shows the FwdObserver federate sending a message out-event of SendInteractionWithRegions to the field artillery federation and BatteryFDC federate receiving the corresponding message in-event of ReceiveInteraction from the federation. Sequencing (i.e., the precedence attribute) of the message transmissions are annotated in the callout boxes. The precedence value of the ACM message event is copied to the initial FAM message event, and its autoincremented value is assigned to the second event. The transformation definition ensures a conflict-free generation of ordering values throughout the FAM LSC. Durable message transformation is defined similarly to OutNonDurableMsg2HLA block, most notably being about three times in size.

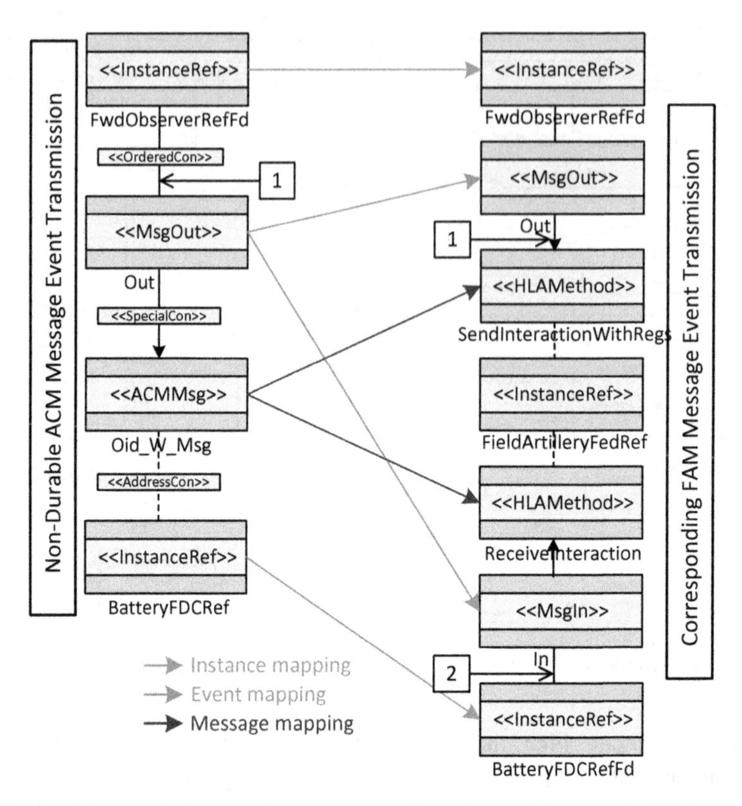

Fig. 7.18 Partial view of non-durable message transformation and its result in FAM

7.4.3 FAM-to-Simulation Code Generation and Execution

Referring back to Fig. 7.4, the content presented up to here constitutes the first phase of the overall transformation process focusing on the ACM-to-FAM transformation. In the second phase, the produced FAM is fed to the code generator to produce simulation code and other useful artifacts such as FDD.

The aspect-oriented programming (AOP) (Elrad et al. 2001) paradigm is adopted in generating distributed simulation code. AOP provides for the separation of crosscutting concerns. In our case, this allows us to generate code that exercises the LSCs in a computation-free manner. The LSC instance is the focal element in code generation. All LSC instance codes are generated in individual class files and are referenced from the diagram code generated from the LSC itself. Using AOP in order to obtain a full-fledged simulation, application-specific computational (and other noncommunication) aspect *advices* are to be crafted by the simulation developer. These advices are then woven onto the generated base code by the aspect-oriented programming environment AspectJ. For example, ballistic computations required by the battery FDC will be handled by the final aspect code provided by the

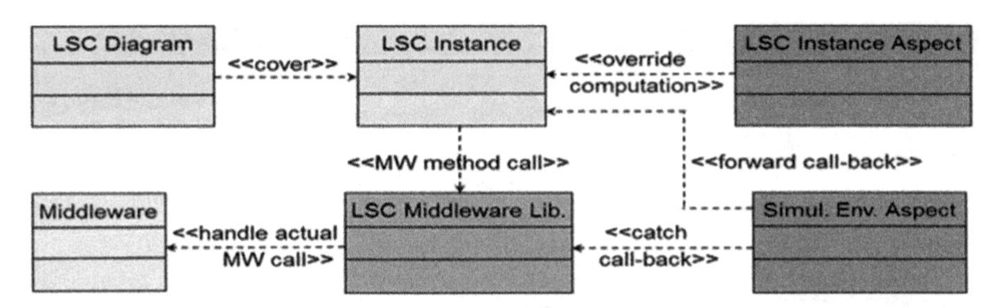

Fig. 7.19 Relationship between the generated source codes for an LSC

developer. On the other hand, HLA-specific portions of the code are automatically woven into the base code generated from the LSC

Figure 7.19 shows the relationship between the generated simulation code files for an LSC. For every LSC message out-event, a simulation middleware (e.g., RTI in the case of HLA) interface method call is made, and for every LSC message in-event, a member application interface method callback is generated. The LSC instance aspect code *intercepts* the middleware interface method calls. It executes developer-written computation code and then redirects the call to the middleware with the computation code in effect. On the middleware side in addition to LSC, aspect code is generated for the overall simulation environment. This aspect code *catches* the middleware callback methods and forwards them to the LSC instance (member application) code. Then in the LSC instance aspect code, the result of the callback (with all arguments) is made available to the developer. The details of the code generator and the code generation process are presented in Adak et al (2010).

The code generator creates an Eclipse project and stores the generated Java and AspectJ codes in the project root folder. AspectJ (Kiczales et al. 2001) is an aspect-oriented extension for the Java programming language. We use an AOP-enabled Eclipse installation to weave the aspects and run the simulation code. A screenshot of the generated code from the FAM of an *Adjustment Followed by Fire For Effect (Adj/FFE)* mission scenario (Özhan et al. 2008) is displayed in Fig. 7.20. The details of code generation for the *AdjFFE* case study are presented in Özhan and Oguztüzün (2011).

After the aspect codes are written and the source is compiled, the simulation is run for execution. Currently, the code generator supports the IEEE-1516 certified RTI implementation developed by Pitch Technologies named pRTI.

7.5 Discussions

This section starts with an informal analysis of this work with respect to several model transformation principles published in literature. Then, the degree of support for the verification of the transformations is explored. Finally, an approach to implement generic model transformations from any conceptual model to an HLA-compliant distributed simulation model is proposed based on our experience.

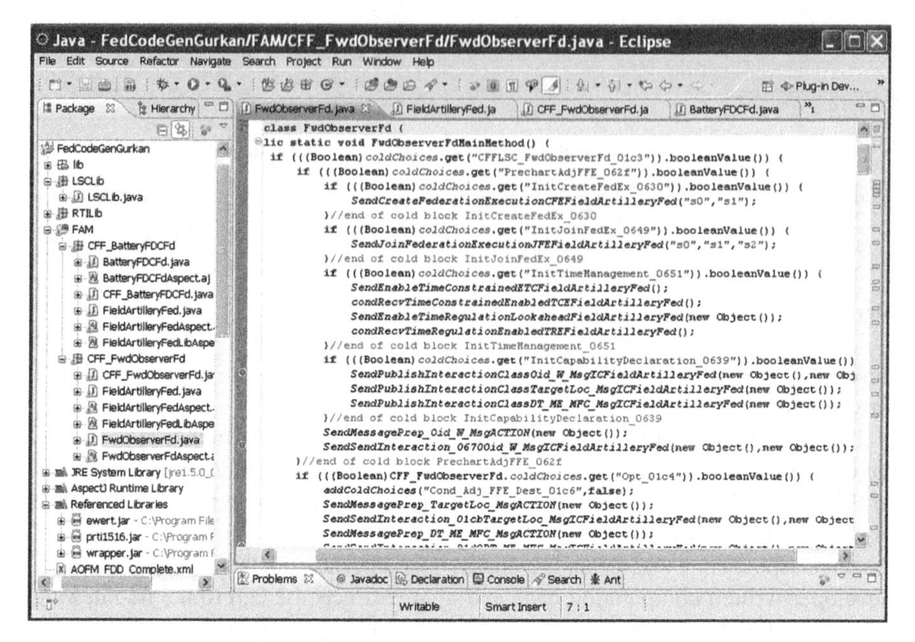

Fig. 7.20 A screenshot of the generated code for the AdjFFE mission in Eclipse

7.5.1 *Analysis of the Transformations*

In this section, we analyze our transformation work from the perspectives of modularity, transformation composition, scope, scalability, and change propagation.

7.5.1.1 Modularity

Modularity is a key factor in developing reusable and maintainable model transformations. Transformation reusability is facilitated if a transformation unit has a specification that describes what is transformed into what but not necessarily how the transformation is carried out (Cuadrado and Molina 2009). Section 7.3.2.1 explains which parts of the ACM data model are transformed into which parts of the FAM data model and similarly for the behavioral models. Figure 7.7 provides an implicit roadmap for the modular breakdown of the ACM2FAM transformation along the LSC structure. Following this breakdown, the whole transformation is defined as a set of hierarchical transformation blocks down to the individual transformation rule level. Moreover, employing expression references for recurring transformation blocks and rules are examples of transformation reuse.

7.5.1.2 Transformation Composition

Kleppe (2006) describes two kinds of transformation compositions, namely, internal and external. Internal transformation composition is defined as the ability of a tool to compose transformation definitions written in the same transformation language. In the ACM2FAM transformation, the data model transformation is performed before the behavior model transformation, and these separately defined transformation modules are internally composed with each other using connection ports between the main blocks. The output of the data model transformation is used by the behavior transformation.

In contrast, external transformation composition is the ability to compose transformation definitions written in different transformation languages. This requires interoperability between different tools and languages. In the case study, the ACM2FAM transformation (phase 1) is realized with the graph-based GReAT tool, and the FAM2Code transformation (phase 2) is realized with the Java-based model interpreter. The transition between the two different transformation phases is facilitated due to the FAM being a common part of the phases, as the target for phase 1 and as the source for phase 2.

The internal and external transformation compositions applied in this work are summarized in Fig. 7.21. Note that GReAT does not provide a composition operator in the sense of Kleppe (2006). The closest constructs would be the connection port used for sequencing and the expression reference used for rule reuse.

7.5.1.3 Scope

Scope is the area of a model (either the source or target) covered by a single transformation step, where a transformation step is usually a single rule application. The pivot of a transformation step is defined as the main source element from which a rule resolves. Four types of transformation steps are identified by van Wijngaardeen and Visser (2003), against which we categorize our rules.

In a local-to-local transformation step, a source element can be directly translated to a target element. All the information needed to create the target element is

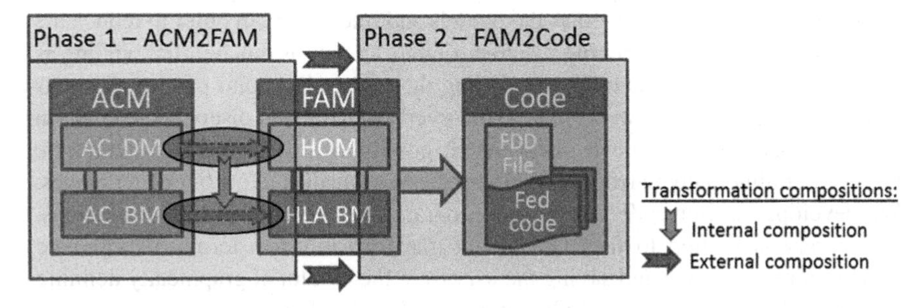

Fig. 7.21 Internal and external transformation compositions

readily available from the source element. Most of our rules transforming LSC components are of this type except for those associated with the payloads of the communicated messages

In a local-to-global transformation step, a source element is transformed into several target elements. Usually, while some of these target elements are part of the target model being generated by the rule, others need to be allocated in different parts of the target model. These target elements are referred to as nonlocal results. A considerable amount of rules in this work are of this type since the transformation of a PIM (i.e., ACM) to a PSM (i.e., FAM) requires the introduction of model elements pertaining to the platform and other target domain–specific aspects. Examples of this category on the data model transformation side include rules that initialize HLA data types, FOM and its subcontainer elements, rules that create and populate interaction and object classes, and rules that create the federation structure elements. On the behavioral side are the rules that create federation initialization and teardown and especially the orderable event transformation rules for durable and nondurable messages.

In a global-to-local transformation step, additional information is required to create a target element from a source element. This additional information is not readily accessible from the source element being transformed (i.e., the pivot), but a complex query is needed. GReAT provides two convenient mechanisms in dealing with them: a *global container* and *cross-links* (Agrawal et al. 2006). Many rules that spread throughout the data and behavior transformations employ cross-links or global containers and thus are examples of global-to-local transformations.

A global-to-global transformation is a combination of the previous two situations. We have tried to avoid such expensive pattern-matching cases as much as possible by dividing the transformation into a number of smaller rules. Still, the orderable event transformation rules for durable and nondurable messages are quite complex and are examples of global-to-global transformations.

7.5.1.4 Scalability and Change Propagation

Scalability and change propagation are two challenging topics in the area of model transformations. Since graph pattern matching is an NP-complete problem, our transformer does not scale well as the models and rules grow. In order to reduce the search space in pattern matching, we first tried to achieve as much initial binding as possible in the rules, effectively discarding those already bound pattern elements from the search. Second, we tried to avoid overly general and disconnected pattern elements that would result in many match possibilities. As a rule of thumb, we favored defining many narrow-scoped specific rules against a few but generic ones. We developed a user code library that programmatically aids in transformations. The library is written to facilitate model transformations in terms of improved execution performance and saving the user from the tedium of graphically defining transformation rules that are slight variations of each other.

The propagation of the changes in the metamodels proved to be challenging. Since the transformation rules are defined in terms of metamodels and rule sequencing follows the structural organizations of the metamodels, any changes in these had profound effects on the transformation definitions.

7.5.2 *Support for the Verification of the Transformations*

Verification of compilers for high-level programming languages has been studied extensively; however, similar studies on model-to-model and model-to-code transformations are less established. For practical purposes, a transformation may be said to have "executed correctly" if a certain instance of its execution produced an output model that preserved certain properties of interest and was endowed with targeted properties.

Narayanan and Karsai (2008b) show that it is both practical and prudent to verify the correctness of every execution of a model transformation as opposed to finding a correctness proof for the transformation specification. This can make the verification tractable and can also find errors introduced during the implementation of a transformation. Our source and target models and the transformation rules conforming to their respective metamodels provide a degree of assurance similar to type checking in programming languages. With the second phase transformation, we are able to generate source code that executes on an RTI. By observing the simulation execution, we are able to conclude whether that transformation instance works as intended.

Narayanan and Karsai (2008a) present a formal verification technique based on the so-called structural correspondence. First, a set of structural correspondence rules specific to a certain transformation are defined. Then, cross-model associations are used to trace source elements with the corresponding target elements, and finally, these associations are used to check whether the structural correspondence rules hold. Their approach is applicable in our work due to the correspondence mappings shown in Table 7.1 and the cross-links between model elements.

Baudry (2009) explores adapting software-testing techniques to validate model transformations especially focusing on the generation and qualification of test data. However, test-based verification is usually not exhaustive, and as systems get more complex, their coverage becomes less and less adequate. Formal verification of model transformations is more desirable since such techniques provide significantly higher confidence of correctness and can even be exhaustive.

7.5.3 *Toward a Domain-Independent CM Transformer for HLA*

The model transformation experience gained in this work has been useful in identifying the key points toward generalizing the transformation perspective

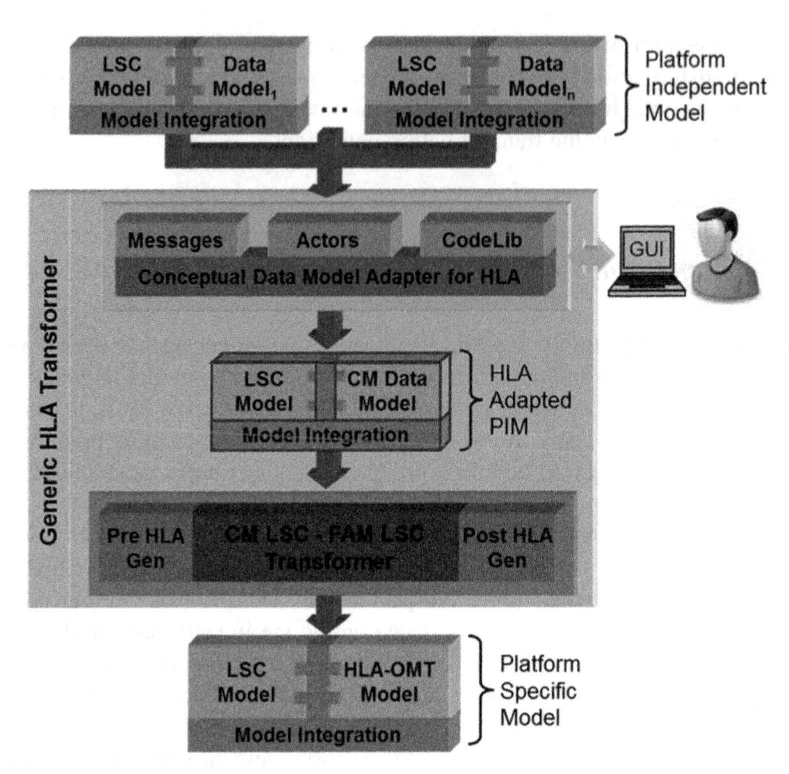

Fig. 7.22 The envisioned domain-independent HLA transformer

from ACM2FAM to (Any) CM2FAM. We claim that if the CM or a part thereof of any given domain can be formulated as entities communicating *durable* and/or *nondurable* (see Sect. 7.3.1.1) data and can be based on the LSC metamodel for behavior representation, then it is viable to implement the model transformation as a source domain–independent LSC-to-LSC transformation.

The key point in obtaining a domain-independent HLA transformer is to devise a mechanism that guides the model transformer in coupling the elements of the source CM with the corresponding elements of FAM (see Table 7.1). Also, the user code library has to be adapted for the parts pertaining to the source domain. Figure 7.22 proposes the high-level architecture of the *domain-independent HLA transformer* which would be used to adapt a given CM for FAM transformation by allowing the user to configure and integrate the transformation's source domain–specific content. Adaptation is accomplished by fitting the source conceptual data model to a so-called abstract CM template as shown in Fig. 7.5. Fitting is used in the sense of extending appropriate user-designated CM elements from the model elements in the template. Note that the CM is assumed to use the LSC formalism for its behavior representation (similar to FAM).

The outcome of the adaptation process is an intermediary model called the *HLA-adapted PIM*, which is a unification of the template and the given CM. This composite model is then fed to the CM2FAM transformer to automatically produce the FAM. In summary, the CM actors and nets are mapped to HLA federates, and the CM data elements are transformed into HLA classes by invoking the configured user code library, which is effectively detached from the transformation rule definitions. The LSC-to-LSC transformations are carried out using the template model elements independent of the CM elements. This can be observed in the way the LSC transformation rules presented in Sect. 7.4.2.3 are defined. The pre- and post-HLA generation parts of the transformer are independent of the source model and only generate the HLA prerequisites, federation initialization, and federation teardown parts of the FAM as explained in Sect. 7.4.2.2.

7.6 Related Work

Although there is work that focuses on behavioral or data model transformations in the literature, reports on transforming a full-fledged conceptual model to an executable model in the spirit of MDE are rare. From our transformation perspective, we treat both data and behavior on an equal basis. In this section, we discuss a selection of the related works and compare them with our work.

Code generation from behavioral specifications in LSC is an ongoing challenge for researchers (Maoz and Harel 2006). There is also a body of literature dealing with transforming LSCs to some executable form, in particular, state charts (Bontemps et al. 2005). We prefer to generate executable code directly from LSC as this approach tends to yield more readable code. Additionally, our metamodeling approach provides the opportunity to extend or tailor the code generator in accordance with the data model by virtue of its data model integration capability.

Kewley and Tolk (2009) specify a systems engineering process for the development of federated simulation models in order to support systems-of-systems analysis. The process borrows principles from the MDA approach to produce models of the simulation system on three different levels, namely, operational, system, and technical levels. Although this work advocates an evolutionary CIM-to-PIM-to-PSM development process and specifies mappings from the upper-level product components to the lower-level ones, it neither takes a model transformation perspective nor any other formal means to derive the lower-level model from the upper-level model.

Similar to this work, Bocciarelli et al. (2012) present a two-stage transformation approach to generate executable code for HLA federations. They also introduce two UML profiles to represent HLA-based information in the models and to support the automated generation of simulation code. They begin with a platform-independent model of the system under study in SysML, while our starting point is a conceptual model conforming to a domain-specific metamodel. On the HLA side, the authors rely on a SysML profile, while our work is based on a purpose-built metamodel, namely, FAMM.

7.7 Conclusion

This chapter presented a comprehensive account of a graph-based model transformation from the field artillery conceptual model (ACM) to HLA federation architecture model (FAM). The extra platform-specific information required for the FAM is provided through the transformation rules. A user code library is developed to simplify rule development and to improve runtime performance. A second phase transformation is applied by a code generator to produce executable simulation code from a FAM. Computation logic has to be woven onto the generated (aspect) code in order to provide legitimate values for the data structures at runtime. Our work can be considered as a step toward validating conceptual models through generation and execution.

Generalizing the experience gained in the case study, we have proposed an outline of a generic model transformer for HLA-based distributed simulation from any conceptual model that is based on LSC for behavioral representation and based on a UML-like language for the data model representation.

In terms of future studies, the conceptual models of other domains can be developed in parallel to building the generic HLA model transformer in the light of the experience reported in this chapter. Another research direction is to investigate the reusability of higher-order transformations, which are declarative rules that allow to capture the recurring patterns of ordinary transformation rules. Another future work direction is the investigation of the applicability of a formal transformation verification approach.

Appendix: Abbreviations and Acronyms

ABM	Artillery Behavior Model
ACM	Artillery Conceptual Model
ACMM	Artillery Conceptual MetaModel
Adj/FFE	Adjustment Followed by Fire For Effect
ADM	Artillery Data Model
ADMM	Artillery Data MetaModel
AOP	Aspect Oriented Programming
API	Application Programming Interface
BatRadNet	Battery Radio Net
BM	Behavioral Model
BMM	Behavioral MetaModel
CIM	Computation Independent Model
CM	Conceptual Model
DIHT	Domain-Independent HLA Transformer
DSAM	Distributed Simulation Architecture Model
DSAMM	Distributed Simulation Architecture MetaModel

FAM	Federation Architecture Model
FAMM	Federation Architecture MetaModel
Fd	Federate
FDC	Fire Direction Center
FDD	FOM Document Data
Fed	Federation
FOM	Federation Object Model
GME	Generic Modeling Environment
GReAT	Graph Rewriting And Transformation
HBM	HLA Behavior Model
HLA	High Level Architecture
HMSC	High Level MSC
HOM	HLA Object Model
HOMM	HLA Object MetaModel
Lib	Library
LSC	Live Sequence Chart
M&S	Modeling and Simulation
JC3IEDM	Joint C3 Information Exchange Data Model
MDA	Model Driven Architecture
MDE	Model-Driven Engineering
MOF	Meta Object Facility
MSC	Message Sequence Chart
Msg	Message
MIC	Model Integrated Computing
OCL	Object Constraint Language
Oid_W	Observer identification & Warning
OMG	Object Management Group
OMT	Object Model Template
PIM	Platform Independent Model
PSM	Platform Specific Model
QVT	Query/View/Transformation
Ref	Reference
RTI	Run-Time Infrastructure
SOM	Simulation Object Model
UDM	Universal Data Model
UML	Unified Modeling Language
UMT	Universal Model Transformer

References

Adak M, Topçu O, Oguztüzün H (2010) Model-based code generation for HLA federates. Softw Pract Exp 40(2):149–175

Agrawal A, Karsai G, Neema S, Shi F, Vizhanyo A (2006) The design of a language for model transformations. Softw Syst Model 5(3):261–288

Bakay A, Magyari E (2004) The UDM framework. Institute for Software-Integrated Systems, Vanderbilt University, Nashville
Baudry B (2009) Testing model transformations: a case for test generation from input domain models, Chapter. In: Model driven engineering for distributed real-time embedded systems. Hermes, Hoboken, NJ, USA
Baudry B (2013) Testing model transformations: a case for test generation from input domain models. In: Babau J-P, Blay-Fornarino M, Champeau J, Robert S, Sabetta A (eds) Model-driven engineering for distributed real-time systems: MARTE modeling, model transformations and their usages. Wiley, Hoboken. doi:10.1002/9781118558096.ch3
Bézivin J (2009) Advances in model driven engineering. In: Jornadas de Ingeniería del Software y Bases de Datos (JISBD), San Sebastian
Bocciarelli P, D'Ambrogio A, Fabiani G (2012) A model-driven approach to build HLA-based distributed simulations from SysML models. In: 2nd international conference on modeling and simulation methodologies, technologies and applications (SIMULTECH), Rome, pp 28–31
Bontemps Y, Heymans P, Schobbens PY (2005) From live sequence charts to state machines and back: a guided tour. IEEE Trans Softw Eng 31(12):999–1014
Cuadrado JS, Molina JG (2009) Modularization of model transformations through a phasing mechanism. Softw Syst Model (SoSyM) 8(3):325–345
Damm W, Harel D (2001) LSCs: breathing life into message sequence charts. Formal Methods Syst Des 19(1):45–80
Elrad T, Akşit M, Kiczales G, Lieberherr K, Ossher H (2001) Discussing aspects of AOP. Commun ACM 44(10):33–38
FM 6–30 (1991) Tactics, techniques, and procedures for observed fire. Headquarters, Department of the Army
FM 6–40 (1996) Tactics, techniques, and procedures for Field Artillery Manual Cannon Gunnery. Headquarters, Department of the Army
Garcia JJ, Tolk A (2013) Executable architectures in executable context enabling fit-for-purpose and portfolio assessment. J Def Model Simul Appl Methodol Tech (JDMS), Online First Version of Record – Jun 24
IEEE 2000 (2000a) IEEE 1516 Standard for modeling and simulation (M&S) high level architecture (HLA) – framework and rules. New York, NY, USA. doi:10.1109/IEEESTD.2000.92296
IEEE 2000 (2000b) Standard for modeling and simulation (M&S) high level architecture (HLA) – federate interface specification (IEEE 1516.1). New York, NY, USA. doi:10.1109/IEEESTD.2001.92421
IEEE 2000 (2000c) Standard for modeling and simulation (M&S) high level architecture (HLA) – object model template specification (IEEE 1516.2). New York, NY, USA. doi:10.1109/IEEESTD.2001.92423
ITU-T (2004) Recommendation ITU-T Z.120, message sequence chart (MSC). New York, NY, USA.
JC3IEDM (2007) The joint command, control and consultation information exchange data model, JC3IEDM – UK – DMWG Edition 3.1a, Greding
Kewley RH, Tolk A (2009) A systems engineering process for development of federated simulations. In: Spring simulation multiconference (SpringSim), San Diego
Kiczales G, Hilsdale E, Hugunin J, Kersten M, Palm J, Griswold W (2001) Getting started with ASPECTJ. Commun ACM 44(10):59–65
Kleppe A (2006) MCC: a model transformation environment. In: 2nd European conference on model driven architecture, LNSC, pp 173–187
Ledeczi A et al (2001a) Composing domain-specific design environments. IEEE Comput 34(11):44–51
Ledeczi A et al (2001b) The generic modeling environment. In: IEEE international Workshop on Intelligent Signal Processing (WISP), Budapest
Maoz S, Harel D (2006) From multi-model scenarios to code: compiling LSCs into AspectJ. In: ACM SIGSOFT Int. symposium on foundations of software engineering, Portland, pp 219–230
Narayanan A, Karsai G (2008a) Verifying model transformation by structural correspondence, Electronic Communications of the EASST, 10:15–29

Narayanan A, Karsai G (2008b) Towards verifying model transformations. Electron Notes Theor Comput Sci (ENTCS) 10:191–200
OMG 2006 (2006) Meta object facility (MOF) core specification. Version 2.0, Available Specification formal/2006-01-01
OMG 2007 (2007) Unified modeling language: infrastructure. Version 2.3, Technical Report formal/2010-05-03
OMG 2011 (2011) Meta object facility (MOF) 2.0 query/view/transformation specification. Version 1.1
Ören TI (1984) Model-based activities: a paradigm shift. In: Ören TI, Zeigler BP, Elzas MS (eds) Simulation and model-based methodologies: an integrative view, vol 10, NATO ASI Series. Springer, Berlin/Heidelberg, pp 3–40
Ören TI (2009) Modeling and simulation: a comprehensive and integrative view. In: Yilmaz L, Ören TI (eds) Agent-directed simulation and systems engineering. Wiley-VCH, Weinheim, pp 3–36
Ören TI (2002) Future of modelling and simulation: some development areas. In: Wallace J, Celano J (eds) 2002 summer computer simulation conference, San Diego, CA, USA, pp 3–8
Ören TI, Zeigler BP (1979) Concepts for advanced simulation methodologies. Simulation 32 (3):69–82
Özhan G, Oguztüzün H (2011) Generating simulation code from federation models: a field artillery case study. In: European simulation interoperability workshop (EuroSIW), 11E-SIW-007, The Hague
Özhan G, Oguztüzün H (2013) Data and behavior decomposition for the model-driven development of an executable simulation model. In: Symposium on theory of modeling & simulation (TMS/DEVS 13), San Diego
Özhan G, Dinç AC, Oguztüzün H (2010) Model-integrated development of field artillery federation object model. In: 2nd international conference on advances in system simulation (SIMUL), Nice, pp 109–114
Özhan G, Oguztüzün H, Evrensel P (2008) Modeling of field artillery tasks with live sequence charts. J Def Model Simul Appl Methodol Tech (JDMS) 5(4):219–252
Schmidt DC (2006) Model-driven engineering. IEEE Comput 39(2):25–32
Sendall S, Kozaczynski W (2003) Model transformation: the heart and soul of model-driven software development. IEEE Softw 20(5):42–45
Topçu O, Adak M, Oğuztüzün H (2008) A metamodel for federation architectures. ACM Trans Model Comput Simul (TOMACS) 18(3):10
van Wijngaardeen JV, Visser E (2003) Program transformation mechanics: a classification of mechanisms for program transformation with a survey of existing transformation systems. Technical Report, Utrecht University

Chapter 8
Using Discrete-Event Cell-Based Multimodels for the Simulation of Evacuation Processes

Gabriel Wainer

8.1 Introduction

Prof. Ören is a pioneer in the fields of Multimodeling and Agent-Based Modeling and Simulation. In Ören (1987), he introduced the concepts of extension and generalization for multimodel formalisms, including formal M&S like DEVS (Discrete Event System Specification) (Zeigler et al. 2000). According to Yilmaz and Ören (2004), a multimodel can be defined as a modular mathematical entity that subsumes multiple submodels that together represent the behavior of the model. Multimodels, introduced in (Zeigler et al. 2000), were extended in Ören (1991) to facilitate generalization of discontinuity in piecewise continuous systems.

The work of Prof. Ören in the area of agent-based simulation is extensive; in particular, we are interested in human behavior modeling, like in Ghasem-Aghaee and Ören (2003). In Yilmaz and Ören (2004), Profs. Ören and Yilmaz define a detailed taxonomy on agent-based multimodeling methodologies. The idea is that agent-based simulation allows defining entities (the agents) that can perceive and reason about their environment and provide responses in order to achieve a goal.

Our objective is to apply these ideas to the field of evacuation simulation processes. In recent years, many simulation models of real systems have been represented as multimodels with agents in the shape of cell spaces (Jalaliana et al. 2011; Weifeng et al. 2003; Vizzari and Bandini 2013). In particular, evacuation processes are important applications and a necessary step in building design. Evacuation simulation is useful for various reasons, such as preventing collapse during evacuation and reducing building evacuation time. Small changes in the building design can result in having important differences; thus, simulation can be used for studying the influence of changing the location of stairways or adding

G. Wainer (✉)
Department of Systems and Computer Engineering, Carleton University, Ottawa, ON, Canada
e-mail: gwainer@sce.carleton.ca

L. Yilmaz (ed.), *Concepts and Methodologies for Modeling and Simulation*,
Simulation Foundations, Methods and Applications,
DOI 10.1007/978-3-319-15096-3_8

sufficient emergency exits to ensure that the building can be evacuated rapidly. In such cases, the models can be built using an agent-based approach for the behavior of the individuals (who need to find the closest exit, might exhibit panic behavior, might need to meet with friends and family, or gather their property; the behavior of the agents should be properly modeled). Using a multimodel approach is important, as different entities at different levels of abstraction exist: individuals, flocks of individuals, buildings, corridors, rooms, and stairs, elevators, obstacles, and even complete city sections. By first creating a virtual version of the building, it is possible to test many different designs to get important measurements such as evacuation time to find the best design. This way, potential problems can be avoided and fixed before construction begins. A multimodel approach and the definition of varied behavior of the agents being evacuated can help in developing a better design.

In this chapter, we will discuss how to address these issues by defining multimodels that can be represented as cell spaces, in which agents represent the behavior of the evacuating agents, and their application in different construction scenarios. We focus on 2D and 3D visualization of the simulation results in order to make the models easier to understand and analyze.

According to the taxonomy in Yilmaz and Ören (2004), our model can be included in the following categories:

- According to the *number of submodels*, it's a multiaspect model, as there are various submodels active at the same time (one per individual at least; we also have models specifying stairs and exits in a building).
- Based on the model's *variability*, it's a static-structure multimodel (representing the building as a cell space and individuals moving to each cell).
- In terms of the *nature of knowledge to activate the models*, it is an adaptive multimodel, as the submodel's behavior is driven by constraints (space, obstacles, building floor plan, panic, etc.).
- The *location of the knowledge* to activate the submodels is within the multimodel; therefore, we can say this is an active multimodel.

8.2 Multimodels in DEVS and Cell-DEVS

A popular multimodel method to describe agents that have spatial properties is called cellular automata (CA), a well-known formalism to describe cell spaces in which individual agents are spatially located in cells in 2D or 3D spaces (Burks 1970; Wolfram 2002). CA is defined as infinite n-dimensional lattices of cells whose values are updated according to a local rule. Cell-DEVS (Wainer and Giambiasi 2000; Wainer 2009) was defined as a combination of cellular automata and DEVS (Discrete Event System specification) (Zeigler et al. 2000). The goal is to improve execution speed building discrete-event cell spaces and to improve their definition by making the timing specification more expressive.

DEVS is a systems theoretical approach that allows the definition of hierarchical modular multimodels. A real system modeled using DEVS can be described as a set of atomic or coupled submodels. The atomic model is the lowest level and defines dynamics, while the coupled are structural models composed of one or more atomic and/or coupled models. DEVS is a formalism proposed to model discrete-event systems, in which a model is built as a composite of basic (behavioral) models called *atomic* that are combined to form *coupled* models. A DEVS atomic model is defined as

$$M = < X, S, Y, \delta_{\mathrm{INT}}, \delta_{\mathrm{EXT}}, \lambda, t_{\mathrm{a}} > \tag{8.1}$$

Where X represents a set of input events, S a set of states, and Y is the output events set. Four functions manage the model behavior: δ_{INT} the internal transitions, δ_{EXT} the external transitions, λ the outputs, and D the duration of a state. Each atomic model can be seen as having an interface consisting of input (X) and output (Y) ports to communicate with other models. Every state (S) in the model is associated with a time advance (t_{a}) function, which determines the duration of the state. Once the time assigned to the state is consumed, an internal transition is triggered. At that moment, the model execution results are spread through the model's output ports by activating an output function (λ). Then, an internal transition function (δ_{INT}) is fired, producing a local state change. Input external events are collected in the input ports. An external transition function (δ_{EXT}) specifies how to react to those inputs.

Once an atomic model is defined, it can be incorporated into a coupled model defined as

$$\mathrm{CM} = < X, Y, D, \{M_i\}, \{I_i\}, \{Z_{ij}\}, \text{select} > \tag{8.2}$$

Each coupled model consists of a set of D basic models M_i. The list of influences I_i of a given model is used to determine the models to which outputs (Y) must be sent and to build the translation function Z_{ij} in charge of converting outputs of a model into inputs (X) for the others. An index of influences is created for each model (I_i). For every j in the index, outputs of model M_i are connected to inputs in model M_j. Coupled models are defined as a set of basic components (atomic or coupled), which are interconnected through the models' interfaces. The models' coupling defines how to convert the outputs of a model into inputs for the others and how to handle inputs/outputs from/to external models. The *select* function decides how to deal with simultaneous events.

Cell-DEVS extended the DEVS formalism, allowing the implementation of cellular models with timing delays. In Cell-DEVS, each cell of a cellular model is defined as an atomic DEVS. Cell-DEVS atomic models are specified as

$$\mathrm{TDC} = < X, Y, S, \theta, N, \text{delay}, d, \delta_{\mathrm{INT}}, \delta_{\mathrm{EXT}}, \tau, \lambda, D > \tag{8.3}$$

Each cell will use N inputs to compute the future state S using the function τ. The new value of the cell is transmitted to the neighbors after the consumption of the delay function. *Delay* defines the kind of delay for the cell and d its duration. The outputs of a cell are transmitted after the consumption of the delay.

Once the cell atomic model is defined, they can be put together to form a coupled model. A Cell-DEVS coupled model is defined by

$$\text{GCC} = < X_{\textbf{list}}, Y_{\textbf{list}}, X, Y, n, \{t_1, \ldots, t_n\}, N, C, B, Z > \tag{8.4}$$

The cell space C defined by this specification is a coupled model composed by an array of atomic cells with size $\{t_1 \times \ldots \times t_n\}$. Each cell in the space is connected to the cells defined by the neighborhood N, and the border (B) can have different behavior. The Z function allows one to define the internal and external coupling of cells in the model. This function translates the outputs of output port m in cell C_{ij} into values for the m input port of cell C_{kl}. The input/output coupling lists (X_{list}, Y_{list}) can be used to interchange data with other models.

The CD++ tool (Wainer and Giambiasi 2000; Wainer 2002, 2009) was developed following the definitions of the Cell-DEVS formalism. CD++ is a tool to simulate both DEVS and Cell-DEVS models. Cell-DEVSs are described using a built-in specification language, which provides a set of primitives to define the size of the cell space, the type of borders, a cell's interface with other DEVS models, and a cell's behavior. The behavior of a cell (the τ function of the formal specification) is defined using a set of rules of the form *VALUE DELAY CONDITION*. When an external event is received, the rule evaluation process is triggered to calculate the new cell value. Starting with the first rule, the CONDITION is evaluated. If it is satisfied, the new cell state is obtained by evaluating the VALUE expression. The cell will transmit these changes after a DELAY. If the condition is not valid, the next rule is evaluated repeating this process until a rule is satisfied.

The specification language has a large collection of functions and operators. The most common operators are included: Boolean, comparison, and arithmetic. In addition, different types of functions are available: trigonometric, roots, power, rounding and truncation, module, logarithm, absolute value, minimum, maximum, G.C.D., and L.C.M. Other available functions allow checking if a number is integer, even, odd, or prime. In addition, some common constants are defined. Figure 8.1 shows the definition of a very simple example of the definition of such models.

The rules in this example say that a cell remains active when the number of active neighbors is 3 or 4 (*truecount* indicates the number of active neighbors) using a transport delay of 10 ms. If the cell is inactive (`(0,0)=0`) and the neighborhood has three active cells, the cell activates (represented by a value of 1 in the cell). In every other case, the cell remains inactive (t indicates that whenever the rule is evaluated, a *True* value is returned).

```
[life]
size:(20,20) delay:transport border:wrapped
neighbors : (-1,-1)(-1,0)(-1,1)(0,-1)(0,0)(0,1)(1,-1) (1,0) (1,1)
localtransition : life-rule

[life-rule]
Rule: 1 10 {(0,0)=1 and (truecount=3 or truecount=4) }
Rule: 1 10 { (0,0) = 0 and truecount = 3 }
Rule: 0 10 { t }
```

Fig. 8.1 Definition of the Life game in CD++

8.3 Agents for Evacuation Processes in Cell-DEVS

In recent years, models for building evacuation have been developed to assist rescue and emergency response crews with proper situation analysis and prompt reaction procedures. The ability to simulate and represent such situations increases training efficiency and creates an opportunity to understand the evacuation process better. The goal is to learn where the bottlenecks can occur and which solutions are effective to prevent congestion during evacuation (Jalaliana et al. 2011; Weifeng et al. 2003; Vizzari and Bandini 2013). The basic idea of the model was to simulate the behavior and movement of every single agent (a person) involved in the evacuation process using a multimodel approach (Brunstein and Ameghino 2003). We defined a Cell-DEVS model with various rules to characterize a person's behavior:

- People try to move toward the closest exit.
- A person in panic might move in the opposite direction to the exit.
- People move at different speeds.
- If the way is blocked, people can decide to move away and look for another way.

We used two planes to represent this spatial model: one for the floor plan of the building and to represent the people moving and the other for the orientation toward the exits, as we can see in Fig. 8.2.

Each cell in the grid represents 0.4 m^2 (one person per cell). The coordinates of each object are divided into two: Boundaries and people, and Objects (i.e., walls, chairs, columns, etc.). The orientation layer contains information that serves to guide persons toward the exits. We assigned a potential distance to an exit to every cell of this layer. The persons will move for the room trying to minimize the potential of the cell in which they are. That is, the people move to a cell with decreasing potential. Each cell can contain different values: exits (value $= -2$), obstacles (value $= -1$), distance to the exit (positive value in the first plane), and information about the individuals in the cell. The individuals' information is represented by a six-digit value, in which each digit represents a different property, as follows.

For instance, a cell with a value of `009121` represents an individual going to the W (first digit $=1$), at a speed of 4.5 km/h (two cells per second, each cell is 0.4 m

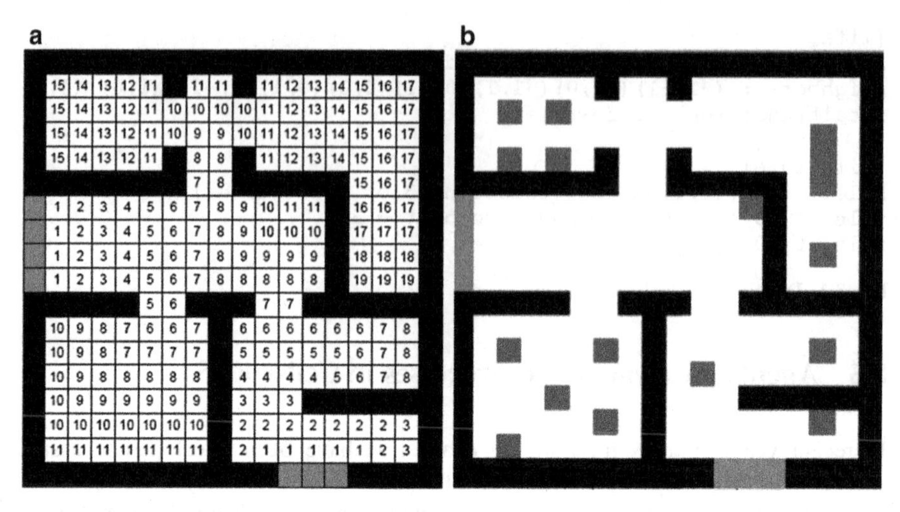

Fig. 8.2 (**a**) Orientation layer: potential value (**b**) Individuals

long), the last movement was W (the individual is keeping the current direction), a stable emotional state, the current panic level is 0 (no panic), and the person will not change the direction of potential. A person moves to decrease the movement potential by decreasing the distance to the exit. If there is no available move that will decrease the potential, a person will try to move to a neighboring cell that has the same potential. If none is available, the person will move further away in an attempt to find another route.

Figure 8.3 shows a subset of the rules used for evacuation models in CD++. We first define the Cell-DEVS multimodel (two layers, 18×18 cells each). The model uses inertial delays (which allows preemption, which is needed because we have to deal with collision behavior). The first set of rules we can see in the figure is used to define the path taken by a person using the orientation plane. The basic idea is to take the direction decreasing the potential of a cell, building a path following the lower value of the neighbors. We use eight different rules to control the people's movement, one for each direction. In all cases, the rules analyze the 8 near neighbors to understand what direction the person should take. We use a random direction (*randint*) when the near neighbors have the same value. The second set of rules model panic: a person in panic will take a wrong path or will not follow the orientation path. In that case, the direction is calculated by taking the path where the cell's potential is increased.

The following figures show different visualizations for the simulation results for this model. Figure 8.4 shows a simple graphical representation of the simulation results. We can see the building shape (with walls in black and two exits: one to the left and one to the bottom right) and people who want to leave the building using the exit doors. The evacuation path is the one previously presented in Fig. 8.2a. As we can see, there is a group of people blocking the left exit because individuals tend to

```
[evacuation]
dim : (18,18,2)        delay : inertial       border : wrapped
localtransition : EvaRule
neighbors : (-1,-1,0) (-1,0,0) (-1,1,0) (0,-1,0) (0,0,0) (0,1,0)
(1,-1,0) (1,0,0) (1,1,0) (-1,-1,1) (-1,0,1) (-1,1,1) (0,-1,1) ...

[EvaRule]
% Rules to govern people movement
rule : {trunc((0,0,0)/10)*10+1} {1000 / remainder(
        trunc((0,0,0) /10),10) } {((0,0,0)>0 AND
         remainder(trunc((0,0,0)/1),10) =0 AND remainder(trunc(
           (0,0,0)/100000),10) =0 AND ((0,-1,0)=0 OR (0,-1,0)=-2)
       AND cellPos(2)=0 )AND (((0,-1,1) <= (1,-1,1) OR (1,-1,0)>0
        OR (1,-1,0)=-1 OR (randint(5)=0)) AND ((0,-1,1)<= (1,0,1)
         OR (1,0,0)>0 OR (1,0,0)=-1 OR (randint(5)=0) ) AND
       ((0,-1,1) <= (1,1,1) OR (1,1,0)>0 OR (1,1,0)=-1 OR
         (randint(5)=0)) AND ((0,-1,1)<= (0,1,1) OR (0,1,0)>0 OR
           (0,1,0)=-1 OR (randint(5)=0)) AND ((0,-1,1)<=(-1,1,1)
             OR (-1,1,0)>0 OR (-1,1,0)=-1 OR (randint(5)=0) ) AND
              ((0,-1,1) <= (-1,0,1) OR (-1,0,0)>0 OR (-1,0,0)=-1
                OR (randint(5)=0) ) AND ((0,-1,1) <= (-1,-1,1) OR
                 (-1,-1,0)>0 OR (-1,-1,0)=-1 OR (randint(5)=0) ))
}  ...

% Rules to control panic behavior
rule : {trunc((0,0,0)/10)*10+1}
        {1000/remainder(trunc((0,0,0)/10),10) } {((0,0,0)>0 AND
          remainder(trunc((0,0,0)/1),10)=0 AND remainder(trunc
           ((0,0,0)/100000),10)>0 AND ((0,-1,0)=0 OR (0,-1,0)=-2)
        AND cellPos(2)=0)AND (((0,-1,1)>= 1,-1,1) OR (1,-1,0)>0
         OR (1,-1,0)=-1) AND ((0,-1,1)>=(1,0,1) OR (1,0,0)>0 OR
           (1,0,0)=-1) AND ((0,-1,1)>= (1,1,1) OR (1,1,0)>0 OR
             (1,1,0)=-1) AND ((0,-1,1)>=(0,1,1) OR (0,1,0)>0 OR
               (0,1,0)=-1) AND ((0,-1,1)>=(-1,1,1) OR (-1,1,0)>0
                 OR (-1,1,0)=-1) AND ((0,-1,1)>=(-1,0,1)
          OR (-1,0,0)>0 OR (-1,0,0)=-1) AND ((0,-1,1)>=(-1,-1,1)
                 OR (-1,-1,0)>0 OR (-1,-1,0)=-1))
}
```

Fig. 8.3 Specification of evacuation model

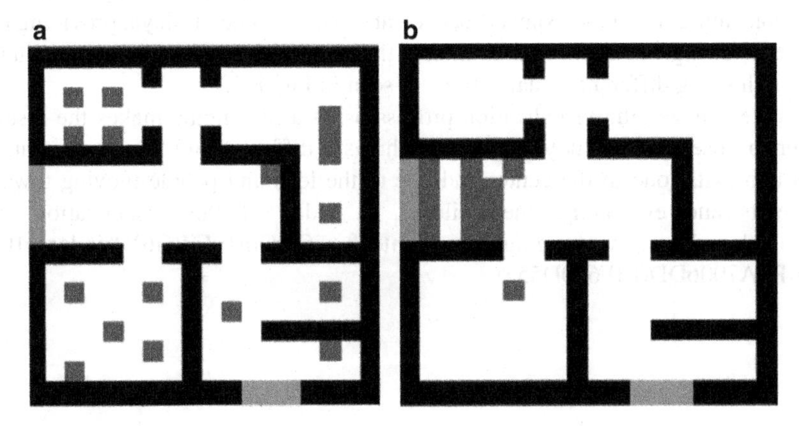

Fig. 8.4 (**a**) People seeking an exit. (**b**) After 15 s, people found the exit

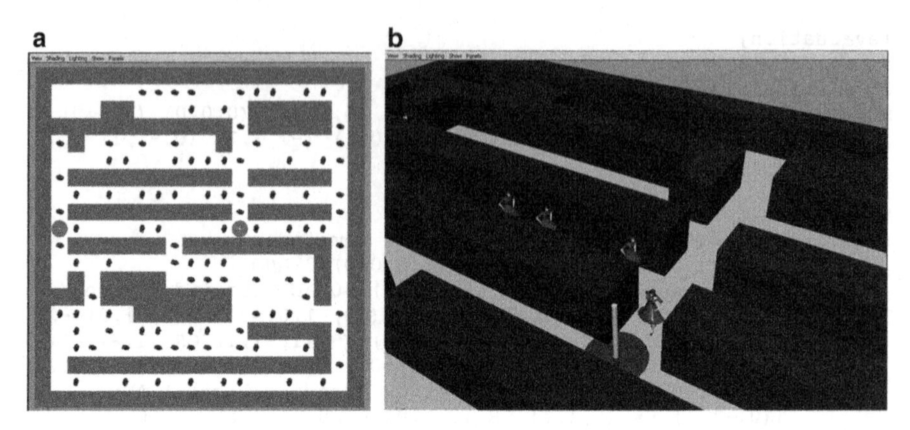

Fig. 8.5 Evacuation Model at time (**a**) 00:00:00:000 (**b**) close up at time 00:00:05:240

move reducing the potential (and based on the original configuration, the closest exit for most people is to the left). Although CD++ provides different visual tools to the ones used to generate the graphical results above, we need to build more sophisticated 3D graphics, which improves data exploration. To overcome these drawbacks and to meet the diverse needs of different users, we have developed mechanisms to integrate the CD++ environment with a variety of both commercial and open-source visualization and rendering techniques, including Autodesk Maya, OpenGL, and Blender (Wainer and Liu 2009).

In this section, we will elaborate on these advanced techniques and demonstrate their capabilities with a wide range of applications. Autodesk Maya is one of the leading commercial software packages for 3D modeling, animation, and visual effects. Maya software interface is fully customizable, and it allows users to extend their functionality by providing access to the Maya Embedded Language (MEL). With MEL, users can tailor the GUI to fulfill their specific needs and to develop in-house tools. The MEL scripting language has been used in our research to create a high-performance 3D visualization engine (Poliakov et al. 2007), allowing for interoperability between a DEVS-based M&S tool and an advanced generic visualization environment like Maya. Users create a static scene in Maya, providing the necessary background for 3D animation of the simulation results. This Maya plugin allows showing different visualizations as seen in Fig. 8.5.

As we can see, the visualization process using a 3D engine makes the results easier to observe and study. Figure 8.5 shows a different building configuration with two exits (one in the center and one to the left) and people moving toward the exits and evacuating the building. A video of this visualization can be found at https://www.youtube.com/watch?v=GOOm1vFWG6Y&index=10&list=PLA7006DDBBF660D55

8.3.1 *Evacuation Example 1: The SAT Building*

The Society for Arts and Technology (SAT) building is located on Blvd. St. Laurent in downtown Montreal. This building is a center devoted to the creation, development, and conservation of digital culture. We have built a model based on existing floor plans to study the evacuation processes in the SAT building. This multimodel also uses various agents with different panic level, considers the distance from exits, etc. The model represents people moving through a room or group of rooms trying to gather their belongings or related persons and to get out through an exit door. Following a similar idea as in the previous section, the agents moving through the cells representing the space of the building use different values to represent different phenomena as follows Table 8.1.

Once the model has been specified as in Fig. 8.7 as above, the simulator generates a log file with the simulation results, as follows (Fig. 8.8).

As we can see, the log file contains the time of the output messages generated by the agents on each cell and the current value representing the combination of digits presented in Table 8.1 for each of the agents. In this case, they represent different individuals moving in different directions, some leaving a cell and others arriving into a new one. For instance, the person in cell (Wainer and Liu 2009; Wainer 2009) abandons the cell (value = 0) and moves to cell (Poliakov et al. 2007; Wainer 2002). The emotional state is 5 (average); it was moving in direction SW and now moves in direction S. It moves at a speed of 4 cells per second.

In the above examples, we show different simulation scenarios showing different agents moving through this building. The first example, presented in Fig. 8.8, shows a simple scenario with eight people distributed throughout the building. The initial values for the cells in the figure are as follows:

`(13,10,0)=5040, (36,24,0)=5010, (45,25,0)=5040, (40,18,0)=5020, (29,5,0)=5040, (35,7,0)=5030, (42,6,0)=5040,` and `(43,4,0)=5060.`

Table 8.1 Values used to represent the agents behavior

Digit	Property
6	Next movement direction. 1:W; 2:SW; 3:S; 4:SE; 5:E; 6:NE; 7:N; 8:NW
5	Speed (cells per second: 1–5)
4	Last movement direction, it can vary from 1 to 8 (as digit #6)
3	Emotional state: the higher this value is the lower the probability that a person gets in panic and changes direction.
2	Number of movements to increase the potential of a cell. If a person moves this number of times, the person, which is now in panic, can move into a different direction in which the potential is increased.
1	Panic Level, representing the number of cells that a person will move in increasing direction of potential.

```
[floor]
type : cell          dim : (49,27,2)        delay : INERTIAL
border : wrapped     initialCellsValue : eva-ej1.val
localtransition : EvaRule

% Neighbors
neighbors : (-1,-1,0) (-1,0,0) (-1,1,0)
neighbors : (0,-1,0)  (0,0,0)  (0,1,0)
neighbors : (1,-1,0)  (1,0,0)  (1,1,0)

% Neighbors in the lower level
neighbors : (-1,-1,1) (-1,0,1) (-1,1,1)
neighbors : (0,-1,1)  (0,0,1)  (0,1,1)
neighbors : (1,-1,1)  (1,0,1)  (1,1,1)

% Rules to control the movement decision of each individual
[EvaRule]
rule : {#pos1+1} {1000/#pos0} {((0,0,0)>0 AND #pos0 =0 ...
rule : {#pos1+3} {1000/#pos0} {((0,0,0)>0 AND #pos0 =0 ...
rule : {#pos1+5} {1000/#pos0} {((0,0,0)>0 AND #pos0 =0 ...
rule : {#pos1+7} {1000/#pos0} {((0,0,0)>0 AND #pos0 =0 ...
rule : {#pos1+2} {1000/#pos0} {((0,0,0)>0 AND #pos0 =0 ...
rule : {#pos1+4} {1000/#pos0} {((0,0,0)>0 AND #pos0 =0 ...
rule : {#pos1+6} {1000/#pos0} {((0,0,0)>0 AND #pos0 =0 ...
rule : {#pos1+8} {1000/#pos0} {((0,0,0)>0 AND #pos0 =0 ...
```

Fig. 8.6 Evacuation rules in CD++

```
...
Message Y / 00:00:00:754 / floor(16,13,0)(893) / out / 5440
Message Y / 00:00:00:755 / floor(15,12,0)(837) / out /  0
Message Y / 00:00:01:005 / floor(16,13,0)(893) / out /  5443
Message Y / 00:00:01:006 / floor(17,13,0)(947) / out /  5340
Message Y / 00:00:01:007 / floor(16,13,0)(893) / out /  0
Message Y / 00:00:01:256 / floor(17,13,0)(947) / out /  5344
...
```

Fig. 8.7 Simulation log files

As we can see, the first two digits on each agent have not been used (as we are not modeling panic). We can also notice different speed levels, which make for a more realistic simulation since not all people move at the same speed or pace.

We use the rules presented in Fig. 8.6, with the agents placed at random inside the building and following the path defined in the second layer to exit the building (no one is in panic). As the level of complexity is small, we could observe that they all followed the exit path. The building is almost empty (which is a normal

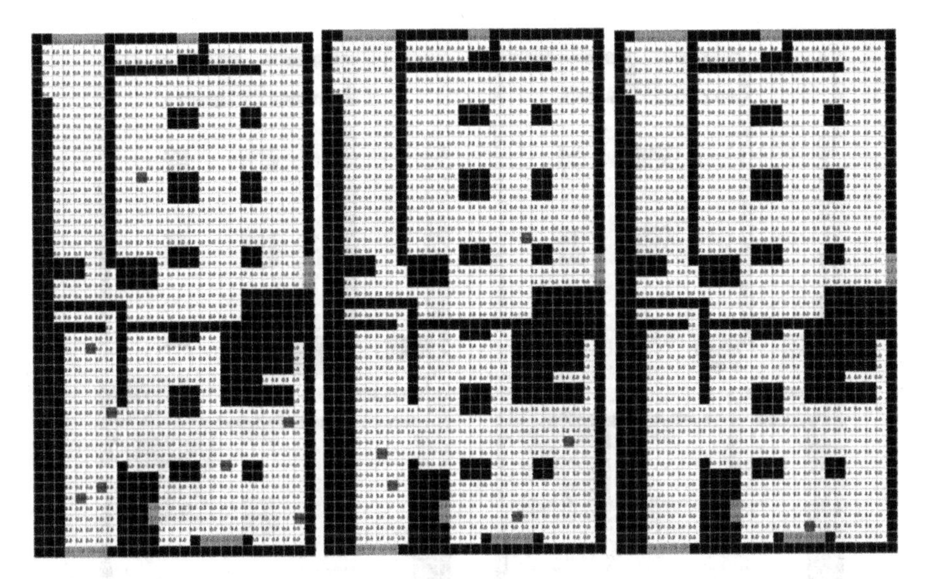

Fig. 8.8 SAT evacuation scenario: eight individuals under normal circumstances

condition for SAT); however, there are people in each sector. This evacuation gives us a general idea of the exit directions people will follow. In this case, no one is in panic, and we did not change the movement potential using a high level of patience. The total evacuation time for this scenario was 13:015 s.

The example presented in Fig. 8.9 also represents eight people; however, they are all located in the bottom left corner of the floor plan. Although we do not include panic behavior and the agents follow an organized evacuation pattern, we can see a bottleneck situation in one of the exits. Although the total evacuation time is short (04:005), this occurs due of the proximity of the people to the exit.

Our following scenario includes panic behavior for one of the agents. If we analyze the execution results on Fig. 8.10, we can notice that this person moves away from the exit because it is blocked. The rest of the individuals leave the building normally. The total evacuation time is 05:004 s (it takes longer because the person in panic finally returns to the main door after the bottleneck disappears). In order to observe the effect of panic on the simulation time, we used the exact same number of people and their positions as specified above.

The following test uses a larger number of people in the building, but they are located closer to the exit on the right, which allows us to study the results of two separate exits close to each other (Fig. 8.11).

In order to observe the effect of panic on the simulation time, we used the exact initial configuration on the left part of the building and more people on the right. This time, however, we introduced the maximum panic level in all the individuals. We noticed an increase in the evacuation time up to three times larger than what was observed in the previous simulation (the total evacuation time was 14:774). We can see people moving away from the exits in any situation where there is a blockage, making the evacuation process much slower than in the previous cases.

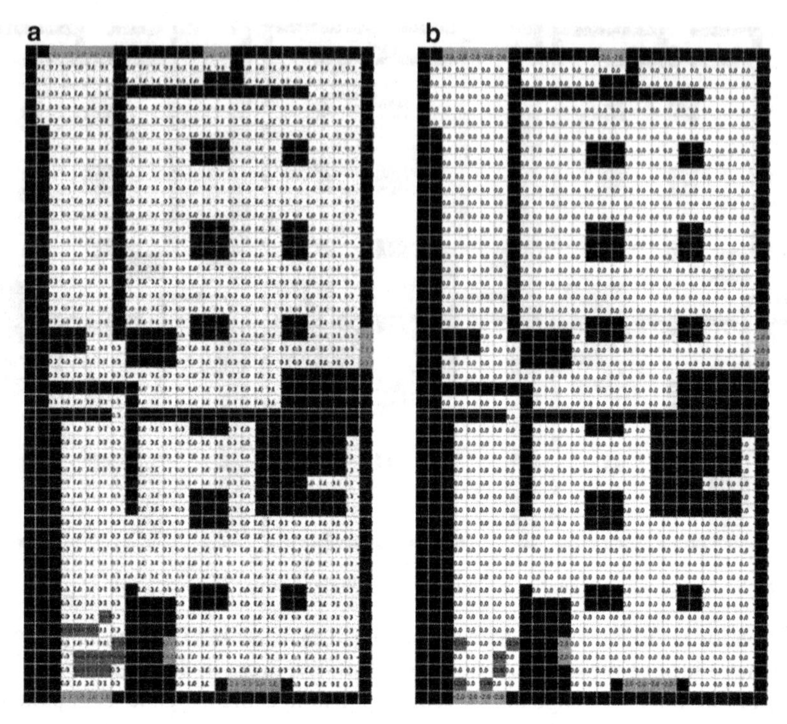

Fig. 8.9 (a) Time: 00:000; (b) time: 01:005

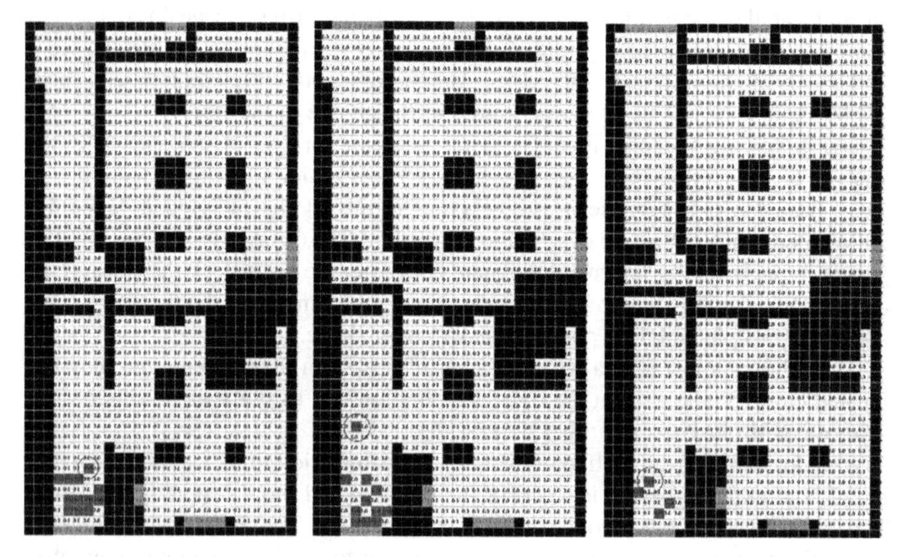

Fig. 8.10 Evacuation with panic (one person)

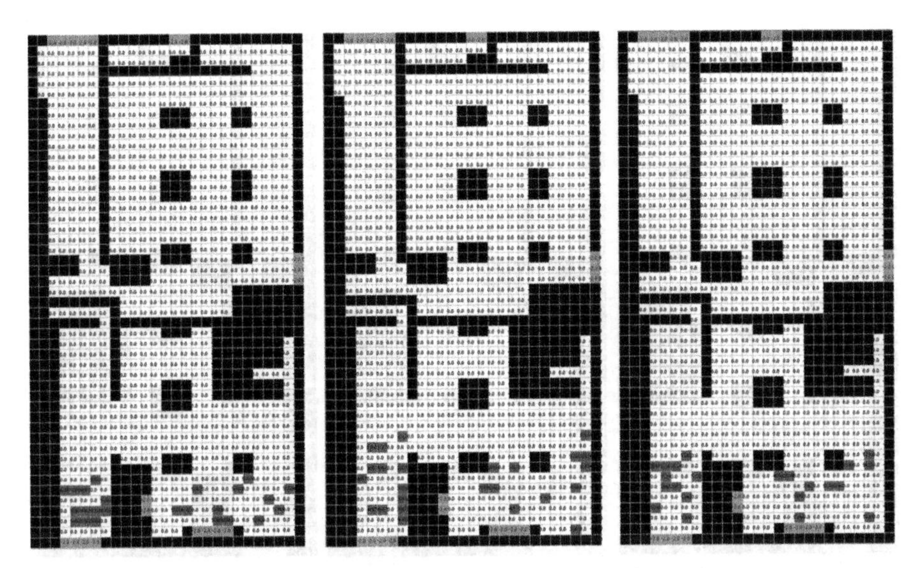

Fig. 8.11 Evacuation with panic (everybody)

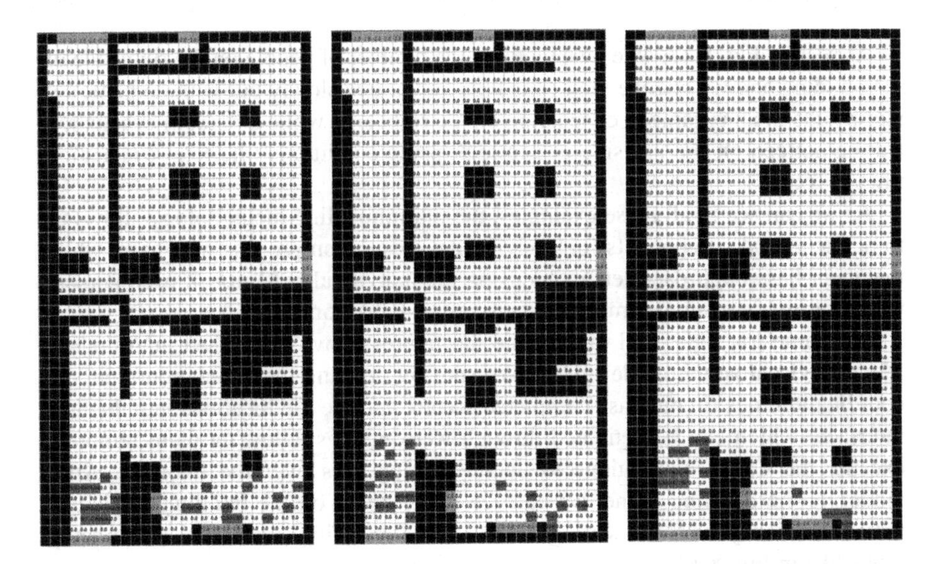

Fig. 8.12 Evacuation with panic (left part of the building)

The following scenario, presented in Fig. 8.12, shows an initial configuration with identical people positions as in Fig. 8.11; however, the panic condition has been removed from the individuals on the right side of the building (representing, for instance, the fact that there is a fire on the left side of the building and the people on the right cannot see it). We can notice a better organized evacuation with movements focused on the exit on that side of the building. Nevertheless, the total evacuation time on the left was 15:607 because it took longer to evacuate the people in panic.

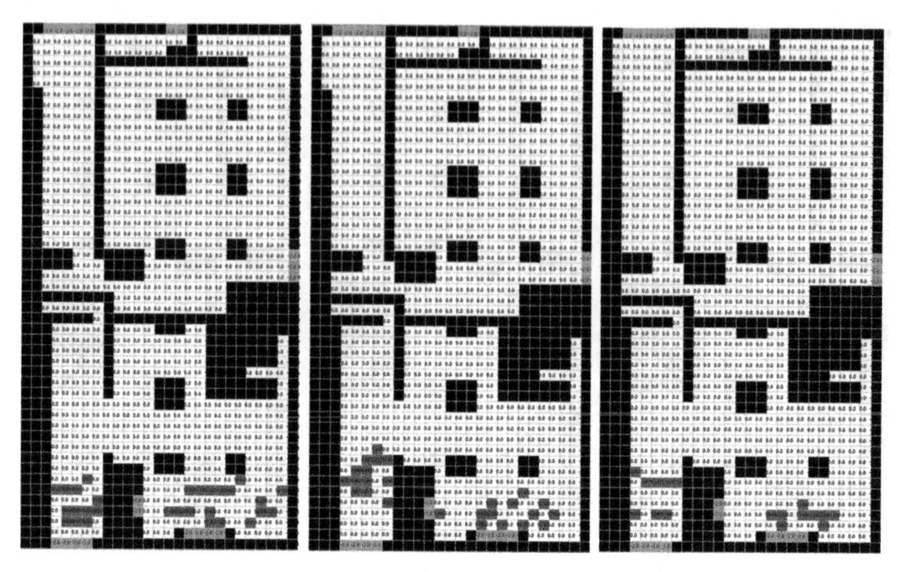

Fig. 8.13 Evacuation with panic (left part of the building)

As seen in Fig. 8.13, when we increase the number of people on the right side of the building and we change their speed (so they move slowly but with no panic), it still results in an organized evacuation on the right side while a random evacuation can be noticed on the left side due to high panic conditions (the total evacuation time is 16:611).

As we can see in all these examples, the multimodeling methodology allows us to define varied components for studying the evacuation process with ease, while the agent-based approach lets us focus on the individual behavior of each person, which results in a simpler mechanism to define behavior.

The same log files used to generate the figures above were used in Autodesk Maya to visualize the model in 3D. We start by defining the simulation type, the coordination files (in our case completed scene), and the file locations into the user interface (which can be activated through web services, allowing us to remotely execute the CD++ simulations to obtain the log files over the internet). After rendering the building scene, we can see better detail on the building to give us better familiarity with the setting. For the SAT building, the initial scene setup looks like in Fig. 8.14.

Once the building floor plan has been loaded and rendered, the CD++/Maya plugin loads the initial values for the cell spaces—in our case people inside the building. Then, we search the log file looking for the Y messages (which, as seen in Fig. 8.7, carries information about the current cell values and locations). The MEL script uses these values and coordinates to relocate the human figures. This organization results in a frame-based motion of the human figures and hence makes an easy-to-see evacuation model. The following are five rendered images of separate frames that show the progressive motion of the human figures toward the dedicated building exits (Fig. 8.15).

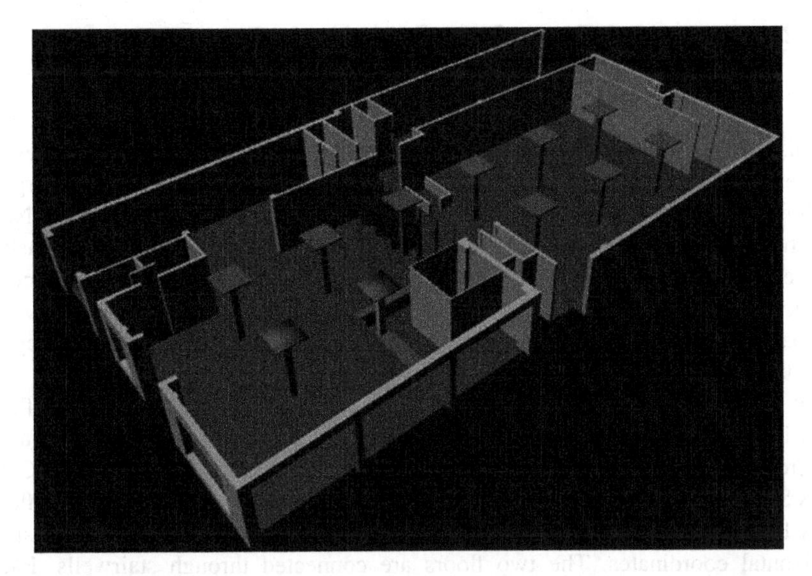

Fig. 8.14 Initial configuration for the SAT building in Maya

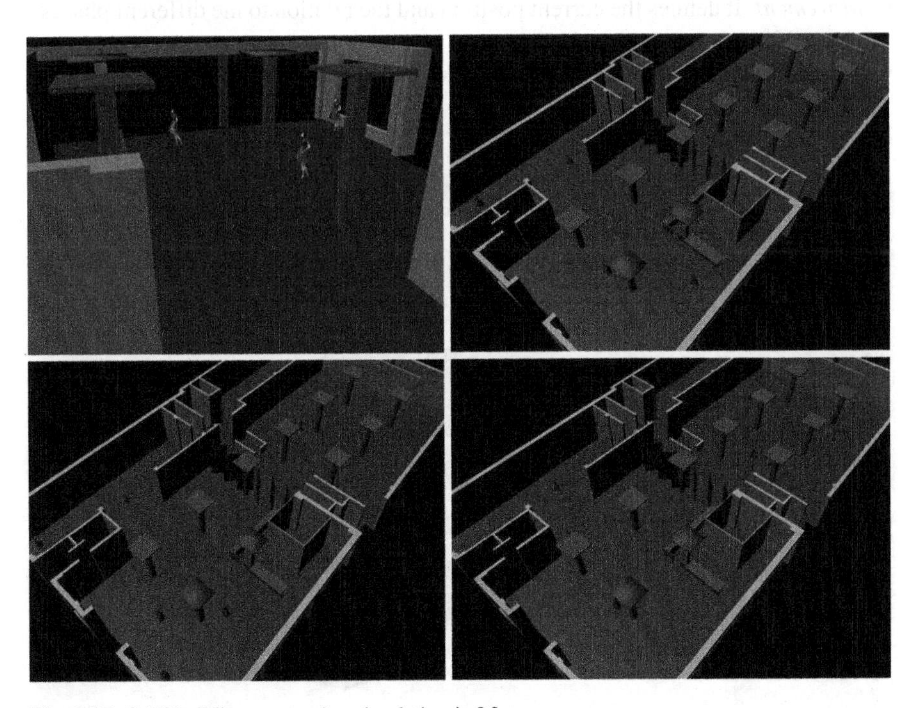

Fig. 8.15 SAT building evacuation simulation in Maya

8.3.2 Evacuation Example 2: Copenhagen Zoo's New Elephant House

In this section, we present a case study focused on analyzing the occupancy levels of the Copenhagen Zoo's New Elephant House. The Copenhagen Zoo is the largest cultural institution in Denmark, attracting over 1.2 million visitors a year. The New Elephant House, which has two floors, tries to create a close visual relationship between the Zoo and the park. Visitors walk in from the main entrance, move downstairs, and leave the house through the exit, moving at random and following the pathway and spending time watching the elephants. The level of occupancy of the building is important in case of needing to evacuate it.

In this case, we have used Autodesk Revit as a tool to input the building floor plan into the simulation model and Autodesk 3Ds Max as the visualization tool. Figure 8.16 shows a view of the building using Autodesk Revit. We used Cell-DEVS to simulate the behavior of the level of occupancy of the building. Each floor uses 10×22 cells, and each cell represents a square place associated with physical horizontal coordinates. The two floors are connected through stairwells. Each individual on each cell uses different state variables to represent the movement:

- *Movement*: It defines the current position and the relation to the different phases, defined below.
- *Phase*: Each movement cycle goes through four phases (Intent, Grant, Wait, Move), to be discussed in detail later.

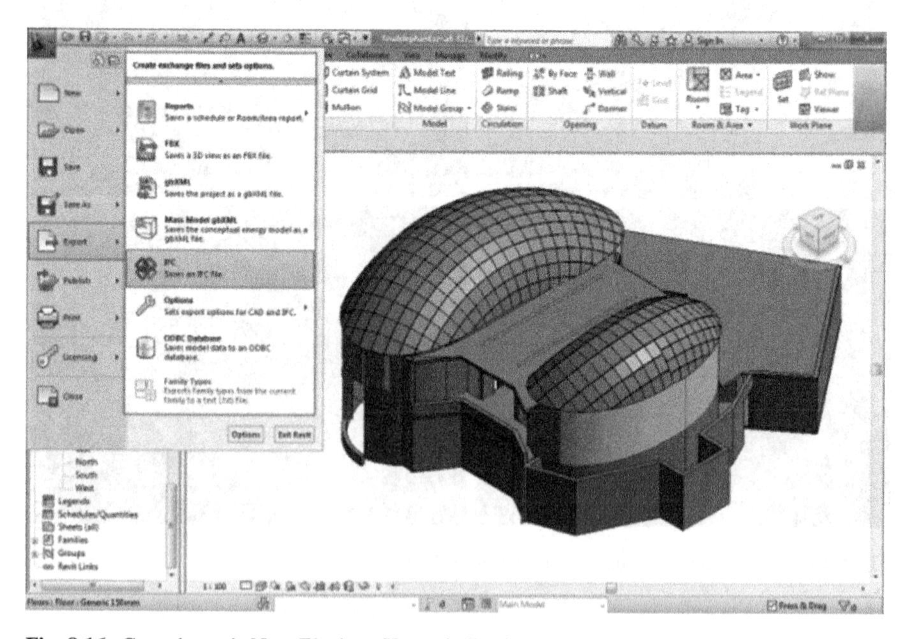

Fig. 8.16 Copenhagen's New Elephant House in Revit

- *Pathway*: Visitors tend to move following the pathway with certain probabilities. Normally, the pathway points to the shortest path toward the exit. In our case, we overlay a Voronoi diagram of the route to an exit or stairwell.
- *Layout*: Each cell can be an empty space, a wall, an entrance, a stairwell, an exit, etc.
- *Hot Zones*: They reflect the popularity levels of certain spots influencing different potential waiting time. The higher value the hot zone is, the higher the probability that a visitor would stay.

In order to implement random movement and random waiting, the movement behavior is divided into four phases (*Intent*, *Grant*, *Wait*, and *Move*). In general, for an occupied cell, a visitor chooses a direction at random during the *intent* phase. If the target cell is available, the visitor state changes to get *grant*; otherwise, it turns to get *rejected*. If granted, the visitor would *wait* for some time at random (according to the hot zone where the visitor is) and then *empty* the cell at the *move* phase. If rejected, the visitor needs to wait. For an *empty* cell, the logic is simpler: it chooses a surrounding cell, which is in the *grant* phase, and changes to *occupied* at the *move* phase.

```
rule : {~movement := 2; ~phase := 1;} 0 {uniform(0,1) <
  #VisitorRate and remainder(time, 4)=1 and (0,0,0)~phase = 0 and
    (0,0,0)~movement=0 and $layout=3}
```

At the beginning of the simulation, visitors go to the main entrances with certain probability (*VisitorRate*) in order to mimic different input flow rates with rush/slash hour during the opening time. In the current implementation, each cycle has 4 s (each phase has 1 s). We check to see if it is the beginning of the cycle (*remainder* (*time*, *4*) *=0*) and then generate a new individual.

During the *intent* phase, the desired direction is determined using the pathway direction and probability.

```
rule : {~movement := 10; ~phase := 2;} 1 { (0,0,0)~phase = 1 and
            (0,0,0)~movement = 1 and $layout = 5}
rule : {~movement := uniform(0,1); ~phase := 1.1;} 0 { (0,0,0)~phase =
1 and (0,0,0)~movement=1 and (0,0,0)~pathway>=5}
...
rule : {~movement := 11; ~phase := 2;} 0 (0,0,0)~phase = 1.1 and
   (0,0,0)~pathway = 5 and (0,0,0)~movement > 0.0 and
        (0,0,0)~movement <= #Front }
...
rule : {~movement := 18; ~phase := 2;} 0 (0,0,0)~phase = 1.1 and
(0,0,0)~pathway =8 and (0,0,0)~movement > #Front + #Left-Front and
       (0,0,0)~movement <= #Front+...+#Right-Front}
```

We first check if the individual is in the intent phase, if the cell is a stair, and if the cell below is empty (intent direction = 10). If the cell is not a stair, we find the probability to move in different directions. We then generate a random number between 0 and 100 and check in which direction that random number is located. Finally, the cell value changes to 10–18, whose unit value corresponds with the intent direction: D(0), E(1), NE(2), N(3), NW(4), W(5), SW(6), S(7), SE(8); e.g., for going up, it should be 13. Note here: we do not care whether the target cell is available; it will be checked in the following phases.

After choosing the intended direction, we need to handle collisions (i.e., to see if more than one person want to enter into the same cell). This phase, called *grant*, is used to choose only one agent to move to a neighbor cell. To do so, it checks neighbors in the *intent* phase and will mark one of the eight reverse directions (i.e., 41 means the current cell accepts the left neighbor to come in). The cells with *intent* direction 10–18 change to 20–28 and phase 3 (*waiting*). The rules for the Grant phase are as follows:

```
rule : {~movement := 40; ~phase := 4;} 1 { (0,0,0)~movement = 0
  and (0,0,-1)~movement = 10 }
rule : {~movement := 41; ~phase := 4;} 1 { (0,0,0)~movement = 0
  and (0,-1,0)~movement = 11 and $layout != 2} ...
rule : {~movement := 48; ~phase := 4;} 1 { (0,0,0)~movement = 0
  and (-1,-1,0)~movement = 18 and $layout != 2}
rule:{~movement := ((0,0,0)~movement+10); ~phase := 3;} 1 {
  (0,0,0)~movement >= 10 and (0,0,0)~movement <= 18 }
```

The *wait* phase defines a random wait. If a person is *granted* to move, they wait for a random amount of time before moving, based on the hot zone where the person is. We implement this by adding different delays in the associated rules, as follows:

```
rule : {~movement := 30; ~phase := 4;} 1 { (0,0,0)~phase = 3 and
  (0,0,0)~movement = 20 and (0,0,1)~movement = 40 }
rule : {~movement := 31; ~phase := 4;} { 1 + 4*randInt($hotzone) }
  { (0,0,0)~phase = 3 and (0,0,0)~movement = 21 and
      (0,1,0)~movement = 41 }
rule : {~movement := 38; ~phase := 4;} { 1 + 4*randInt($hotzone) }
    { (0,0,0)~phase = 3 and (0,0,0)~movement = 28 and
        (1,1,0)~movement = 48 }
rule : {~movement := 39; ~phase := 4;} 1 { (0,0,0)~phase = 3 and
      (0,0,0)~movement >=20 and (0,0,0)~movement <= 28 }
```

Now, every individual that intended to move has a value of 30–38 (the movement was *granted*) or 39 (the movement was *rejected*). A granted individual can move to the target cell. To finish the moving for the next cycle, we empty the intended cells (value = 0) and change the rejected ones to 1.

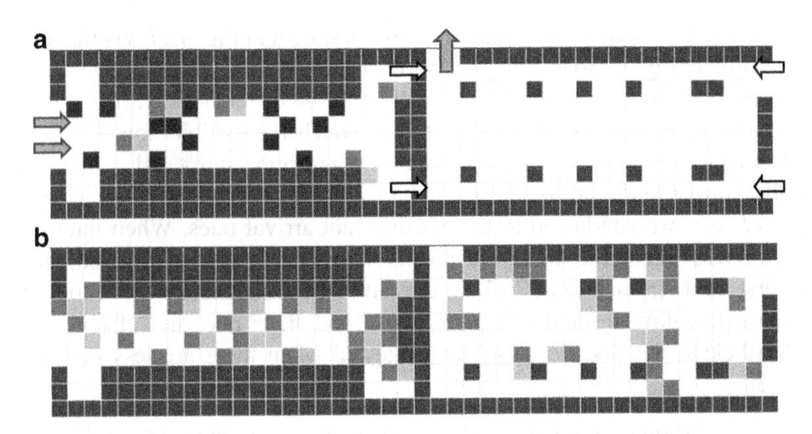

Fig. 8.17 Simulation results of basic properties at different simulation times. (**a**) At 2.5 min. (**b**) At 7.5 min

```
rule : {~movement := 0; ~phase := 1;} 100 { (0,0,0)~phase = 4 and
  (0,0,0)~movement = 30 and (0,0,1)~movement = 40 } ...
rule : {~movement := 1; ~phase := 1;} 100 { (0,0,0)~phase = 4 and
  (0,0,0)~movement = 48 and (-1,-1,0)~movement = 38}
```

As seen in the previous section, 3D visualization provides a more intuitive way to observe simulation results, enabling the designers to check the building performance and people behaviors under different properties. Most authoring tools support full-featured 3D visualization of buildings. Among them, Autodesk 3ds Max is a powerful tool for 3D animation and rendering. We have developed an advanced visualization tool in 3ds Max, providing options for hiding building floors for visibility and filtering models. We include arrow models with key framing ability and humanoids to animate real body movement (using the Motion Mixer plugin). The simulation results presented in Fig. 8.17 show the basic behavior of the visitors under normal conditions. We can see the two floors in the building and visitors arriving in the building and moving around the floors (they arrive using the main doors on the left). Then, they move to the first floor downstairs (following the white arrows) and leave the building. Each visitor goes through the four cycles discussed above.

In Wang et al. (2013), we presented different simulation scenarios for this building, showing the impact of door location/stairs number in terms of occupancy, and simulated two modifications to the original design. It was found that arrival rate and stairs affect the occupancy level more significantly than other properties do. In order to evaluate different options, the following parameters were modified:

1. *Hot zones*: we decreased the probability of people waiting, representing people moving faster. The result showed that the occupancy level decreased relatively obviously, which indicates the influence of people movement speed to the occupancy.

2. *Movement direction*: in the simulation results shown in Fig. 8.17, visitors have a 70 % probability to move forward. We changed this probability to 50 %, giving visitors freedom to move in other directions. The differences with the original simulations were minimal. People stop to watch the exhibits longer than any influence in their moving direction, and they reach the stairs and exits at a rate similar to that in the original case.
3. *Arrival rate*: we conducted tests with different arrival rates. When interarrival interval was longer, the simulation results showed a decrease in the occupancy of the first floor (from 38.9 to 26.7 %) but only a small change in the second floor because the flow from the first to the second floor does not change much. Nevertheless, the first floor is less congested when there are less individuals arriving.

Figure 8.18 shows the results of the occupancy simulation using our 3D visualization tool. As we can see, we can combine the simulation results with the original 3D floor plan in Revit, which is used to generate initial conditions for the simulation. Then, we use 3Ds Max to visualize the results of the simulation.

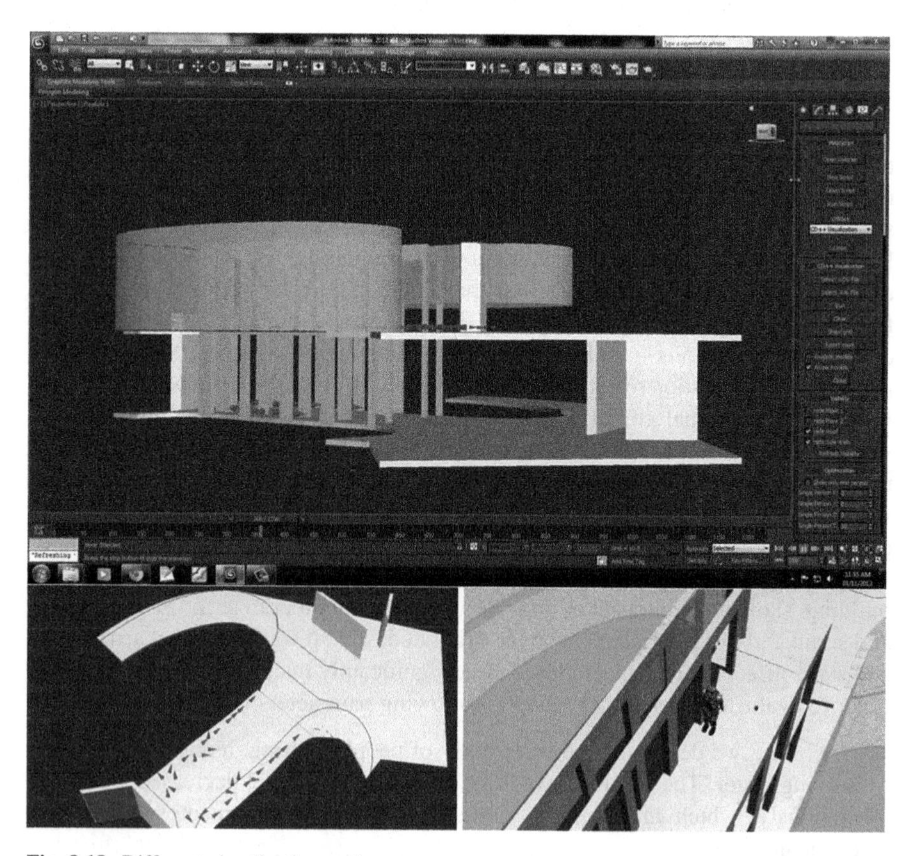

Fig. 8.18 Different visualization options

The figure shows some visualization results of different options. We can see the whole building and the two floors and different individuals represented as cones in the direction of the movement of the individuals. The complete visualization of this simulation can be seen at https://www.youtube.com/watch?v=ciA5mtXdHIA.

8.4 Conclusion

Multimodeling can help builders of complex models and their simulations to organize their work better, address each of the problems at the right level of abstraction, and resolve the problems quicker and easier. Prof. Ören's invention allowed us to address these complex problems with ease. The application of multimodels into agent-based simulation provides a good combination for solving complex simulation problems. Here, we showed how to use these concepts combined with the DEVS formalism proposed by Prof. Zeigler and cellular models to describe the phenomena using a spatial-based notation. We showed different modeling and simulation examples focusing on evacuation and occupation of buildings. We defined a solution based on building information modeling, mixing the results of buildings and simulations in Cell-DEVS. We also presented new methods to view advanced 3D visualization in 3Ds Max. We showed two different case studies: one for the SAT building in Montreal and another one for Copenhagen's New Elephant House.

The models are public domain and can be easily modified to be applied for other purposes. The tools can be found at http://cell-devs.sce.carleton.ca. The different models can be found at http://www.sce.carleton.ca/faculty/wainer/wbgraf.

These techniques can benefit building designers and engineers to understand better some issues related to the buildings under construction (e.g., doors location, stairs number, rush/slash hours, different movement probabilities of directions, etc.), allowing them to better manage the design and to provide suggestions for improvements.

Acknowledgments Numerous authors participated in the research reported in this chapter (see the references for complete citations), including J. Ameghino, M. Braunstein, V. Freire, A. Khan, Q. Liu, E. Poliakov, V. Rajus, and S. Wang. This work was partially supported by NSERC, Autodesk Research, and MITACS.

References

Brunstein M, Ameghino J (2003) Modeling evacuation processes using Cell-DEVS. Internal report. Computer Science Department. Universidad de Buenos Aires

Burks AW (1970) Von Neumann's self-reproducing automata. In: Burks AW (ed) Essays on cellular automata. University of Illinois Press, Champaign, pp 3–64

Ghasem-Aghaee N, Ören TI (2003) Towards Fuzzy agents with dynamic personality for human behavior simulation. In: Proceedings of the 2003 summer computer simulation conference, Montreal

Jalaliana A, Chalupa S, Ostwald M (2011) Architectural evaluation of simulated pedestrian spatial behavior. Archit Sci Rev 54(2):132–140

Ören TI (1987). Model update: a model specification formalism with a generalized view of discontinuity. In: Proceedings of the summer computer simulation conference, Montreal

Ören TI (1991) Dynamic templates and semantic rules for simulation advisors and certifiers. In: Fishwick PA, Modjeski RB (eds) Knowledge-based simulation: methodology and application. Springer, Berlin/Heidelberg/New York/Tokyo

Poliakov E, Wainer G, Hayes J, Jemtrud M (2007) A busy day at the SAT Building. In: Proceedings of AIS 2007, artificial intelligence, simulation and planning, Buenos Aires

Vizzari G, Bandini S (2013) Studying pedestrian and crowd dynamics through integrated analysis and synthesis. IEEE Intell Syst 28(5):56–60

Wainer G (2002) CD++: a toolkit to define discrete-event models. In: Software, practice and experience, vol 32, no. 3. Wiley, pp 1261–1306

Wainer G (2009) Discrete-event modeling and simulation: a practitioner's approach. CRC Press/ Taylor and Francis, Boca Raton

Wainer G, Giambiasi N (2000) Timed Cell-DEVS: modelling and simulation of cell spaces. In: Discrete event modeling & simulation: enabling future technologies. Springer-Verlag, New York.

Wainer G, Liu Q (2009) Tools for Graphical Specification and Visualization of DEVS Models. SIMULATION: Transactions of the Society for Modeling and Simulation International 85 (3):131–158

Wang S, Wainer G, Rajus V, Woodbury R (2013) Occupancy analysis using building information modeling and Cell-DEVS simulation. In: Proceedings of 2013 SCS/ACM/IEEE symposium on theory of modeling and simulation, TMS/DEVS'13, San Diego

Weifeng F, Lizhong Y, Weicheng F (2003) Simulation of bi-direction pedestrian movement using a cellular automata model. Physica A: Statistical Mechanics and its Applications 321 (3):633–640

Wolfram S (2002) A new kind of science. Wolfram Media, Champaign

Yilmaz L, Ören TI (2004) Dynamic model updating in simulation with multimodels: a taxonomy and a generic agent-based architecture. In: Proceedings of SCSC 2004 – summer computer simulation conference, San Jose

Zeigler B, Praehofer H, Kim T (2000) Theory of modeling and simulation: integrating discrete event and continuous complex dynamic systems. Academic Press, San Diego

Part III
Quality Assurance and Reliability of Simulation Studies

Chapter 9
Quality Indicators Throughout the Modeling and Simulation Life Cycle

Osman Balci

9.1 Introduction

I met Dr. Tuncer Ören in April 1980 at a meeting in New York City where he presented a paper entitled "Concepts and Criteria to Assess Acceptability of Simulation Studies: A Frame of Reference," which was later published in the *Communications of the ACM* (Ören 1981). During that time, I was focusing on my Ph.D. dissertation research in the area of simulation model validation. I also presented a paper at that meeting entitled "A Methodology for Cost-Risk Analysis in the Statistical Validation of Simulation Models," which was later published right after his paper in the same special issue of the *Communications of the ACM* (Balci and Sargent 1981).

Dr. Ören's seminal paper in the *Communications of the ACM* has expanded my horizon in assessing the acceptability of modeling and simulation applications. I am honored to contribute this chapter presenting quality indicators that can be used for such acceptability assessment. Dr. Ören has published and presented more than 85 articles just on the topic of reliability, quality assurance, and failure avoidance in modeling and simulation since his seminal paper. He has been an internationally recognized leading authority not only in this topic but also in the whole modeling and simulation discipline. Dr. Ören's linguistic ability is beyond my comprehension. He is the only person I know who can deliver a very technical speech in Turkish without using a single English word! That is unbelievable!

As the saying goes, "Quality is Job 1!" Quality is a critically important issue in almost every discipline. Whether we manufacture a product, employ processes or provide services, quality often becomes a major goal. Achieving that goal is the

O. Balci (✉)
Mobile/Cloud Software Engineering Lab, Department of Computer Science, Virginia Polytechnic Institute and State University (Virginia Tech), Blacksburg, VA, USA
e-mail: balci@vt.edu

L. Yilmaz (ed.), *Concepts and Methodologies for Modeling and Simulation*, Simulation Foundations, Methods and Applications,
DOI 10.1007/978-3-319-15096-3_9

challenge. Many associations have been established worldwide for quality, e.g., American Society for Quality (http://www.asq.org), Australian Organization for Quality (http://www.aoq.asn.au), European Organization for Quality (http://www.eoq.org), and Society for Software Quality (http://www.ssq.org). Manufacturing companies have quality control departments, business and government organizations have Total Quality Management programs, and software development companies have Software Quality Assurance departments to be able to meet the quality challenge.

Quality can be generically defined for anything X as: Quality of X is the degree to which the X possesses a desired set of characteristics.

The ultimate goal of a modeling and simulation (M&S) project is to develop an M&S application with sufficient quality characteristics. M&S quality assurance (QA) refers to the planned and systematic activities that are established throughout the M&S life cycle to substantiate adequate confidence that an M&S application possesses a set of characteristics required for a set of intended uses.

M&S applications are mostly made up of software or are software based. Software is inherently complex and very difficult to engineer. Under the current state of the art, we continue to face serious technical challenges in developing a reasonably large and complex software product with acceptable accuracy. *Accuracy* refers to the transformational and representational/behavioral correctness and is considered just one of dozens of quality characteristics of an M&S application. M&S accuracy is judged by conducting M&S verification and validation (V&V). As advocated by Balci et al. (2002), we can increase our confidence in the accuracy of large-scale and complex M&S applications by employing a quality-centered evaluation approach.

The purpose of this chapter is to present quality indicators throughout the M&S life cycle that can be used for such a quality-centered evaluation approach. After providing background information and an introduction, we present a life cycle for modeling and simulation, which is applicable for any kind of M&S project. We describe the quality indicators throughout the M&S life cycle based on the experience and knowledge gained by the author in the U.S. Department of Defense large and complex M&S application verification, validation, and accreditation projects. Concluding remarks are stated to conclude this chapter.

9.2 Modeling and Simulation Life Cycle

A life cycle for M&S is presented in Fig. 9.1. This life cycle is a different representation of the same life cycle described by Balci (2012). The author has developed this life cycle based on his many years of experience and knowledge gained in more than a dozen U.S. Department of Defense large and complex M&S application development projects.

An M&S life cycle is a framework for organization of the processes, work products, quality assurance activities, and project management activities required

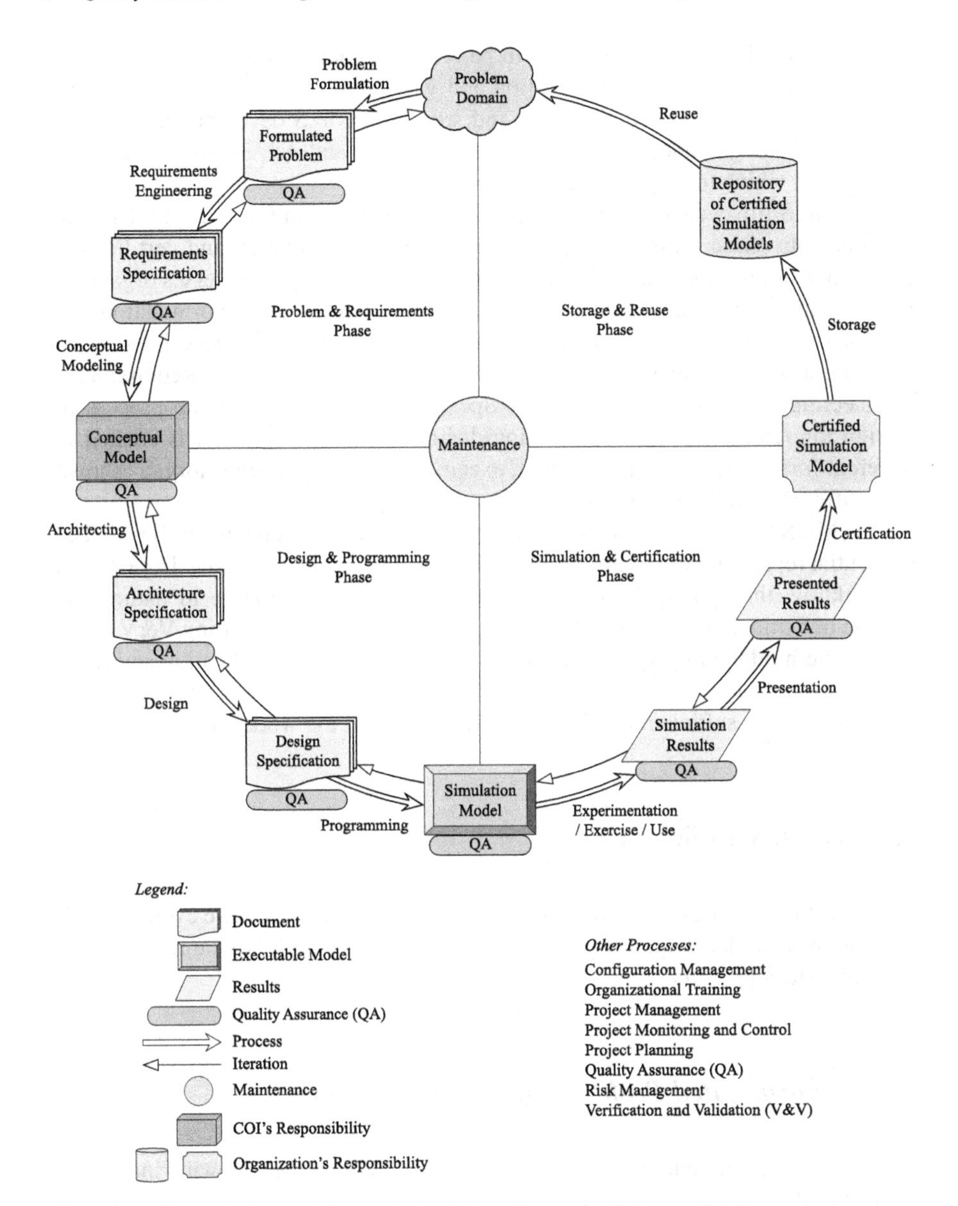

Fig. 9.1 A life cycle for modeling and simulation (Copyright © Osman Balci)

to develop, use, maintain, and reuse an M&S application from birth to retirement. The M&S life cycle is created to modularize and structure an M&S application development and to provide guidance to an M&S developer (engineer), manager, organization, and community of interest (COI).

The M&S life cycle presented in Fig. 9.1 enables to view M&S development from four perspectives (or Ps): Process, Product, People, Project. The M&S life cycle (a) specifies the *work* products to be created under the designated *processes* together with the integrated verification and validation (V&V) and quality assurance (QA) activities, (b) modularizes and structures M&S development and provides valuable guidance for *project* management, and (c) identifies areas of expertise in which to employ qualified *people.*

The M&S life cycle consists of four phases as depicted in Fig. 9.1: problem and requirements phase, design and programming phase, simulation and certification phase, and storage and reuse phase. It consists of eleven major processes organized in a logical order, as depicted in Fig. 9.1, starting with Problem Formulation and culminating with Reuse. A process, represented by a double-line arrow, is executed to create a work product. For example, we execute the process of Requirements Engineering to create a Requirements Specification document or the process of Design to create a Design Specification document. A work product is created in different forms, i.e., document, model, executable, results, or repository, as shown with different symbology in Fig. 9.1.

The M&S life cycle should not be interpreted as strictly sequential or linear. The sequential representation of the double-line arrows is intended to show the direction of workflow throughout the life cycle. The life cycle is iterative in nature and reverse transitions are expected. For example, an error identified during V&V of the executable model may require changes in the requirements specification and redoing the earlier work. We typically bounce back and forth between the processes until we achieve sufficient confidence in the quality of the work products.

9.3 Quality Indicators

In this section, we present quality indicators throughout the M&S life cycle that can be employed under a quality-centered assessment approach for large-scale and complex M&S projects.

9.3.1 Formulated Problem Quality Indicators

Formulated problem quality can be assessed by employing the following indicators (Balci 2012):

1. What are the chances that the *real* problem is not completely identified due to the possibility that
    1.1. People might have personalized problems?
    1.2. Information showing that a problem exists might not have been revealed?
    1.3. The problem context is too complex for the analyst to comprehend?
    1.4. Root problems might have arisen in contexts with which people have had no experience?

1.5. Cause and effect may not be closely related within the problem context?
1.6. The analyst might have been unable to distinguish between facts and opinions?
1.7. The analyst might have been misguided deliberately or accidentally?
1.8. The level of extraction of problem context was insufficiently detailed?
1.9. The problem boundary was insufficient to include the entire real problem?
1.10. Inadequate standards or definition of desired conditions exist?
1.11. The root causes might be time dependent?
1.12. A root cause might have been masked by the emphasis on another?
1.13. Invalid information might have been used?
1.14. Invalid data might have been used?
1.15. Assumptions might have concealed root causes?
1.16. Resistance might have occurred from people suspicious of change?
1.17. The problem was formulated under the influence of a solution technique?
1.18. The real objectives might have been hidden accidentally, unconsciously, or deliberately?
1.19. Root causes might be present in other unidentified systems, frameworks, or structures?
1.20. The formulated problem may be out of date?

2. Stakeholders and Decision Makers

2.1. Do you know or can you think of any stakeholders and decision makers, other than the ones identified by the analyst, who might be aided by the solution of the problem?
2.2. Are all active stakeholders (e.g., users of the solution system, administrators of the solution system, trainers of the solution system users) identified? (An *active stakeholder* is the one who will actively interact with the solution system once it is operational and in use.)
2.3. Are all passive stakeholders (e.g., developers, decision makers about the use of the solution system, logistics personnel, manufacturer, owners/sponsors if they do not use/operate the solution system) identified? (A *passive stakeholder* is the one who will not actively interact with the solution system once it is operational and in use.)

3. Constraints

3.1. Do you know or can you think of any other constraints, which should have been identified by the analyst?
3.2. Are there any incorrect or irrelevant constraints?
3.3. Are there any constraints that make the formulated problem infeasible to solve?

4. Objectives

4.1. How well are the objectives stated?
4.2. Do you believe any objectives to be inconsistent, ambiguous, or conflicting in any way?

4.3. How realistic are the objectives?
4.4. Are there any priorities specified for the case where only some of the objectives are achievable?
4.5. Do you know or can you think of any relevant stakeholders and decision makers whose objectives are conflicting with any of those specified?
4.6. In case of multiple objectives, do you agree with the way the objectives are weighted?
4.7. Do you agree that the stated objectives are the real objectives of the stakeholders and decision makers involved?
4.8. Do you know or can you think of any associated objective, which is disguised or hidden accidentally, unconsciously, or deliberately?
4.9. How often could the stated objectives change?

5. Data and Information

5.1. Are there any sources of data and information used by the analyst that you believe to be unreliable?
5.2. Are there any data and information used by the analyst that you believe to be out of date or need to be updated?
5.3. Are there any data and information, which you believe to be not sufficiently accurate?

6. Assumptions

6.1. How well are the assumptions stated?
6.2. Are there any invalid assumptions based on which the problem is formulated?
6.3. Are there any invalid inferences or conclusions drawn by the analyst?

9.3.2 Requirements Quality Indicators

M&S requirements quality can be assessed by employing the following indicators:

1. *M&S Requirements Accuracy* is the degree to which the requirements possess sufficient transformational (verity) and representational (validity) correctness.

1.1. *M&S Requirements Verity* is assessed by conducting M&S requirements verification. M&S requirements verification is substantiating that the M&S requirements are transformed from higher levels of abstraction into their current form with sufficient accuracy judged with respect to the M&S intended uses. M&S requirements verification addresses the question of "Are we creating the M&S requirements *right*?"
1.2. *M&S Requirements Validity* is assessed by conducting M&S requirements validation. M&S requirements validation is substantiating that the M&S requirements represent the *real* needs of the application sponsor with

sufficient accuracy. M&S requirements validation addresses the question of "Are we creating the *right* M&S requirements?"

2. *M&S Requirements Clarity* is the degree to which the M&S requirements are unambiguous and understandable.

   2.1. *M&S Requirements Unambiguity* is the degree to which each statement of the requirements can only be interpreted one way.

   2.2. *M&S Requirements Understandability* is the degree to which the meaning of each statement of the requirements is easily comprehended by all of its readers.

3. *M&S Requirements Completeness* is the degree to which all parts of a requirement are specified with no missing information, i.e., each requirement is self-contained. For example, "radar search pulse rate must be 10" is an incomplete requirement because it is missing the "per second" part. The requirement "missile kill assessment delay must follow the Uniform probability distribution" is incomplete because it is missing the range parameter values. Also use of the placeholder "TBD" (to be determined or to be defined), "TBR" (to be resolved), "TBP" (to be provided), and use of the phrases such as "as a minimum," "as a maximum," and "not limited to" are indications of incomplete requirements specification.
4. *M&S Requirements Consistency* is the degree to which (a) the requirements are specified using uniform notation, terminology, and symbology, and (b) any one requirement does not conflict with any other.
5. *M&S Requirements Feasibility* is the degree of difficulty of (a) implementing a single requirement, and (b) simultaneously meeting competing requirements. Sometimes it may be possible to achieve a requirement by itself, but it may not be possible to achieve a number of them simultaneously.
6. *M&S Requirements Modifiability* is the degree to which the requirements can easily be changed.
7. *M&S Requirements Stability* is (a) the degree to which the requirements are changing while the M&S application is under development, and (b) the possible effects of the changing requirements on the project schedule, cost, risk, quality, functionality, design, integration, and testing of the M&S application.
8. *M&S Requirements Testability* is the degree to which the requirements can easily be tested. A testable requirement is the one that is specified in such a way that pass/fail or assessment criteria can be derived from its specification. For example, the following requirement specification is not testable: "The probability of kill should be estimated based on the simulation output data." The following requirement specification is testable: "The probability of kill should be estimated by using a 95 % confidence interval based on the simulation output data."
9. *M&S Requirements Traceability* is the degree to which the requirements related to a particular requirement can easily be found. Requirements should be specified in such a way that related requirements are cross-referenced. When it is

necessary to change a requirement, those requirements affected by the changed requirement should be easily identified by using the cross-references.

9.3.3 Conceptual Model Quality Indicators

The quality of a conceptual model created for a particular problem domain can be assessed by employing the following indicators (Balci et al. 2011):

1. How well does the conceptual model assist in designing not just one simulation model but also many in a particular problem domain?
2. How well does the conceptual model assist in designing any type of simulation model?
3. How well does the conceptual model assist in achieving *reusability* in simulation model design? (Balci et al. 2011)
4. How well does the conceptual model assist in achieving *composability* in simulation model design? (Balci et al. 2011)
5. How well does the conceptual model enable effective communication among the people involved in a large-scale M&S project such as stakeholders, potential users, managers, analysts, and M&S developers?
6. How well does the conceptual model assist in overcoming the complexity of designing large-scale complex simulation models in a particular problem domain?
7. How well does the conceptual model provide a multimedia knowledge base covering the areas of expertise needed for designing large-scale complex simulation models in a particular problem domain?
8. How well does the conceptual model help a subject matter expert (SME) involved in an M&S project to understand another SME's work?
9. How well does the conceptual model facilitate the collaboration among the SMEs for designing a large-scale complex simulation model in a particular problem domain?
10. How well does the conceptual model assist in verification, validation, and certification (VV&C) of an M&S application in a particular problem domain?
11. How well does the conceptual model support effective and efficient VV&C of an M&S application in a particular problem domain?
12. How well does the conceptual model assist in the specifications of test designs, test cases, and test procedures for an M&S application in a particular problem domain?
13. How well does the conceptual model assist in proper formulation of *intended uses* (objectives) for an M&S application in a particular problem domain?
14. How well does the conceptual model assist in the generation of new M&S requirements?
15. How well does the conceptual model provide significant economic benefits through its repeated use?

9.3.4 Architecture Quality Indicators

The following indicators can be employed for assessing how well a specified architecture such as High Level Architecture (HLA) (IEEE 2000) enables an M&S application to possess a desired set of quality characteristics under a set of indented uses (Balci and Ormsby 2008):

1. *Adaptability* is the degree to which the architecture enables the M&S application to be easily modified to satisfy changing requirements.
2. *Compliance with standards* is the degree to which the architecture enables the M&S application to comply with required standards.
3. *Dependability* is the degree to which the architecture enables the M&S application to (a) deliver services when requested, (b) deliver services as specified, (c) operate without catastrophic failure, and (d) protect itself against accidental or deliberate intrusion.
   - 3.1. *Availability* is the degree to which the architecture enables the M&S application to function according to its requirements at a given point in time. Availability refers to the ability of the M&S application to deliver services when requested.
   - 3.2. *Reliability* is the degree to which the architecture enables the M&S application to perform its required functions without failure under prescribed conditions in a specified period of time for a specific purpose. Reliability refers to the ability of the M&S application to deliver services as specified.
   - 3.3. *Safety* is the degree to which the architecture enables the M&S application to operate, normally or abnormally, without threatening people or the environment. Safety refers to the ability of the M&S application to operate without catastrophic failure.
   - 3.4. *Security* is the degree to which the architecture enables the M&S application to provide protection and authentication of information in transit or stationary, as well as the confidentiality of sensitive information. Security refers to the ability of the M&S application to protect itself against accidental or deliberate intrusion.
4. *Deployability* is the degree to which the architecture enables the M&S application to be easily transformed to run on more than one hardware, software, or network environment.
5. *Extensibility* is the degree to which the architecture enables the M&S application to (a) be capable of growing by including more and a greater diversity of subsystems, and (b) facilitate the extension of its capabilities by modifying current features or adding new features.
6. *Interoperability* is the degree to which the architecture enables the M&S application to exchange data with other systems or subsystems and be able to use the data that has been exchanged.

7. *Maintainability* is the degree to which the architecture enables the M&S application to facilitate changes for: (a) adaptations required as the system's external environment evolves (*adaptive maintenance*), (b) fixing bugs and making corrections (*corrective maintenance*), (c) enhancements brought about by changing customer requirements (*perfective maintenance*), and (d) preventing potential problems or for reengineering (*preventive maintenance*).
8. *Modifiability* is the degree to which the architecture enables the M&S application to be easily changed.
9. *Openness* is the degree to which the architecture enables the M&S application to possess interface specifications of its components or subsystems that are fully defined, publicly available, nonproprietary, and maintained by recognized standards bodies.
10. *Performance* is the degree to which the architecture enables the M&S application to execute its work in a speedy, efficient, and productive manner.
11. *Scalability* is the degree to which the architecture enables the M&S application to continue to function correctly as its workload (e.g., number of users, size of network, and amount of processing) is increased within anticipated limits.
12. *Survivability* is the degree to which the architecture enables the M&S application to satisfy and continue to satisfy specified critical requirements (e.g., security, reliability, real-time responsiveness, and accuracy) under adverse conditions.
13. *Testability* is the degree to which the architecture enables the M&S application to facilitate the creation of test criteria and conduct tests to determine whether those criteria have been met.
14. *Usability* is the degree to which the architecture enables the M&S application to be easily employed for its intended uses.

9.3.5 Simulation Model Design Quality Indicators

The following indicators can be employed for assessing the quality of a simulation model design:

1. *Adaptability* is the degree to which the simulation model design can accommodate changing requirements.
2. *Complexity* is the degree to which the simulation model design can be understood without difficulty and can easily be communicated to others.
3. *Composability* is the degree to which the simulation model design is capable of being constituted by combining modules, parts, or elements.
4. *Correctness* is the degree to which the simulation model design possesses sufficient transformational, representational, and behavioral accuracy.
5. *Detailedness* is the degree to which the simulation model design possesses sufficient level of detail to enable its programming into an executable model.

6. *Efficiency* is the degree to which the simulation model design enables the M&S application to fulfill its purpose without waste of resources.
7. *Flexibility* is the degree to which the simulation model design accommodates modifications.
8. *Integrity* is the degree to which the simulation model design enables the M&S application to be capable of controlling access to sensitive information by unauthorized persons or other applications.
9. *Interoperability* is the degree to which the simulation model design enables the M&S application in a distributed environment to exchange data with other applications and to be able to use the data that has been exchanged.
10. *Maintainability* is the degree to which the simulation model design facilitates changes for: (a) adaptations required as the model's external environment evolves (*adaptive maintenance*), (b) fixing bugs and making corrections (*corrective maintenance*), (c) enhancements brought about by changing customer requirements (*perfective maintenance*), and (d) preventing potential problems or for reengineering (*preventive maintenance*).
11. *Modularity* is the degree to which the simulation model design has the highest level of cohesion and the lowest level of coupling.
12. *Cohesion* is the degree to which the elements included within a simulation model design component are highly related to each other.
13. *Coupling* is the degree to which the simulation model design components depend on each other in terms of their internal logic.
14. *Portability* is the degree to which the simulation model design can easily be transformed to enable the M&S application to run on more than one hardware or software platform.
15. *Reusability* is the degree to which the simulation model design facilitates the reuse of its components in the development of other simulation model designs.
16. *Testability* is the degree to which the simulation model design facilitates the creation of test criteria and conducting tests to determine whether those criteria have been met.

9.3.6 M&S Application Quality Indicators

M&S application quality can be assessed by employing the following indicators (Balci 2004; Pressman 2010; Sommerville 2011):

1. *M&S Application Dependability* is the degree to which the M&S application (a) delivers services when requested, (b) delivers services as specified, (c) operates without catastrophic failure, and (d) protects itself against accidental or deliberate intrusion.
   1.1. *M&S Application Availability* is the probability that the M&S application functions according to its requirements at a given point in time. Availability

refers to the ability of the M&S application to deliver services when requested.

1.2. *M&S Application Reliability* is the degree to which the M&S application performs its required functions without failure under prescribed conditions in a specified period of time for a specific purpose. M&S application reliability refers to the ability of the M&S application to deliver services as specified.

1.2.1. *M&S Application Accuracy* is the degree to which the M&S application possesses sufficient transformational and representational/behavioral accuracy.

1.2.1.1. *M&S Application Verity* is assessed by conducting M&S application verification, which is substantiating that the M&S application is transformed from one form into another with sufficient accuracy. M&S application verification addresses the question of "Are we building the M&S application *right*?"

1.2.1.2. *M&S Application Validity* is assessed by conducting M&S application validation, which is substantiating that the M&S application possesses sufficient representational and behavioral accuracy. M&S application validation addresses the question of "Are we building the *right* M&S application?"

1.2.2. *M&S Application Mean Time to Failure* (MTTF) is the average time between observed M&S application failures. MTTF = 300 h means that, on the average, one failure can be expected to occur every 300 h.

1.2.3. *M&S Application Mean Time to Restore* (MTTR) is the average time it takes to restore the M&S application after failure.

1.2.4. *M&S Application Recoverability* is the degree to which the M&S application provides mechanisms to enable users to recover from errors.

1.3. *M&S Application Safety* is the ability of the M&S application to operate, normally or abnormally, without threatening people or the environment. M&S safety refers to the ability of the M&S application to operate without catastrophic failure. The safety may be an issue particularly for training simulations.

1.4. *M&S Application Security* is the degree to which the M&S application provides protection and authentication of information in transit or stationary, as well as the confidentiality of sensitive information. M&S security refers to the ability of the M&S application to protect itself against accidental or deliberate intrusion.

2. *M&S Application Functionality* is the degree to which the M&S application completely captures all of the desired functional modules that need to be present.

2.1. *M&S Application Capabilities* is the degree to which the M&S application is capable of performing its feature set, e.g., capability of simulating a particular combat at the soldier level of granularity.
2.2. *M&S Application Detailedness* is the degree to which the M&S application is characterized by abundant use of detail or thoroughness of treatment.
2.3. *M&S Application Feature Set* is the degree to which the M&S application provides the set of features that need to be present, e.g., simulating a particular combat.
2.4. *M&S Application Generality* is the degree to which the M&S application can be used for a wide range of intended uses.

3. *M&S Application Performance* is the degree to which the M&S application executes its work in a speedy, efficient, and productive manner.

3.1. *M&S Application Algorithmic Efficiency* is the degree to which the algorithms used in the M&S application provide the optimal execution time.
3.2. *M&S Application Architectural Efficiency* is the degree to which the M&S application architecture enables the optimal execution time.
3.3. *M&S Application Communication Efficiency* is the degree to which the M&S application fulfills its purpose of communicating with its user over a network without waste of resources. Communication efficiency is influenced by the communication protocol (e.g., HTTP or RMI) used by the M&S application, encryption/decryption of the communication, or the existence of a firewall.
3.4. *M&S Application Resource Use Efficiency* is the degree to which the M&S application fulfills its purpose without waste of resources such as CPU, main memory, and hard disk space.

4. *M&S Application Supportability* is the degree to which the M&S application can be supported.

4.1. *M&S Application Compatibility* is the degree to which the M&S application can be integrated into or used with other M&S applications, products, or systems.
4.2. *M&S Application Configurability* is the degree to which the M&S application can easily be set up or configured for a particular application or intended use.
4.3. *M&S Application Conformity* is the degree to which the M&S application adheres to standards and conventions.
4.4. *M&S Application Installability* is the degree to which the M&S application can easily be prepared for use.
4.5. *M&S Application Interoperability* is the degree to which the M&S application in a distributed environment (e.g., federation of models) can exchange data with one or more other M&S applications and be able to use the data that has been exchanged.
4.6. *M&S Application Localizability* is the degree to which the M&S application can easily be adopted, preferably via preferences or options, (a) to satisfy

the needs of languages other than English, and (b) to local standards such as decimal separator, currency symbol, time zone, calendar, etc.

4.7. *M&S Application Maintainability* is the degree to which the M&S application facilitates changes for: (a) adaptations required as the M&S application's external environment evolves (*adaptive maintenance*), (b) fixing bugs and making corrections (*corrective maintenance*), (c) enhancements brought about by changing customer requirements (*perfective maintenance*), and (d) preventing potential problems or for reengineering (preventive maintenance or software reengineering).

   4.7.1. *M&S Application Adaptability* is the degree to which the M&S application can accommodate changes to its external environment.
   4.7.2. *M&S Application Correctability* is the degree to which the M&S application facilitates changes for fixing bugs and making corrections.
   4.7.3. *M&S Application Extensibility* is the degree to which the M&S application capabilities can be extended by modifying current features or adding new features.
   4.7.4. *M&S Application Preventability* is the degree to which the M&S application facilitates changes for preventing potential problems or for reengineering.

4.8. *M&S Application Portability* is the degree to which the M&S application can easily be transformed to run on more than one hardware or software platform.
4.9. *M&S Application Testability* is the degree to which the M&S application facilitates the creation of test criteria and conducting tests to determine whether those criteria have been met.

5. *M&S Application Usability* is the degree to which the M&S application can easily be employed for its intended use.

   5.1. *M&S Application Documentation Quality* is the degree to which the M&S application external documentation (e.g., user manuals, reference guides, online help) possesses a desired set of characteristics.
   5.2. *Ease of Experimentation or Exercise Specification* is the degree to which a simulation experiment (for analysis) or a simulation exercise (for training) can easily be specified.
   5.3. *Ease of M&S Application Input Specification* is the degree to which the input conditions and data of the M&S application are easily specified under a set of prescribed intended uses.
   5.4. *M&S Application Ease of Learning* is the ease with which the M&S application can be learned.
   5.5. *M&S Application Output Understandability* is the degree to which the meaning of the M&S application output is easily comprehended by its users under a set of prescribed intended uses.

9.3.7 Simulation Results Quality Indicators

The simulation results make up the solution to the problem (for problem solving), show effectiveness of simulation-based training (for training purposes), or indicate some benefit in using the simulation model (e.g., for research). The quality of the results obtained from a simulation model by way of experimentation (for problem solving), exercise (for training purposes) or otherwise use can be assessed by employing the following indicators:

1. How reliable is the random number generator used as judged by the community?
2. How theoretically accurate are the algorithms used for random variate generation?
3. How accurately are the random variate generation algorithms translated into executable code?
4. How well are the simulation experiments designed to gather the desired information at minimal cost and to enable the analyst to draw valid inferences?
5. How accurately are the designs of experiments translated into executable code?
6. How appropriate are the statistical techniques used for the analysis of simulation output data?
7. How well are the assumptions underlying the statistical techniques used satisfied?
8. How appropriately is the problem of the initial transient (or the start-up problem) addressed?
9. How correctly are identical experimental conditions created for each of the alternative operating policies compared?

9.3.8 Presented Results Quality Indicators

The life cycle process of *presentation* consists of (a) interpretation of the simulation results, (b) documentation of the simulation results, and (c) communication of the simulation results to the decision makers. Simulation results must be interpreted because all simulation models are descriptive in nature. A *descriptive model* is a model that describes the behavior of a system without any value judgment on the "goodness" or "badness" of such behavior. For example, a simulation result can be "average waiting time is 5 minutes" without indicating how good or bad the value 5 is. That value must be interpreted and judged before presenting it to the decision makers. Due to the complexity of some simulation results, failing to properly interpret, document, and communicate the simulation results may lead to wrong decisions in spite of the fact that the simulation results are sufficiently credible.

The quality of the presented results can be assessed by employing the following indicators:

1. How accurately are the simulation results interpreted?
2. How properly are the simulation results documented?

3. How correctly are the simulation results communicated to the decision makers?
4. How accurately are the simulation results converted from the technical jargon into a language the decision makers can understand?
5. How accurately are the simulation output data transformed into visualizations, spreadsheets, tabulations, and/or graphical representations for effective presentation?

9.4 Concluding Remarks

Accuracy undoubtedly stands out to be the most important quality indicator of an M&S application. It is assessed by conducting verification, validation, and testing (VV&T) (Balci 2003). Tremendous amount of literature exists on VV&T. More than 75 VV&T techniques have been described in the published literature (Balci 1998).

Assessment of accuracy alone, however, is not sufficient for judging the acceptability of a large-scale and complex M&S application. An M&S application may be sufficiently accurate, but it may not satisfy other quality indicators such as the ones described in this chapter. Gaining an acceptable level of confidence in the accuracy of a large-scale and complex M&S application may not be feasible due to the complexity. Therefore, a total quality-centered assessment approach should be used to gain sufficient level of confidence in certifying the acceptability of a large-scale and complex M&S application.

Quality assessment activities must be tied to a well-structured M&S life cycle (Balci 2012). Quality assessment is not a stage but a continuous activity carried out hand in hand throughout the entire M&S life cycle. The use of an effective M&S life cycle is critically important for success in gaining sufficient confidence in M&S application acceptability.

References

Balci O (1998) Chapter 10. Verification, validation, and testing. In: Banks J (ed) The handbook of simulation. Wiley, New York, pp 335–393

Balci O (2003) Verification, validation, and certification of modeling and simulation applications. In: Proceedings of the 2003 Winter Simulation Conference (New Orleans, LA, Dec. 7–10). IEEE, Piscataway, pp 150–158

Balci O (2004) Quality assessment, verification, and validation of modeling and simulation applications. In: Proceedings of the 2004 Winter Simulation Conference (Washington, DC, Dec. 5–8). IEEE, Piscataway, pp 122–129

Balci O (2012) A life cycle for modeling and simulation. Simulation Trans Soc Model Simul Int 88 (7):870–883

Balci O, Ormsby WF (2008) Network-centric military system architecture assessment methodology. Int J Syst Syst Eng 1(1–2):271–292

Balci O, Sargent RG (1981) A methodology for cost-risk analysis in the statistical validation of simulation models. Commun ACM 24(4):190–197

Balci O, Nance RE, Arthur JD, Ormsby WF (2002) Expanding our horizons in verification, validation, and accreditation research and practice. In: Proceedings of the 2002 Winter Simulation Conference (San Diego, CA, Dec. 8–11). IEEE, Piscataway, pp 653–663

Balci O, Arthur JD, Ormsby WF (2011) Achieving reusability and composability with a simulation conceptual model. J Simul 5(3):157–165

IEEE (2000) IEEE standard for modeling and simulation (M&S) high level architecture (HLA) – framework and rules, IEEE Standard No. 1516–2000. IEEE, Piscataway

Ören TI (1981) Concepts and criteria to assess acceptability of simulation studies: a frame of reference. Commun ACM 24(4):180–189

Pressman RS (2010) Software engineering: a practitioner's approach, 7th edn. McGraw-Hill, New York

Sommerville I (2011) Software engineering, 9th edn. Addison-Wesley, Reading

[illegible]

Chapter 10
Verification, Validation, and Replication Methods for Agent-Based Modeling and Simulation: Lessons Learned the Hard Way!

S.M. Niaz Arifin and Gregory R. Madey

10.1 Introduction

Reproducible modeling and simulation research has been identified as one of the Modeling and Simulation (M&S) Grand Challenge activities (Taylor et al. 2013). Recently, uncertainty quantification has seen a renewed emphasis (National Research Council 2012). While methods for verification and validation (V&V) have been widely developed for discrete-event simulations, newer simulation approaches such as the agent-based, agent-directed, and multi-agent simulation approaches introduce new V&V challenges. The active elements in these newer approaches have greater heterogeneity, e.g., every agent can be unique, with complex attributes and behaviors. Those behaviors can result in actions based on interaction with other agents, the environment, and even the outcome of simulated or artificial intelligence. The simulation spaces are often less constrained, e.g., rather than a network of servers and queues, the space can be continuous 2D Euclidian space with multiple associated geographic information systems (GIS) layers influencing the behavior of the actors.

Over the last decade, a multitude of techniques has been used in agent-based modeling and simulation (ABMS) to perform V&V as well as replication and reproducibility (R&R) of the models. In this chapter, we review and summarize some important papers by Tuncer Ören and his colleagues and describe the influence of some of the early works by Ören. We present an overview of other contributions in V&V, quality assurance (QA), and R&R of simulation studies, with special focus on ABMS. We also discuss the lessons learnt in V&V and replication from a series of simulation experiments using agent-based models (ABMs).

S.M.N. Arifin (✉) • G.R. Madey
Department of Computer Science and Engineering,
University of Notre Dame, Notre Dame, IN, USA
e-mail: sarifin@nd.edu

L. Yilmaz (ed.), *Concepts and Methodologies for Modeling and Simulation*,
Simulation Foundations, Methods and Applications,
DOI 10.1007/978-3-319-15096-3_10

Verification and validation (V&V), accreditation, quality assurance (QA), certification, replication and reproducibility (R&R), acceptability assessment and uncertainty quantification are critical steps in the modeling and simulation (M&S) process. Without properly conducting these steps there can be little trust in the insights and predictions provided by a simulation study. For these steps in M&S studies, Tuncer Ören recognized both their pragmatic importance and their ethical implications for the modelers conducting such studies.

In this chapter, we review and summarize some important papers by Ören and his colleagues and describe the influence of some of the early works by Ören. We present an overview of other contributions in V&V, QA, and R&R of simulation studies, with special focus on agent-based modeling and simulation. We also discuss the lessons learnt in V&V and replication from a series of simulation experiments using agent-based models (ABMs).

The V&V and R&R experiences we present in this chapter originate from using several ABMs of the population dynamics of one of the most efficient mosquito species for transmitting malaria, *Anopheles gambiae*. Malaria is the third most important pathogen-specific cause of human morbidity and mortality in the world today. The populations of sub-Saharan Africa experience the highest burden of the disease with an estimated two million deaths per year. Agent-based modeling of malaria plays important roles to quantify the effects of malaria-control interventions and to answer interesting research questions (Arifin et al. 2013, 2014).

From the philosophical and methodological viewpoints, we also discuss the connection of the following to the M&S grand challenges:

- Challenges faced during V&V of multiple ABMs, all of which originated from the same conceptual model
- Unique challenges encountered during replication of published ABMs
- Major sources from which model differences may arise and/or the process of replication may become more time-consuming and challenging, etc.

Some commonly used methodologies and techniques used for the assessment of accuracy of M&S research described in this chapter are listed in Table 10.1.

The rest of this chapter is organized as follows. In the remainder of this section, we review some of the earliest works in V&V by Ören and others in advanced simulation methodologies, assessing the acceptability of simulation studies, categorizations and taxonomies of M&S, and M&S applications. In Sect. 10.2, we summarize the contributions of Ören and others in the areas of V&V and QA. Section 10.3 discusses some important works in R&R. In Sect. 10.4, we briefly introduce the malaria entomological model and the ABMs we used to perform the V&V and R&R works. Sections 10.5 and 10.6 describe the V&V and R&R challenges we faced, discuss some important V&V issues from which model differences may arise and offer recommended guidelines for ABM modelers. Finally, Sect. 10.7 concludes with a summary.

Table 10.1 Methodologies and techniques commonly used for the measurement and assessment of accuracy in M&S research. Methodologies are ordered alphabetically

	Definition	Alternative or synonymous terms	References
Acceptability assessment	The official certification that a model, simulation, or federation of models and simulations is acceptable for use for a specific purpose.	Accreditation, certification, credibility assessment	Ören (1981), Balci (1989, 1998a, 2003)
Docking	A form of V&V that tries to align multiple models in order to investigate whether they yield similar results.	Alignment, model-to-model comparison	Axtell et al. (1996), Arifin et al. (2010a, b)
Quality assurance	Ensuring that an M&S application possesses a desired set of characteristics to match a desired degree of quality.	Quality assessment	Ören (1984), Balci (2004)
Replication	To allow independent researchers to address scientific hypotheses in a model and produce evidence for or against them in order to judge the scientific claims presented by the model.	Alignment, cross-model validation, model-to-model comparison	Peng (2011), Jasny et al. (2011)
Reproducibility	The ability to independently verify the prior findings reported by an established model. The ability to independently replicate, reproduce and, if needed, extend computational artifacts associated with published work.		Santer et al. (2011) Fomel and Hennenfent (2009)
Uncertainty quantification	The process of quantitative characterization and reduction of uncertainties in M&S applications. It tries to determine how likely certain outcomes are if some aspects of the system are not exactly known.	Calibration, parameter estimation, parameter fitting, sensitivity analysis	National Research Council (2012), Walker et al. (2003), Thiele et al. (2014)
	To learn about the relative importance of the various mechanisms represented in the model and how robust the model output is to parameter uncertainty by exploring the sensitivity of model output to changes in parameters.		

(continued)

Table 10.1 (continued)

	Definition	Alternative or synonymous terms	References
Verification	The assessment of transformational accuracy of the model by addressing the question of "are we creating the model right?"		Balci (1998a, 2003)
	Ensuring that the computer program of the computerized model and its implementation are correct.		Sargent (2001, 2004)
Validation	The assessment of behavioral or representational accuracy of the model by addressing the question of "are we creating the right model?"		Balci (1998a, 2003)
			Sargent (2001, 2004)

10.1.1 Advanced Simulation Methodologies

One of the earliest works in this area includes the 1979 seminal paper by Ören and Zeigler (1979) on advanced simulation methodologies, which is still one of the 50 most-frequently cited articles in *Simulation* (5/50) as of 3 Dec 2014. This has been considered a thought-provoking article, earning 177 citations so far. In this paper, a set of comprehensive concepts is proposed for the design and implementation of advanced simulation methodologies. Later, it influenced the development of several special-purpose simulation systems and general-purpose languages, including the concepts of functional separation of model development, entity description, scenario specification, output analysis, etc. (Ketcham et al. 1984, 563). It also presented the concept of hierarchical/structured modeling and experimental frames, changing the emphasis of simulation software development from then on (An et al. 1989, 500; Wang and Li 1991, 1062). The described approach can be connected to general systems theory, a field to which both authors of this paper have made frequent and significant contributions (Karplus 1979). It also pointed out several weaknesses and drawbacks in current simulation languages (Shannon 1986, 150; Lin 1990; Pegden et al. 1995), inspired the development of tools covering the whole trajectory from systems analysis through model construction (Elzas 1988, 49), new approach to definitions (Wittmann 1992), etc.

10.1.2 Assessing the Acceptability of Simulation Studies

In assessing the acceptability of simulation studies, Ören proposed a framework that permits the discussion of the concepts and criteria related to the acceptability of various components of a simulation study, including simulation results, real world and simulated data, parametric model and the values of the model parameters, specification of the experimentation, representation and execution of the computer program, and modeling, experimentation, simulation, and programming methodologies or techniques used, etc. (Ören 1980, 1981). The acceptability issues, which are closely related to credibility of simulation studies, validation of models, and verification of simulation programs, are described in a highly cited 1981 paper (Ören 1981) with 82 citations (according to Google Scholar) so far. These issues are discussed with respect to the goal of the simulation study, the structure and data of the real system, the parametric model, the model parameter set, the specification of the experimentation and the existing or conceivable norms of modeling methodology, experimentation technique, simulation methodology, software engineering etc. (Ören 1981).

10.1.3 Categorizations and Taxonomies of Modeling and Simulation (M&S)

The detailed classification, categorizations, and taxonomies of M&S tools proposed by Ören influenced and guided the research in multiple related topics, including identification of regions that require immediate additional efforts (Garzia 1979), simulation model management (Nance et al. 1981), simulation software development (Standridge and Pritsker 1982), general systems models (Troncale 1985) etc. The related works by Ören and Zeigler (1979) provided a framework for preparing taxonomy of expert systems (ES) and artificial intelligence (AI) techniques related to simulation studies (Zeigler 1982; Ketcham 1986).

10.1.4 M&S Applications

Ören and others later reported results of the ideas introduced in the papers discussed above in several applications. For example, to deal with various types of errors that can be generated by simulations of nuclear fuel waste management programs, Ören et al. (1985) offered M&S frameworks to clarify issues of model reliability and software quality assurance. They identified potential problems with reference to the main areas of concern for reliability and quality, e.g., experimental issues, decomposition, scope, fidelity, verification, requirements, testing, correctness, and robustness, and provided a list of 80 most common computerization errors, as well as software tools and techniques to detect and to correct the errors (Ören et al. 1985).

In an attempt to promote the credibility and integrity of simulation as a field itself, Ören (2000) pointed out the existence of professional *codes of ethics* in domains that are relevant to simulation, including science, engineering, business, computerization, software engineering, AI, software agents, the Internet, and defense industries. The article discuses a set of important philosophical questions related to simulation and ethics, for example: the need for a code of ethics for simulation, where to start to develop such a code, whose responsibility and responsibility to whom, etc. (Ören 2000). While addressing quality assurance problems in simulation, Ören also introduced and defined the concept of *normative assessment* of an element of a simulation study as its evaluation with respect to some norms of a value system, which can be pragmatic or ethical (Smit 1999). He also strongly advised the development of a code of conduct for simulation professionals, and presented a procedure of composing such a code of conduct (Kettenis 2001).

10.2 Verification and Validation (V&V) and Quality Assurance (QA)

In general, verification, validation, docking, testing, accreditation, certification, quality assurance, and credibility assessment activities primarily deal with the measurement and assessment of accuracy of M&S (Balci 1998a, b; Ören and Yilmaz 2009; Arifin et al. 2010a, b).

Verification involves transformational accuracy of the model artifacts in model development, in order to ensure that the implementation is a correct realization of the conceptual model. Validation, on the other hand, involves substantiation that a model within its domain of applicability possesses a satisfactory range of accuracy consistent with its intended application.

Regarding the quality assurance of system design for complex problems, Ören recognized the needs to critically examine the nature and problem-solving paradigms of complex interrelated problems, and listed a set of questions and metaquestions relevant to the task under the six categories of: (1) problem and problem solving paradigms, (2) goals, (3) performance measures, (4) decision and its components, (5) system, and (6) resources (Ören 1983). This work was soon followed by presenting a basis for the taxonomy of concepts related with quality assurance in M&S, which included three major parts: (1) criteria for assessment, (2) types of assessments, and (3) a comprehensive categorized list of related terms (Ören 1984).

Ören and Yilmaz (2009) pointed out various sources of failures in order to take necessary precautions to minimize the risks associated with agent-based modeling (ABM) and agent-directed simulation (ADS), and suggested new failure avoidance (FA) paradigms for V&V as well as quality assurance (QA) purposes for M&S. They recognized the significant repercussions of any failure to the application area in the use of M&S, provided a five-part representative list of common mistakes in M&S projects, described three groups of concepts (i.e., types, criteria, and elements) involved in an assessment paradigm, and discussed various V&V, QA, and FA paradigms for successful M&S projects (Ören and Yilmaz 2009). They also noted the importance of lessons learned and best practices from earlier mistakes – portraying the collection and analyses of lessons learned from studies in V&V, QA, and FA to be useful in benefiting from the shared experiences of many simulation professionals (Ören and Yilmaz 2009).

Guided by the pioneering work of Ören and others (1985), Balci (1998a, b) proposed the need for a simulation quality assurance (SQA) group, portraying it as critically essential for the success of a simulation study. The SQA is a managerial approach whose job is to ensure quality management by working closely with the simulation project managers in planning, preparing, and administering QA activities throughout the simulation study (Balci 1998a). He also presented guiding principles for conducting verification, validation, and accreditation (VV&A) of M&S applications (Balci 1998a, b).

Some of the other previous works involving V&V, docking, alignment, and model-to-model comparison are described below.

Axtell et al. (1996) described the docking process of Axelrod and Sugarscape models and concluded that by comparing independently built simulations using different tools, docking (or alignment) may discover bugs, misinterpretation of model specification, and inherent differences in toolkit implementations. Sargent (2001) described different approaches to decide validity by two different paradigms that relate V&V to the model development process, with the use of graphical data statistical references for operational validity. Carley (2002) described the importance of docking in computational social and organizational science. North and Macal (2002) implemented the Beer game using Mathematica programming, Repast and Swarm ABMS, reproducing (i.e., docking) published results with these new implementations. Burton (2003) argued that docking provides a guide in use of different laboratories to address organization questions, and computational and non-computational models can be docked to broaden the understanding of organization science. Xu et al. (2003) discussed the results of docking a Repast simulation and a Java/Swarm simulation of four social network models of the Open Source Software (OSS) community.

Edmonds and Hales (2003) replicated a published model involving co-operation between self-interested agents in two independent implementations to align the results and the conceptual design, revealing a host of minor bugs and ill-defined implementation issues that otherwise appeared unnoticed. By using model-to-model comparison (docking), Xiang and others demonstrated the V&V processes for a natural organic matter simulation model (Xiang et al. 2005; Huang et al. 2005). Kennedy and others (Kennedy et al. 2005, 2006) presented V&V results involving two different case studies (a scientific and an economic model), identifying general guidelines on the best approach to new simulation experiments and drawing conclusions on effective V&V techniques. They argued that one of the goals of docking is to try to align multiple models in order to investigate whether they produce similar results (Xiang et al. 2005; Kennedy et al. 2006).

Yilmaz (2006) presented a process-centric perspective for the V&V of agent-based computational organization models. He emphasized the importance of V&V in assessing not only predictive capability but also explanation accuracy of formal models in terms of the degree of realism, and presented a framework for the V&V of multi-agent organizations and a set of formal validation metrics to substantiate the operational validity of emergent macro-level behavior (Yilmaz 2006).

We performed a docking case study on agent-based models of malaria epidemiology, and showed how docking helped in increasing confidence to the conceptual model (from which several agent-based implementations were built), revealing conceptual and/or programming errors, and eliminating dubious assumptions (Arifin et al. 2010a, b). To examine the effects of spatial heterogeneity, we later extended the non-spatial ABMs to have explicit spatial representations, and performed V&V between several non-spatial and spatial ABMs, all of which originated from the same conceptual model (Arifin et al. 2012).

In many cases, ABMs can produce large volumes of textual outputs, potentially in the range of hundreds of gigabytes. In general, these outputs contain inherent logical structures that can be naturally expressed in terms of abstract mathematical notions such as graphs, relations, etc. The modeler is often interested in visualizing these structures in the forms of data plots, time-series analysis, visual graphs, and the like. Thus, to understand the results and patterns exhibited by the agents, it is crucial to be able to effectively analyze the voluminous textual outputs, and to produce the desired visualization with ease. Appropriate analysis and visualization also play important roles in V&V of the ABMs. To meet these goals, we developed a software module called P-SAM (Post-Simulation Analysis Module) for post-simulation output analysis and visualization (Arifin et al. 2010c). As a case study, we described its application to a biological simulation model named LiNK, which models pathogen transmission amongst long-tailed macaque monkeys on the island of Bali in Indonesia. P-SAM allowed the modelers to analyze and visualize the logical structures inherent in LiNK outputs, thus helping in V&V of the model (Arifin et al. 2010c).

10.3 Replication and Reproducibility (R&R)

Replication and reproducibility (R&R) cover a wide spectrum of issues related to M&S, and fall under the broader subject area of V&V. In M&S, replication is also known as model-to-model comparison, alignment, or cross-model validation. In general, many replication tasks may be viewed as a *weaker* form of docking, in which the goal of the modeler is to *qualitatively* (as opposed to *quantitatively*) replicate the results of previously published models.

Replication is treated as the scientific gold standard to judge scientific claims. It allows independent researchers to address a scientific hypothesis and produce evidence *for* or *against* the hypothesis (Peng 2011; Jasny et al. 2011). Replication confirms reproducibility, which refers to the independent verification of prior findings, and is at the core of the spirit of science (Santer et al. 2011). Reproducibility, as a fundamental principle of the scientific method, refers to the ability to independently replicate, reproduce and, if needed, extend computational artifacts associated with published work (Fomel and Hennenfent 2009). Replication exercises can also provide unique opportunities for V&V processes (Zhong and Kim 2010).

Although computational science has led to exciting new developments, the nature of the work has also exposed shortcomings in the general ability of the research community to evaluate published findings (Peng 2011). As the use of computer simulation is becoming increasingly important, lack of proper documentation, validation, and distribution of models may hamper reproducibility, causing a *credibility gap* (Yilmaz 2011; Yilmaz and Ören 2013). In order to minimize this credibility gap, replicability of the *in silico* experiments and simulations performed by various models bear special importance (Arifin et al. 2013). In the following, we describe some of the other previous works involving R&R.

Will and Hegselmann (2008) showed the importance of replication by reporting a failure to replicate the results presented on a published model. Obtaining the source code from the original authors, Will (2009) later found that the simulation unintentionally implemented an assumption that was never mentioned in the original model, and showed that the model crucially depended on that dubious assumption, and its removal leaded to dramatically different results. Others (Rand and Wilensky 2006; Wilensky and Rand 2007) presented a case study that replicated the Axelrod-Hammond model and described the challenges in recreating the model and in determining whether the replication was successful. Pavón et al. (2008) proposed the use of agent-based graphical modeling languages to specify social systems as multi-agent systems, which allow replication of simulations on different platforms. Rouchier et al. (2008) described advancements in model-to-model analysis, and categorized comparative modeling research into a number of areas. Olaru et al. (2009) reported the docking experience and validation stages performed when replicating a fuzzy logic model with an agent-based model in innovation business networks.

In a "Letter from the Editor" for the *Simulation* journal, Yilmaz (2011) emphasized that since science is a collective phenomenon, progress in simulation-based science requires the ability of scientists to create new knowledge, elaborate and combine prior computational artifacts, and establish analogy and metaphor across models. Thus, models that are not designed and disseminated to be discovered, extended, or combined with other models may hinder scientific progress (Yilmaz 2011, 2012a, b). He argues that reproducibility should become the responsibility of the broader scientific community, and suggested a set of guidelines for the authors, publishers, funding agencies, journals, and the broader scientific community. Some of these guidelines include the following to be made available, along with the simulation code, by the modelers:

- Links to data, source code, standard documentation, and experimental conditions (pertaining to testing and validating the model)
- Open source repositories which may help facilitate making data available to others and allow replication through the use of openly accessible simulations
- A version control system which may help to uniquely identify and cite each version of published models and data

- Automated batch scripts that can produce tables, data summaries, and figures to support data analysis
- Automated model documentation tools that can be integrated with model development processes
- Annotation technologies and standard notation for metadata which may help presenting comments, notes, explanations, or other types of external remarks for each shared model

In addition, to promote reproducibility, Yilmaz (2011) suggested the following: (1) the funding agencies and grant reviewers can give preference to highly qualified proposals with fully implemented, transparent and online scientific workflows, (2) the reviewers can provide incentives to research groups to develop shared artifacts for reproducibility, and (3) a classification system of reproducibility categories (e.g., verifiable, verified, annotated, shared etc.) be used.

Yilmaz (2012a) also classified the major areas of focus in reproducible research under three dimensions: (1) scholarly communication, (2) methodology of scientific practice, and (3) technical infrastructure. He identified issues, strategies, and implications for each of these dimensions for simulation model development. To promote the reproducibility of computational research, he also proposed a multi-dimensional strategy, a software framework as a technical solution for establishing and maintaining open M&S research platforms, and suggested a set of systematic guidelines with regard to legitimization, dissemination, and access dimensions for authors and institutional environments (Yilmaz 2012a, b). Yilmaz and others also introduced the *e-Portfolio* concept as an ensemble of integrated active documents that encompass published manuscript, computer code, data, and scientific workflow specification (Yilmaz 2011, 2012a, b; Yilmaz and Ören 2013).

10.4 The Models

Our V&V and R&R experiences originate from using several agent-based models (ABMs), all of which were developed from a core entomological model (hereafter referred to as the *core model*) of the population dynamics of *Anopheles gambiae*, which is regarded as one the most efficient malaria-transmitting mosquito species. Details about the core model and the ABMs can be found in (Arifin et al. 2013, 2014). In this section, we present a brief overview of the models.

The *An. gambiae* mosquito life cycle consists of aquatic and adult phases. The aquatic phase consists of three aquatic stages (egg, larva, and pupa). The adult phase consists of five adult stages (immature adult, mate seeking, blood meal seeking, blood meal digesting, and gravid). The development and mortality rates in all stages of the life cycle are described in the core model in terms of the aquatic and adult mosquito populations.

The core model assumes simplistic, homogeneous aquatic habitats (which serve as breeding sites for female adult mosquitoes). Each aquatic habitat is set with a carrying capacity, which regulates the density-dependent egg-laying (oviposition) mechanism and the age-adjusted biomass that the habitat can sustain. Age-specific mortality rates are used for the larva stage, and for all the adult stages in the core model. Female adult mosquito abundance and potentially infectious female (PIF, which denotes the number of female adult mosquitoes that are potentially capable of transmitting malaria) abundance are treated as the primary outputs of the model. In order to compare language-specific dependencies as well as other V&V features, the core model was implemented as ABMs in two programming languages (Java and C++). Over the last few years, several versions evolved in both languages to explore various related research problems. For example, one specific Java version extended the core model to have an explicit spatial representation by appending spatial properties to the mosquito agents and the underlying landscapes (Arifin et al. 2013).

10.5 Verification and Validation Challenges

The V&V techniques we used form a subset of the taxonomy of more than 77 V&V techniques listed by Balci (1998b). These techniques, and the new dynamic techniques of *phase-wise docking* and *compartmental docking*, are listed in Table 10.2. However, among these, *face validation* was by far the more frequently used than the rest. Note the absence of any *formal* category, due to the population dynamics-based ABM abstractions of the models.

Most of the V&V were performed by using model-to-model comparison, i.e., *docking*, between the agent-based implementations. The docking workflows, performed in two separate episodes, are depicted in Fig. 10.1. In both episodes, different versions were developed in the two languages (Java and C++). For *Episode 1*, they are noted as *Java1*, *Java2*, *Java3*, and *CPP*. For *Episode 2*, they are noted as *Java1*, *Java2*, *CPP1*, and *CPP2*. All versions are hereafter referred to by using these notations.

Table 10.2 V&V Techniques We Used

Primary category	Secondary category
Informal	**Face validation**, inspections, **reviews**
Static	**Data analysis**, **failure analysis**
Dynamic	**Phase-wise docking***, **compartmental docking***, bottom-up testing, comparison testing, **debugging**, execution testing, functional (black-box) testing, **graphical comparisons**, partition testing, predictive validation, **sensitivity analysis**, statistical techniques, structural (white-box) testing, submodel testing, top-down testing, **visualization**
Formal	*(none)*

The more frequently used techniques are marked in bold. The new techniques of phase-wise docking and compartmental docking are marked with *

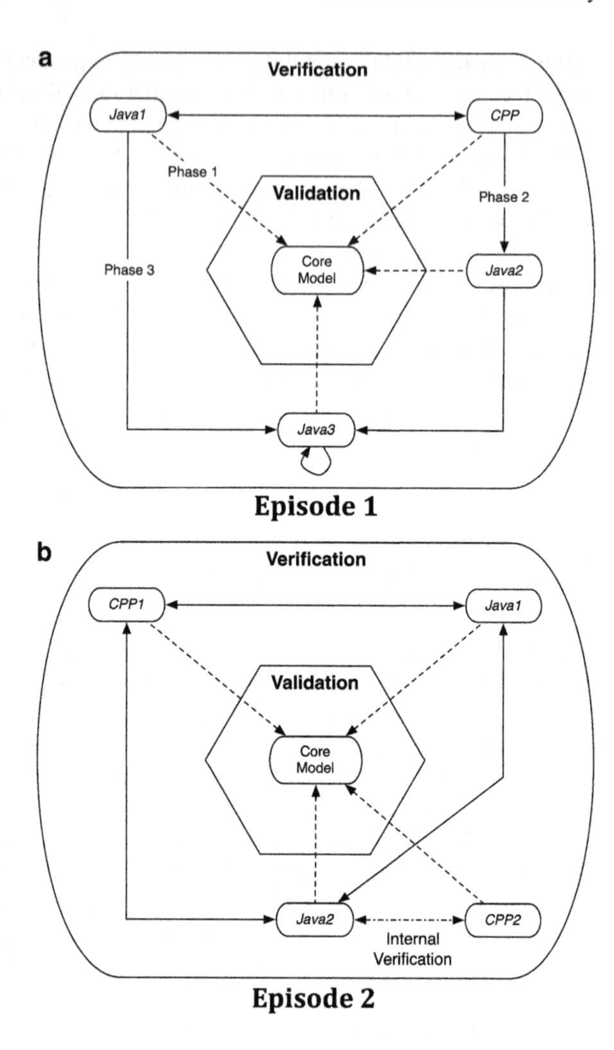

Fig. 10.1 *The V&V workflows*. (**a**) The *phase-wise docking* workflow reported in (Arifin et al. 2010a). (**b**) The *compartmental docking* workflow reported in (Arifin et al. 2010b). The *solid arrows* indicate verification relationships and immediate successorship between the ABMs. The *dashed arrows* indicate validation relationships with the core model, as well as internal verifications between the ABMs. In both episodes, different versions were developed in the two languages (Java and C++), as abbreviated in the figure

10.5.1 Episode 1: Phase-Wise Docking

As depicted in Fig. 10.1a, Episode 1 consists of the phase-wise docking workflows. Not surprisingly, the initial V&V results obtained in this episode showed substantial differences, highlighting different conceptual (mental) images of the core model among the developers, and thus necessitating further investigation. We identified the following major differences in Episode 1:

1. Both the adult and aquatic mosquito populations differed significantly
2. Proportion of the older female adult mosquitoes (with an age of 12 days or higher) was significantly lower in *CPP*
3. The aquatic population sizes were consistently higher in *CPP*

Before analyzing the above differences, we addressed some minor issues. First, we verified that the initial parameter settings and constants used were identical for both models. Next, we logged all randomly-generated numbers with the specified distribution parameters, and ensured that proper distributions were being generated, ruling out any potential differences due to the use of different random number generator libraries. We also explicitly typecasted all floating point arithmetic into decimal arithmetic to ensure that the former was being correctly computed in both models.

To address the first difference as listed above, we verified the age-specific mortality rate functions for the adult mosquitoes. Comparisons of the specific routines to calculate the age-specific mortality functions revealed that although it was intended to eliminate all adult mosquitoes in a given age group, *CPP* placed an artificial bound to kill only 80 % mosquitoes in the group. This modification, once applied to *Java1*, had little impact on the results. However, it revealed another critical error: in the models, an agent entered the simulation with an age of 0 days, and in each simulated day the age was increased by 1, which was supposed to continue without any resetting of the age when the agent transitioned into the first adult stage. At this transition, the mosquito's age was reset to 0 in *CPP*, but inadvertently, not in *Java1* or *Java2*. This also partially explained the second difference, i.e., why a larger proportion (approximately half) of the female adult mosquito population was $\geq$12 days old in *Java1*. While resetting the mosquito's age reduced the oscillation in the proportion of older females, there were still significant differences in terms of average number of mosquitoes (*CPP* still had much higher abundances).

To address the third difference (which primarily dealt with the aquatic mosquito population), we noted that the number of eggs in *Java1* was consistently less than *CPP*, causing *Java1* to eventually inject fewer agents into the system. This, in conjunction with the second difference, suggested that younger mosquitoes were killed more rapidly than the older ones. However, at least on the surface, this insight was counterintuitive, since the age-specific mortality rate functions were designed to kill the older mosquitoes at a higher rate.

We further verified the age-specific mortality for the aquatic stages. This revealed that *Java1* and *CPP* calculated an important age-adjusted biomass parameter (the 1-day old equivalent larval population, see Arifin et al. 2014) differently: *CPP* computed it once at the start of each day, using the previous day's aquatic populations. *Java1*, on the other hand, recomputed it for every egg-laying event, creating a selection bias for female mosquitoes trying to lay eggs ahead of others within the same simulation day, and thus causing more repulsive force to the females trying later. This issue, when resolved, yielded some impact to increase the number of eggs being laid in *Java1*. It also revealed a transition logic variance in *CPP* for the egg development time: eggs took 2 days (instead of 1 day as intended). When resolved, this further reduced the difference in number of adult mosquitoes.

However, at this point, the Java (*Java1*, *Java2*, and *Java3*) and *CPP* versions were not completely *docked*, especially in terms of the aquatic mosquito populations. To achieve a complete docking, we used the technique of *compartmental docking* in Episode 2, as described in the following.

10.5.2 Episode 2: Compartmental Docking

Episode 2 consists of the compartmental docking workflows, as depicted in Fig. 10.1b. The results of the separate implementations revealed that seemingly insignificant differences leaded to significant mismatch in overall model outputs. Hence, for V&V purposes, we decided to separate the artificial simulation world into *isolated compartments*, so that errors in one specific compartment were not propagated. Thus, they could not influence the discovery (and correction) of errors in other compartments. Following the *divide and conquer* paradigm, we compartmentalized the mosquito world into the adult and aquatic populations. Then, we first ensured that the compartments were working as intended, and later combined them to perform more complex V&V experiments.

In this episode, we made the following simplifying assumptions in all implemented versions:

- All sources of randomness were removed
- The carrying capacity of each aquatic habitat was fixed as a constant
- The mate seeking state was omitted altogether (for simplicity)
- The egg-laying mechanism was simplified
- All male mosquito agents were omitted

The compartmental docking was performed by first isolating the adult mosquito populations from the aquatic mosquito populations. In order to ensure the theoretical expected measures for age-structure and age-specific mortality rates for the adult stages, some additional simplifying assumptions were made, which included using fixed number of days for some stages in the adult development, a simplified egg-laying mechanism, etc.

The next phase dealt with the isolated aquatic mosquito populations. To ensure the theoretical expected measures for age-structure and age-specific mortality rates of the aquatic populations, the simplifying assumptions in this phase included using fixed number of female eggs per oviposition, fixed carrying capacity for each aquatic habitat, etc.

The last phase combined the isolated adult and aquatic mosquito populations, verified the transitions from aquatic to adult stages, the egg-laying mechanism, the actual number of eggs laid in aquatic habitats and the durations of stages. The four-fold outputs from the two Java (*Java1* and *Java2*) versions and the two C++ versions (*CPP1* and *CPP2*) were compared, and potential misinterpretations were analyzed and fixed. The major issues discovered in different phases during compartmental docking are listed in Table 10.3.

Once all the issues listed in Table 10.3 were addressed, the results produced a complete four-fold dock between the separate versions.

Table 10.3 The major issues discovered in different phases during *compartmental docking*

Issue	Resolution
In *CPP1* and *Java1*, the adult populations were slightly less in number than those in *Java2* and *CPP2*. Also, the aquatic populations in *CPP1* and *Java1* were killed at a higher rate than suggested by the theoretical expected measures.	Resolved after all versions used the same carrying capacities for the aquatic habitats.
CPP1 and *Java1* had female mosquitoes lay all of their eggs on the first egg-laying attempt. This created an egg-laying pattern where bursts of eggs were laid on the same day, followed by a few days when no eggs are laid, then another burst, and so on. In *Java2* and *CPP2*, however, the eggs were laid over a period of successive days.	This issue suggested a difference in the egg-laying code. It was partially resolved after all versions ensured to use the same formulas for calculating how many eggs a female could lay in different egg-laying attempts.
On the first egg-laying attempt, eggs were laid one day sooner in *Java1* and *CPP1* than in *Java2* and *CPP2*.	It was revealed that in *CPP1* and *Java1*, eggs were laid at the end of the blood meal digesting stage, rather than waiting until the first day of the gravid stage. Consequently, they were updated to ensure that eggs were only laid while in the gravid stage.
In *Java1*, after all female mosquitoes laid all their eggs in the gravid stage, they did not transition back to the blood meal seeking stage on the same day.	*Java1* was updated to ensure that the females transitioned back to the blood meal seeking stage from the gravid stage on the same day once all eggs are laid.
CPP1 and *Java1* allowed the female mosquitoes to lay all their eggs on the first egg-laying attempt during day 5 and 6. However, as suggested by the density-dependent egg-laying mechanism, laying all the eggs in a single attempt was possible only if the aquatic habitat was empty.	It was ensured that the egg-laying was indeed complete after three attempts in *CPP1* and *Java1*, and we made a simple adjustment delay before the female mosquitoes could leave the gravid stage.
CPP2 and *Java2* placed an upper bound of 80 % on the larval mortality rate.	*CPP1* and *Java1* were ensured to use the same value.
In calculating the maximum number of eggs a female could lay, *CPP2* and *Java2* erroneously used another parameter: the number of eggs remaining to lay.	*CPP2* and *Java2* were ensured to use the same parameter.
In *Java1*, gravid females incorrectly laid all eggs, and the biomass of the aquatic habitat did not affect the number of eggs actually laid.	In the calculation of the number of eggs allowed to lay, some double-precision values were erroneously converted to integer values, making the expressions evaluating to 0 in these instances. This, in turn, affected the related variables. *Java1* was updated to use double-precision values by explicit typecasting.
After 14 simulation days, *CPP1* and *Java1* killed different number of adult mosquitoes.	After analyzing the relevant code, a rounding error was found. Also, *Java1* erroneously used an extra parameter in the adult age-specific mortality rate function. These were corrected.

10.6 Replication and Reproducibility (R&R) Challenges

Our R&R experiences originate from replicating the results and extending some assumptions of several published malaria models (both mathematical and agent-based), including Gu and Novak (2009a, b), Yakob and Yan (2009), Chitnis et al. (2010) and Eckhoff (2011). We emphasize that the goal of replication was to achieve *qualitative* (not *absolute* or *quantitative*) matches between the respective models.

Critical examination of the first two studies (Gu and Novak 2009a, b), which explored the impact of applying several mosquito control interventions, revealed that despite providing reasonably plausible results, the models adopted two major dubious assumptions regarding: (1) the number of replicated simulation runs, and (2) the boundary type of the landscapes.

In general, any simulation model which involves substantial stochasticity should conduct sufficient number of replicated runs (with identical parameter settings but different random seeds), and the average and/or aggregate results of these replicated runs should be reported, as opposed to reporting results from a single run. Sufficient number of replications is required to ensure that, given the same input, the average response can be treated as a deterministic number, and not as random variation of the results. This allows modelers to obtain a complete statistical description of the model variables. The same principle also applies to a set of stochastic (Monte Carlo) simulation models in other domains (e.g., traffic flow, financial problems, risk analysis, supply chain forecasting, etc.), where, in most cases, the standard practice is to report the averages and standard deviations of the measures of interest (also known as the measures of effectiveness, or MOEs).

We argued that since most epidemiology models (including ABMs) involve substantial stochasticity in the forms of probability-based distributions and equations, performing sufficient number of replicated runs is also important for validation of the results. In malaria ABMs, agents' decisions are often simulated using random draws from certain distributions. These sources of randomness are used to represent the diversity of model characteristics, and the behavior uncertainty of the agents' actions, states, etc., with the goal to mimic/simulate the reality as closely as desired. For example, in our ABM, when a host-seeking mosquito searches for a blood meal in a house covered by insecticide-treated bed nets, a 50 % mortality by bed nets would mean that it may die with a probability of 0.5, which can be simulated using random draws from a uniform distribution. As another example, the number of eggs in each egg-batch of a gravid mosquito is simulated using random draws from a normal distribution with mean (average) = 170 and standard deviation = 30. As we showed, these sources of randomness can have significant impact on the results of the simulation, and different simulation runs can therefore produce significantly different results, due to a different sequence of pseudo-random numbers drawn from the distributions (Arifin et al. 2013).

The second issue, the use of a specific boundary type, may greatly impact the movement process of the agents (mosquitoes). In general, three different boundary types are commonly used in ABMs: absorbing, non-absorbing, and reflecting. With an absorbing boundary, mosquitoes are permanently removed (effectively killed) when they hit an edge of the landscape's boundary. On the other hand, with a non-absorbing boundary, when mosquitoes hit an edge, they re-enter the landscape from the edge directly opposite of the exiting edge (and thus are not killed due to hitting the edge). Unless the underlying landscape reflects a completely isolated geographic location (e.g., an island far away from the mainlands), in reality, when mosquitoes hit an edge, logical approaches are either to reflect the mosquito back from the same edge (reflecting boundary), or to coerce the mosquito to re-enter from the opposite edge (non-absorbing boundary). We argued that a non-absorbing boundary might capture the mosquito population dynamics more realistically. This is especially true when the resource (e.g., the houses and aquatic habitats) densities are high, and the resources are more evenly distributed across the landscape. Gu and Novak (2009a, b) used an absorbing boundary for all landscapes. In replicating their results, we used a non-absorbing boundary for all landscapes, which were modeled topologically as 2D torus spaces.

We also replicated the results reported by Yakob and Yan (2009), which examined the combined impact of several mosquito control interventions. A systematic comparison of some R&R features and assumptions of several recent malaria models is given in Table 10.4. Note that the features refer to the particular results presented in the cited publications, and should not be treated as limitations of the respective models. For example, for the feature "average of multiple simulation runs", both EMOD (Eckhoff 2011) and OpenMalaria (Chitnis et al. 2010) are capable of reporting the averages of multiple simulation runs; however, in the cited publications, they were not explicitly reported.

Figure 10.2 shows some important R&R results depicting the effects of performing sufficient number of replicated simulation runs and the type of boundary used for the landscapes. The importance of performing multiple simulation runs (instead of a single run) can be seen by comparing the simulation output (mosquito abundance) for the maximum, the minimum, and the average cases.

As we found (Arifin et al. 2013), in 94 % cases, use of an absorbing boundary yielded less mosquito abundance than that with a non-absorbing boundary. Also, with an absorbing boundary, even before applying any control intervention, abundances were too low when compared to those with a non-absorbing boundary. Due to using an absorbing boundary, more mosquitoes died out because of the additional unrealistic killing effect imposed by the absorbing boundary. This suggested the importance of using the proper boundary type in the ABM in order to avoid any potential bias created by a specific boundary type.

Our R&R works posed some unique challenges which include: (1) the unavailability of source codes of the original models inhibited us from performing direct model-to-model comparison (docking) and (2) the structural characteristics of the ABMs, which are fundamentally different from, for example, equation-based mathematical models, also ruled out the possibility of systematic R&R of model features. These experiences, many of which were reported and addressed by the earlier works of Ören and others (Ören 1983, 1984; Ören et al. 1985; Ören and

Table 10.4 Summary of R&R feature comparisons from several malaria models

Model feature	Malaria models				
	Gu and Novak (2009a, b)	Yakob and Yan (2009)	EMOD (Eckhoff 2011)	OpenMalaria (Chitnis et al. 2010)	Arifin et al. (2013)
Model type	Agent-based	Mathematical	Individual-based	Hybrid (agent-based/ mathematical)	Agent-based
Automation of landscape generation (e.g., using separate tools)	No	N/A	No	N/A	Yes
Boundary type of landscape*	Absorbing	N/A	N/A	N/A	Non-absorbing
Average of multiple simulation runs	No	No	No	No	Yes
Time-step resolution	Daily	Daily	Daily	Daily	Hourly
Age-specific mortality	No	No	N/A	No	Yes
Daily mortality rate (aquatic stages)	Fixed (0.2)	Fixed (0.15)	Temperature-dependent	N/A	Age-specific (for larvae)
Daily mortality rate (adult stages)	Fixed (0.2)	Fixed (0.15)	Adult life expectancy of 10 days	N/A	Age-specific
Fecundity (eggs/ oviposition)	Fixed (80)	N/A	Fixed (100)	N/A	*N*(170, 30)
Variability in daily temperature	No	No	Yes	Yes	Yes
Length of individual simulation run	200 days or 300 days	N/A	>6 years	N/A	≥1 year
Interventions modeled	LSM, ITNs	LSM, ITNs	IRS, ITNs, larvicides, space spraying	ITNs, IRS	LSM, ITNs
Time-step of intervention application	Day 100 for LSM, day 150 for ITNs	N/A	N/A	N/A	Day 100
Explores combined interventions	No	Yes	Yes	Yes	Yes
Variability in human populations	No	Yes	No	Yes	Yes

(continued)

Table 10.4 (continued)

Model feature	Malaria models: Gu and Novak (2009a, b)	Yakob and Yan (2009)	EMOD (Eckhoff 2011)	OpenMalaria (Chitnis et al. 2010)	Arifin et al. (2013)
Coverage scheme used for ITNs*	Proportion of households with bed nets	Proportion of populations sleeping under bed nets	Proportion of populations sleeping under bed nets	Proportion of populations sleeping under bed nets	Partial and complete coverage schemes
Comparison of coverage schemes for ITNs	No	No	No	No	Yes

Adapted and edited from Arifin et al. (2013)
The features refer to the particular results presented in the cited publications, and should not be treated as limitations of the respective models. Each row represents a specific model feature. Each column represents a specific malaria model. Features marked with * were either modeled with extensions, or were treated as new (not modeled earlier by other studies). Text in the cells represent whether the feature was implemented/available in the model, including simple yes/no, or other comments. N/A means not applicable or not available. *N* indicates a normal distribution with *mean* and *standard deviation*. LSM, ITNs, and IRS mean larval source management, insecticide-treated nets, and indoor residual spraying, respectively

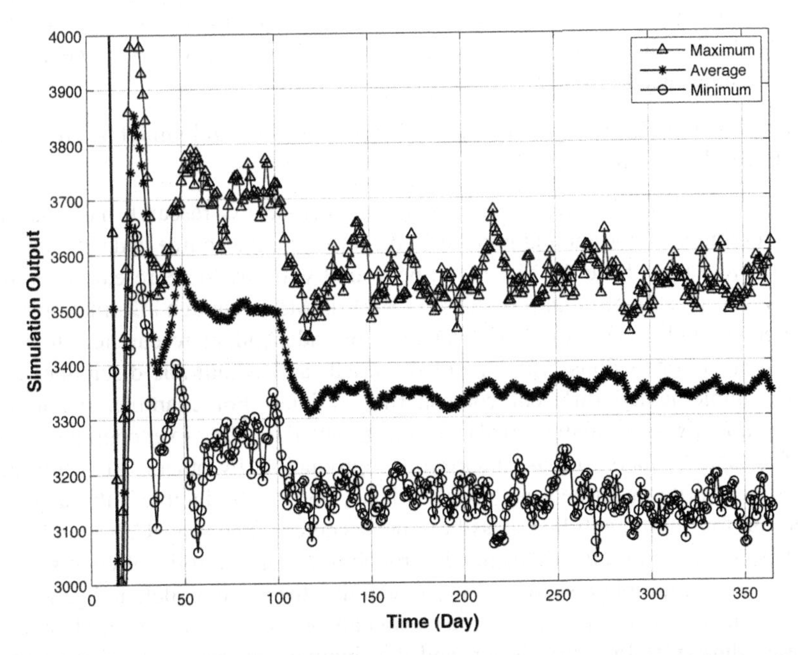

Fig. 10.2 Smoothing out the simulation stochasticity effects by performing sufficient number of replicated simulation runs (Data obtained from (Arifin et al. 2013). The results are derived from 50 replicated runs. The maximum, the minimum, and the average time-series plots represent the maximum, the minimum, and the average output (mosquito abundance in this case) values, respectively, obtained across all 50 replicated runs in each time-step)

Yilmaz 2009; Balci 1998a, b), drew some important V&V issues from which model differences may arise, and/or the process of R&R may become more time-consuming and challenging (Arifin et al. 2013):

- *Conceptual image of the model*: the intended logical view of the ABM may be perceived differently by different modelers, thus creating different conceptual, mental images of the logical view.
- *Choice of tools*: selection of programming languages and tools (e.g., C++ vs. Java) from the numerous options offered these days may be another potential source. The availability and limitations of a particular programming language, the use of specific data structures and other language constructs, and even the coding style of individual modelers, can compound the differences.
- *Availability of additional resources*: in some cases, additional resources used by the model (in the forms of artificial maps, object-based landscapes, etc.), if not defined or made explicitly available, pose subtle challenges. Although the importance of these resources may seem somewhat arbitrary in the broader context, goals, and outputs of the original models, their precise specification still remains important for R&R. For example, in replicating the landscapes, the absence of a listing of the spatial coordinates of the objects (which may be provided as supplementary materials) not only forces future modelers who try to replicate the landscapes to spend a significant amount of time in reproducing the landscapes (some parts of which inevitably rely on best guesses, due to the lack of additional information), it also increases the possibility of judgment errors being introduced in this phase.

Based on the above experiences, we also proposed several guidelines for future ABM modelers (Arifin et al. 2013):

- *Code and data sharing*: As the trends of open-access research have become increasingly important and popular in recent years, the source code and executable programs of the ABMs should be shared with the M&S research community. To ensure a minimum standard of R&R in M&S, enough information about methods and code should be available for independent researchers to reach consistent conclusions. Many reputed journals across multiple disciplines have also implemented different code-sharing policies. For example, the journal *Biostatistics* (Biostatistics 2014) has implemented a policy to encourage authors of accepted papers to make their work reproducible by others: based on three different criteria termed as "code", "data", and "reproducible", the associate editor for reproducibility classifies the accepted papers as C, D, and/or R, respectively, on their title pages. Reproducibility, in the form of code sharing, may allow multiple research groups examining the same model, and generating new data or output may lead to robust conclusions. However, for certain reasons (e.g., during preliminary design and development phases, exploratory feature testing phases, etc.), it may not always be the ideal case to share the source code. In these cases, we recommended that for ABM-based studies which are accepted for publication, at least the associated executable programs and/or other tools be made available as supplementary materials with detailed instructions.

- *Relevant documentation*: Modelers who share the source code and/or executable programs of their ABMs should also provide well-written documentation. Documentation is an important part of software engineering. The journal *PLOS Computational Biology*, which publishes articles describing outstanding open source software, emphasizes that the source code must be accompanied with documentation on building and installing the software from the source, including instructions on how to use and test the software on supplied test data (PLoS Computational Biology Guidelines for Authors 2014). An ABM documentation may include statements describing the attributes, features, and characteristics of the agents and environments of the ABM, the overall architecture or design principles of the code, algorithms and application programming interfaces (APIs), manuals for end-users, interpretation of additional materials (e.g., object-based landscapes), etc. Free and commercial software tools are also available which can help automating the process of code annotation, code analysis, and software documentation.
- *Standardized models*: The general workflow of the ABM, including the input/output requirements, program logic, etc. should follow a standardized approach. The need for standardization becomes more important when the broader utility of the model is considered by multiple research groups, and within an integrated modeling platform. For example, two malaria models, OpenMalaria (OpenMalaria 2014) and EMOD (Institute for Disease Modeling (IDM) 2014), are currently being integrated within the open-access execution environment of the Vector-Borne Disease Network (VecNet) (VecNet 2014). The proposed VecNet cyberinfrastructure (VecNet CI), within a shared execution environment, establishes three modes of access sharing for model developers: (1) shared data, (2) shared execution, and (3) shared software. Once integrated, these models can utilize other components of the VecNet CI, including the VecNet Digital Library, web-based user interfaces (UI), tools for visualization, job management, query and search, etc. in order to, for example, import and use malaria-specific data to run specific scenarios or campaigns of interest, and display their outputs using the visualization and/or the UI tools of the VecNet CI. It is envisaged that most malaria ABMs, in future, will be accommodated within similar integrated modeling frameworks. Hence, to expedite the integration process, future malaria ABMs should plan and follow a well-defined integration path from the early phases of model development.

10.7 Conclusions

In this chapter, we reviewed some of the earliest works in V&V by Ören and others, which include advanced simulation methodologies, assessing the acceptability of simulation studies, categorizations and taxonomies of M&S, and M&S applications. We also summarized the contributions of Ören and others in the areas of V&V, QA, and R&R.

We then described our V&V and R&R experiences from using several agent-based models, all of which were developed from a core entomological model of the malaria-transmitting mosquito species *An. gambiae*. In general, the V&V works were influenced by previous works of Ören, Balci, Yilmaz, and others, and performed primarily using docking (model-to-model comparison) between the agent-based versions.

For V&V, with the different language-specific implementations (versions), we showed how docking helped to verify the different implementations, to validate these against the core model, and to reveal incorrect assumptions and errors, which, being unnoticed, initially led to erroneous results. Our results showed that seemingly insignificant differences in separate versions may lead to significant mismatch in overall model outputs, suggesting that docking should be iterative and should involve well-planned feedback from earlier versions. Using compartmental docking and following the divide and conquer paradigm, we also showed how the V&V produced incremental agreements in model outputs.

For R&R, as our results indicated, replicability of the experiments and simulations performed by malaria models published earlier bear special importance. Due to several factors (including new tools and technologies, massive amounts of data, interdisciplinary research, etc.), the task of replication may become complicated. We summarized the challenges, insights, and experiences gained from the R&R works of several models and offered several guidelines for future ABM modelers. Overall, our V&V and R&R experiences, learned the hard way from these exercises, served the dual purpose of increasing confidence to the core model as well as revealing conceptual errors in different versions.

References

An Y, Xiong G, Xiao T, Wang X (1989) IPSOS – an integrated simulation system for industrial processes. In: Proceedings of the 1989 Beijing simulation conference. Pergamon Press, London, p 500

Arifin SMN, Davis GJ, Zhou Y (2010a) Verification & validation by docking: a case study of agent-based models of *Anopheles gambiae*. In: Summer computer simulation conference (SCSC), Ottawa, ON, Canada, July 2010

Arifin SMN, Davis GJ, Kurtz SJ, Gentile JE, Zhou Y (2010b) Divide and conquer: a four-fold docking experience of agent-based models. In: Winter simulation conference (WSC), Baltimore, MD, USA, Dec 2010

Arifin SMN, Kennedy RC, Lane KE, Fuentes A, Hollocher H, Madey GR (2010c) P-SAM: a post-simulation analysis module for agent-based models. In: Summer computer simulation conference (SCSC), Ottawa, ON, Canada, July 2010

Arifin SMN, Davis GJ, Zhou Y (2012) A spatial agent-based model of malaria: model verification and effects of spatial heterogeneity. In: Zhang Y (ed) Theoretical and practical frameworks for agent-based systems. IGI Global, Hershey, PA, USA, pp 221–240

Arifin SMN, Madey GR, Collins FH (2013) Examining the impact of larval source management and insecticide-treated nets: replicating recent models and exploring the combined impact using a spatial agent-based model of *Anopheles gambiae* and a landscape generator tool. Malar J 12:290

Arifin SMN, Zhou Y, Davis GJ, Gentile JE, Madey GR, Collins FH (2014) An agent-based model of the population dynamics of *Anopheles gambiae*. Malar J 13:424

Axtell R, Axelrod R, Epstien JM, Cohen MD (1996) Aligning simulation models: a case study and results. Comput Math Organ Theory 1:123–141

Balci O (1989) How to assess the acceptability and credibility of simulation results. In: Proceedings of the 21st conference on winter simulation, Washington, DC. ACM, New York, pp 62–71

Balci O (1998a) Verification, validation and accreditation. In: Proceedings of the 1998 winter simulation conference, Washington, DC, 13–16 Dec 1998. IEEE, Piscataway, pp 41–48

Balci O (1998b) Verification, validation, and testing. In: Banks J (ed) Handbook of simulation. Engineering & Management Press, pp 335–393

Balci O (2003) Verification, validation, and certification of modeling and simulation applications. In: Proceedings of the 35th conference on winter simulation: driving innovation, New Orleans, pp 150–158

Balci O (2004) Quality assessment, verification, and validation of modeling and simulation applications. In: Proceedings of the 2004 winter simulation conference, Washington, D.C., USA, vol 1. IEEE

Biostatistics (2014) http://biostatistics.oxfordjournals.org/. Accessed 2 Dec 2014

Burton RM (2003) Computational laboratories for organization science: questions, validity and docking. Comput Math Organ Theory 9(2):91–108

Carley KM (2002) Computational organizational science and organizational engineering. Simul Model Pract Theory 10(5):253–269

Chitnis N, Schapira A, Smith T, Steketee R (2010) Comparing the effectiveness of malaria vector-control interventions through a mathematical model. Am J Trop Med Hyg 83:230–240

Eckhoff P (2011) A malaria transmission-directed model of mosquito life cycle and ecology. Malar J 10:303

Edmonds B, Hales D (2003) Replication, replication and replication: some hard lessons from model alignment. J Artif Soc Soc Simul 6:4

Elzas MS (1988) Expert simulation systems in practice. In: Vichnevetsky R et al (eds) Proceedings of the 12th World congress on scientific computation, vol 5, Paris, 18–22 July 1988, pp 47–50

Fomel S, Hennenfent G (2009) Reproducible computational experiments using Scons. In: IEEE international conference on acoustics, speech, and signal processing, 4, Taipei, Taiwan, pp 1257–1260

Garzia RF (ed) (1979) Simulation week report. Modelling (IEEE Computer Society Simulation Technical Committee Newsletter) Issue 5, May 1979, p 5

Gu W, Novak RJ (2009a) Agent-based modelling of mosquito foraging behaviour for malaria control. Trans R Soc Trop Med Hyg 103:1105–1112

Gu W, Novak RJ (2009b) Predicting the impact of insecticide-treated bed nets on malaria transmission: the devil is in the detail. Malar J 8:256

Huang Y, Xiang X, Madey G, Cabaniss S (2005) Agent-based scientific simulation. Comput Sci Eng 7(1):22–29. doi:10.1109/MCSE.2005.7. http://ieeexplore.ieee.org/stamp/stamp.jsp?tp=&arnumber=1377072&isnumber=30051

Institute for Disease Modeling (IDM) (2014) http://idmod.org. Accessed 2 Dec 2014

Jasny BR, Chin G, Chong L, Vignieri S (2011) Again, and again, and again. Science 334:1225

Karplus WJ (1979) Review of the article by Ören TI, Zeigler BP (1979) Concepts for advanced simulation methodologies. Simulation 32(3):69–82

Kennedy RC, Xiang X, Madey GR, Cosimano TF (2005) Verification and validation of scientific and economic models. In: Proc. Agent, Chicago, pp 177–192

Kennedy RC, Xiang X, Cosimano TF, Arthurs LA, Maurice PA, Madey GR, Cabaniss SE (2006) Verification and validation of agent-based and equation-based simulations: a comparison. In: 2006 agent-directed simulation (ADS), Huntsville

Ketcham MG (1986) Expert systems and user decisions in simulation studies. In: Cellier FE (ed) Languages for continuous system simulation. SCS, San Diego, pp 44–49

Ketcham MG, Fowler JW, Phillips DT (1984) New directions for the design of advanced simulation systems. In: Sheppard S, Pooch U, Pegden D (eds) Proceedings of the 1984 winter simulation conference, Dallas. SCSI, San Diego, pp 563–568
Kettenis DL (2001) Ethical issues in modeling and simulation. Guest Editor's Introduction to the special issue. Simul-T Soc Mod Sim 17(4):162
Lin M-J (1990) Automatic simulation model design from a situation theory based manufacturing system description. Ph.D. Dissertation, Texas A&M University, College Station, Texas
Nance RE, Mezaache AL, Overstreet CM (1981) Simulation model management: resolving the technological gaps. In: Proceedings of the 1981 winter simulation conference, Atlanta, GA, USA, pp 173–179
National Research Council (2012) Assessing the reliability of complex models: mathematical and statistical foundations of verification, validation, and uncertainty quantification. The National Academies Press, Washington
North MJ, Macal CM (2002) The Beer Dock: three and a half implementations of the beer distribution game. In: Sixth annual swarm users meeting (SwarmFest), University of Washington, Seattle, 29–31 Mar 2002
Olaru D, Purchase S, Denize S (2009) Using docking/replication to verify and validate computational models. In: 18th World IMACS/MODSIM Congress 2009, Cairns, Australia
OpenMalaria (2014) A simulator of malaria epidemiology and control. https://code.google.com/p/openmalaria. Accessed 2 Dec 2014
Ören TI (1981) Concepts and criteria to assess acceptability of simulation studies: a frame of reference. Commun ACM 24(4):180–189
Ören TI (1983) Quality assurance of system design and model management for complex problems. In: Adequate modeling of systems. Springer, Berlin/Heidelberg, pp 205–219
Ören TI (1984) Quality assurance in modelling and simulation: a taxonomy. Simulation and model-based methodologies: an integrative view. Springer, Berlin/Heidelberg, pp 477–517
Ören TI (2000) Responsibility, ethics, and simulation. Transactions 17(4)
Ören TI (1980) Assessing acceptability of simulation studies. In: Proceedings of 1980 winter simulation conference, vol 2, Orlando, FL, USA
Ören TI, Yilmaz L (2009) Failure avoidance in agent-directed simulation: beyond conventional V&V and QA. In: Agent-directed simulation and systems engineering, Systems engineering series. Wiley, Berlin
Ören TI, Zeigler BP (1979) Concepts for advanced simulation methodologies. Simulation 32 (3):69–82
Ören TI, Elzas MS, Sheng G (1985) Model reliability and software quality assurance in simulation of nuclear fuel waste management systems. ACM SIGSIM Simulation Digest 16(4):4–19
Pavón J, Arroyo M, Hassan H, Sansores C (2008) Agent-based modelling and simulation for the analysis of social patterns. Pattern Recogn Lett 29(8):1039–1048
Pegden CD, Shannon RE, Sadowski RP (1995) Introduction to simulation using SIMAN. McGraw-Hill, New York
Peng RD (2011) Reproducible research in computational science. Science 334:1226–1227
PLoS Computational Biology Guidelines for Authors (2014) http://www.ploscompbiol.org/static/guidelines.action#software. Accessed 2 Dec 2014
Rand W, Wilensky U (2006) Verification and validation through replication: a case study using Axelrod and Hammond's ethnocentrism model. In: 14th annual conference of the North American Association for Computational Social and Organization Sciences (NAACSOS), Notre Dame, 22–23 June 2006
Rouchier J, Cioffi-Revilla C, Polhill JG, Takadama K (2008) Progress in model-to-model analysis. J Artif Soc Soc Simul 11(2):8
Santer BD, Wigley TML, Taylor KE (2011) The reproducibility of observational estimates of surface and atmospheric temperature change. Science 334:1232–1233
Sargent RG (2001) Verification and validation: some approaches and paradigms for verifying and validating simulation models. In: Proceedings of the 33rd conference on winter simulation. IEEE Computer Society, Washington, DC, pp 106–114

Sargent RG (2004) Validation and verification of simulation models. In: Proceedings of the 2004 winter simulation conference, Washington, D.C., USA, vol 1. IEEE
Shannon RE (1986) Intelligent simulation environments. In: PA Luker, Adelsberger HH (eds) Proceedings of the conference on intelligent simulation environments, Society for Computer Simulation, San Diego, 23–25 Jan 1986
Smit W (1999) A question of ethics. In: The book edited to honor Prof. Ir. M.S. Elzas, University of Wageningen, Dept. of Informatics, Wageningen, pp 30–33
Standridge CR, Pritsker AAB (1982) Using data base capabilities in simulation. In: Cellier FE (ed) Progress in modelling and simulation. Academic, London, pp 347–365
Taylor SJE, Khan A, Morse KL, Tolk A, Yilmaz L, Zander J (2013) Grand challenges on the theory of modeling and simulation. In: Proceedings of the symposium on theory of modeling & simulation 2013, San Diego, California, USA, 34:1–8
Thiele JC, Kurth W, Grimm V (2014) Facilitating parameter estimation and sensitivity analysis of agent-based models: a cookbook using NetLogo and 'R'. J Artif Soc Soc Simul 17(3)
Troncale LR (1985) The future of general systems research: obstacles, potentials, case studies. Syst Res 2(1):43–84
Vector-Borne Disease Network (VecNet) (2014) https://www.vecnet.org. Accessed 2 Dec 2014
Walker WE, Harremoës P, Rotmans J, van der Sluijs JP, van Asselt MBA, Janssen P, Krayer von Krauss MP (2003) Defining uncertainty: a conceptual basis for uncertainty management in model-based decision support. Integr Assess 4(1):5–17
Wang Z-Z, Li B-H (1991) Computer simulation: the past, present and future. In: Proc. of the 1991 summer computer simulation conference, 22–24 July 1991, Baltimore. The Society for Computer Simulation International, San Diego, pp 1059–1065
Wilensky U, Rand W (2007) Making models match: replicating an agent-based model. J Artif Soc Soc Simul 10(4):2
Will O (2009) Resolving a replication that failed: news on the Macy & Sato Model. J Artif Soc Soc Simul 12:4
Will O, Hegselmann R (2008) A replication that failed: on the computational model in 'Michael W. Macy and Yoshimichi Sato: Trust, Cooperation and Market Formation in the U.S. and Japan. In: Proceedings of the National Academy of Sciences, May 2002'. J Artif Soc Soc Simulat 11(3):3
Wittmann J (1992) Model and experiment – a new approach to definitions. In: Luker P (ed) Proceedings of the 1992 summer computer simulation conference, 27–30 July 1992, Reno. The Society for Computer Simulation International, San Diego, pp 115–119
Xiang X, Kennedy R, Madey G (2005) Verification and validation of agent-based scientific simulation models. In: Agent-directed simulation conference, San Diego
Xu J, Gao Y, Madey G (2003) A docking experiment: swarm and repast for social network modeling. In: Seventh annual Swarm researchers meeting (Swarm2003), Notre Dame, Indiana USA
Yakob L, Yan G (2009) Modeling the effects of integrating larval habitat source reduction and insecticide treated nets for malaria control. PLoS One 4:e6921
Yilmaz L (2006) Validation and verification of social processes within agent-based computational organization models. Comput Math Organ Theory 12:283–312
Yilmaz L (2011) Reproducibility in modeling & simulation research. Simulation, Editorial 87 (1):3–4
Yilmaz L (2012a) Reproducibility in M&S research: issues, strategies, and implications for model development environment. J Exp Theor Artif Intell 24(4):457–474
Yilmaz L (2012b) Scholarly communication of reproducible modeling and simulation research using e Portfolios. In: Proceedings of the 2012 international summer computer simulation conference, Genoa, 8–11 July 2012, pp 241–248
Yilmaz L, Ören T (2013) Toward replicability-aware modeling and simulation: changing the conduct of M&S in the information age. In: Ontology, epistemology, and teleology for modeling and simulation. Springer, Berlin/Heidelberg, pp 207–226

Zeigler BP (1982) Subject plan for an encyclopedia: a taxonomy for modelling and simulation. ACM Simuletter 13:1–4, pp 55–62

Zhong W, Kim Y (2010) Using model replication to improve the reliability of agent-based models. In: Chai S-K, Salerno JJ, Mabry PL (eds) Proceedings of the third international conference on social computing, behavioral modeling, and prediction (SBP'10). Springer, Berlin/Heidelberg, pp 118–127

Chapter 11
Comparisons of Validated Agent-Based Model and Calibrated Statistical Model

Il-Chul Moon and Jang Won Bae

11.1 Introduction

Modeling and simulation are an abstracted generation of a part of the real world, and their validity on the generation is the holy grail of the modeling and simulation study. In spite that this is a key factor in the credibility of simulation, still there is a large unchartered area in the validation of modeling and simulation. Prof. Tuncer Ören acknowledged this challenge in his article (Sheng et al. 1993; Yilmaz and Ören 2009). He claims that there are three major reasons which make the validation a challenge: philosophical, definitional, and theoretical reasons. He points out the philosophical problem by quoting Thomas Kuhn and Karl Popper, and he contrasts two arguments: "....theories are confirmed or refuted on the basis of critical experiments designed to verify the consequences of theories" from Kuhn and "... as scientists we can never validate a hypothesis, only invalidate it" from Popper. If we take Kuhn's positivism argument, it is imperative to validate a simulation model by the consequence of the model execution, yet Popper might agree that a simulation model will be only invalidated when a real-world case opposing to the simulation result is found.

Prof. Tuncer Ören also points out the various definitions of model validation. For instance, a model validation can be measured by the standard of either correct representation or acceptable representation. This definition difference inherits the Kuhn's view and the Popperian view, respectively. When we narrow down the scope to the computer simulation, the Society of Computer Simulation (SCS) defines the validation as "substantiation that a computerized model within its domain of applicability possesses a satisfactory range of accuracy consistent with the intended application of the model." This definition seems to be a mixture of

I.-C. Moon (✉) • J.W. Bae
Department of Industrial and Systems Engineering, KAIST, Daejeon, South Korea
e-mail: icmoon@kaist.ac.kr; repute82@kaist.ac.kr

L. Yilmaz (ed.), *Concepts and Methodologies for Modeling and Simulation*,
Simulation Foundations, Methods and Applications,
DOI 10.1007/978-3-319-15096-3_11

arguments from correct representation and acceptable representation. *The satisfactory range of accuracy* suggests that the correct representation with a certain level of quantitative confidence is required, yet *the intended application of the model* asserts the importance of the acceptance of the model.

Prof. Tuncer Ören argues that one distinction of the validation study in the modeling and simulation is the relative validity concept from the experiments of simulation models. Since a simulation is a limited representation of the real world, and because the limited representation is determined by the modeler, the modeler is able to adjust the scope to make the model more valid. The simulations often used as a part of virtual experiments which is a part of the typical scientific method. The virtual experiments consist of (1) a simulation model, (2) generated/collected/observed data, and (3) the experiment framework. The scope of the model depends on the model features as well as the used dataset in the experiment design.

Though time has passed since Prof. Tuncer Ören's works on validation, this community still struggles to have better theoretical frameworks as well as case studies on validation. Furthermore, when the users of simulation models call for the predictive results, simulation models are considered to be weaker solutions than statistical models only designed for predictions. Such perception is in the line of the continuing argument of *correctness vs. acceptance* of models. However, some simulation models might be better at the prediction as well as the replication to a certain scope at a certain problem set. This chapter introduces such case to see whether there is a chance for a simulation model to be good at the two aspects simultaneously.

11.2 Background

Agent-based model consists of multiple agents, environments, and interactions among the agents and environments (Yilmaz and Ören 2009; Bonabeau 2002). Such model structure reflects a holistic view where the aggregate of a system is different from the sum of its components. This view has been recognized as an efficient method to analyze complex systems that contain a large number of components and interactions among them. From this reason, agent-based model has been applied to understanding, replicating, and resolving problems in various domains, such as sociology (Schelling 1971), economy (Tesfatsion 2003), biology (Grimm et al. 2005), etc. Agent-based modes are generative models so that they can provide unrecorded, yet important information to model users. For example, in the disaster event (Lee et al. 2014), how the disaster response organizations efficiently reacted is hard to be evaluated, because their efficiency is rarely recorded in such immediate events. However, agent-based model, as a generative model, can help us to estimate such efficiency using several partial data which seem not to be related to the efficiency.

Although agent-based model provides such invaluable information, outsider as well as insider of the agent-based model community often considers about its accuracy for the prediction (Moss 2008; Brown et al. 2005; Windrum 2007).

To compare with the model for accurate prediction, such as statistical model, agent-based model might have less accuracy in the prediction. Statistical model describes how a set of variables are related to the other set of variables mathematically. From such mathematical bases, statistical models often represent the accurate predictions. This accuracy depends on the proper data for the prediction, yet finding such data in the real world would be another problem.

Hence, this chapter intends to investigate the differences of the model for regeneration of the real world, i.e., agent-based model, and the model for accurate predictions of the real world, i.e., statistical model. The investigations include quantitative comparisons of the two types of models. As a case study, an agent-based model for city commerce was compared to the corresponding statistical model. This particular comparison, again, casts light on the trade-off of different contributions from different models.

11.3 Validation of Agent-Based Models

The validation of simulation models can be completed at the various levels and through diverse methodologies (Moss 2008). These different types of validations range from the qualitative assessments of models, i.e., quality assurance of models, to the quantitative validations, i.e., correlation between the simulated world and the real world. Some models in a certain virtual experiment can be verified by the quantitative analyses, while other models are only able to be qualitatively validated.

Let's imagine that there is a traffic simulator designed for urban areas. This traffic simulator might be validated for daily traffic estimation which can be validated by historic records of the simulated area. Since it is daily estimation, the validation data would be sufficient, and the details of the model would be easily expected by the modelers. On the other hand, this traffic simulator might be used for the traffic estimation on city-scale evacuation. This case would be very difficult to be numerically validated because it would be impossible to obtain sufficient datasets to be compared with. Moreover, the modeled details of the evacuation simulator should be different from the models for the daily traffic estimations. These hypothetical scenarios reveal that some models might be quantitatively validated by the support of the experimental framework (see Fig. 11.1), and other models are only subject to qualitative quality assurance of the model.

When the validation of a simulation model is perceived as quality assurance, there are two different approaches to view the simulation model. The first perspective is treating the simulation model as software artifact, which is true because the simulation model to be validated needs to be implemented as a software executable (Yilmaz and Ören 2009). Then, we can apply diverse software quality assurance techniques, for instance, CMMI (Chrissis et al. 2003). This is a comprehensive quality checking over the modelers, the simulation model before implementation, and the simulation software after implementation. The other perspective is adopting a qualitative simulation validation approach. Such qualitative assessment often

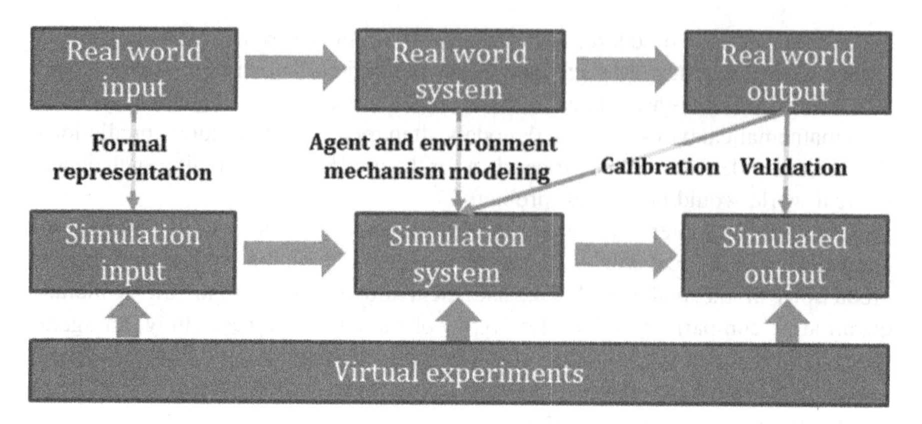

Fig. 11.1 Simulation-based experiment flow and the interaction between the real-world data and the simulated data

relies on the knowledge of subject-matter experts (Sargent 2005). The experts receive the simulation results in various formats, i.e., visualization of simulation progresses, response surface analysis (Inman et al. 1981), etc. With the provided materials from simulations, the experts decide to accredit the simulation model for its validity.

Similar to the levels of quality assurance, the quantitative validation has diverse levels of validation. The fundamental of the quantitative validation is the comparison between the simulated data and the real-world data with statistical approaches. However, the comparison points would be different from the intended acceptance level, which we and Prof. Tuncer Oren discussed in the Background section. A certain virtual experiment might be only interested in the change point of the target system, which is known as tipping point (Davis et al. 2007). Also, another experiment setting would concentrate on the performance and the state dynamics at the end of the target period or at the convergence of the system states. These quantitative validations aim at the point validation, which limits the validity of the simulation to a certain point of the simulated period. The point validation can be simply performed by checking the difference between the simulated point and the real-world point with a null hypothesis. Often, confidence intervals, T-tests, F-tests, etc. are the statistical tools to be used. The experiments with point validation have lower intended acceptance level than the validation on the complete timeline of the simulated period because the point validation does not ensure the generation process of the simulated world to be close to the real world. From this perspective, the point validation is quite analogical to the statistical models for predictions. The statistical models might not produce the generation process of the target system, and the statistical models are only *trained* to be close to the prediction point of the system.

When modelers need to validate the generative process of the simulations, it is necessary to validate the matching between the simulated data and the real-world data, and this process is often called as trend validation (Kleijnen 1995; Barlas 1996). This is different from the point validation which requires a correlation on a single time-step. Because the trend validation requires match the trend, not a point, this validation needs a further calibration than the before. The calibration now includes adjusting the parameters controlling the temporal flow of the simulation. The trend validation can be performed by multiple statistical methods. Firstly, each point over the simulated period might be tested with the point validation technique, such as *T*-test. Secondly, the overtime data from the real world is fitted to an auto-regression model, either linear or nonlinear; then we can see the fitness of the simulated data to the auto-regression model. Finally, there are techniques for temporal data comparisons, such as dynamic time warping. The dynamic time warping technique selects the most similar temporal flow between two overtime data, so this supplements the temporal flow discrepancy between a simulation and the real world.

11.4 Prediction with Agent-Based Model

In this section, we presented city commerce models for a quantitative comparison of validated agent-based model and calibrated statistical model.

11.4.1 Case Study: City Commerce Models

The agent-based model was developed in our previous paper (Lee et al. 2013) to estimate the impact on the city commerce by relocation policy. Relocation policy, which moves city functions from the centralized city to newly developed city, has been implemented in several countries such as the UK, Ireland, and Germany (Marshall 2007). In Korea, the government recently executed a relocation policy to resolve problems from overpopulation in Seoul, which is the capital of Korea, and achieve the balanced regional development. By the relocation policy, some of government branches from Gwacheon city, near Seoul, moved to a newly built city, located about 100 km south from Seoul.

While the government cares about the positive effect of the relocation policy, on the other side, several researchers concern about potential negative impacts associated with relocation, such as depressed economy in the region which people left. For investigating such negative impacts from the relocation policy from Korea, we developed an agent-based model for Gwacheon city.

To evaluate the changes, we established an assumption: the change of city commerce might have a positive correlation with the mobility of the population in Gwacheon city. So, the agent-based model describes a daily movement of the citizen in Gwacheon city. More specifically, the daily movement indicates a traffic flow in a day with respect to the job types of peoples in the city and daily time schedules associated with the jobs. To represent such characteristics in the agent-based model, we applied three kinds of real data: micropopulation data, time-use data, and GIS data.

Micropopulation data are detailed statistics about population in a certain region. In the United States, the micropopulation data are called the Public Use Microdata Sample (PUMS); in Korea, the data are named the Micro Data Service System (MDSS). MDSS contains a collection of attributes, such as the individual's address, occupation, family composition, education level, and so on. Because there is a concern about privacy violation, the dataset is usually provided as an anonymized sample. We could obtain a 5 % sample of population data, which contains 1,189 population data out of 23,780 populations in Gwacheon city, from MDSS, and use them to identify jobs of each agent (see the top of Fig. 11.2).

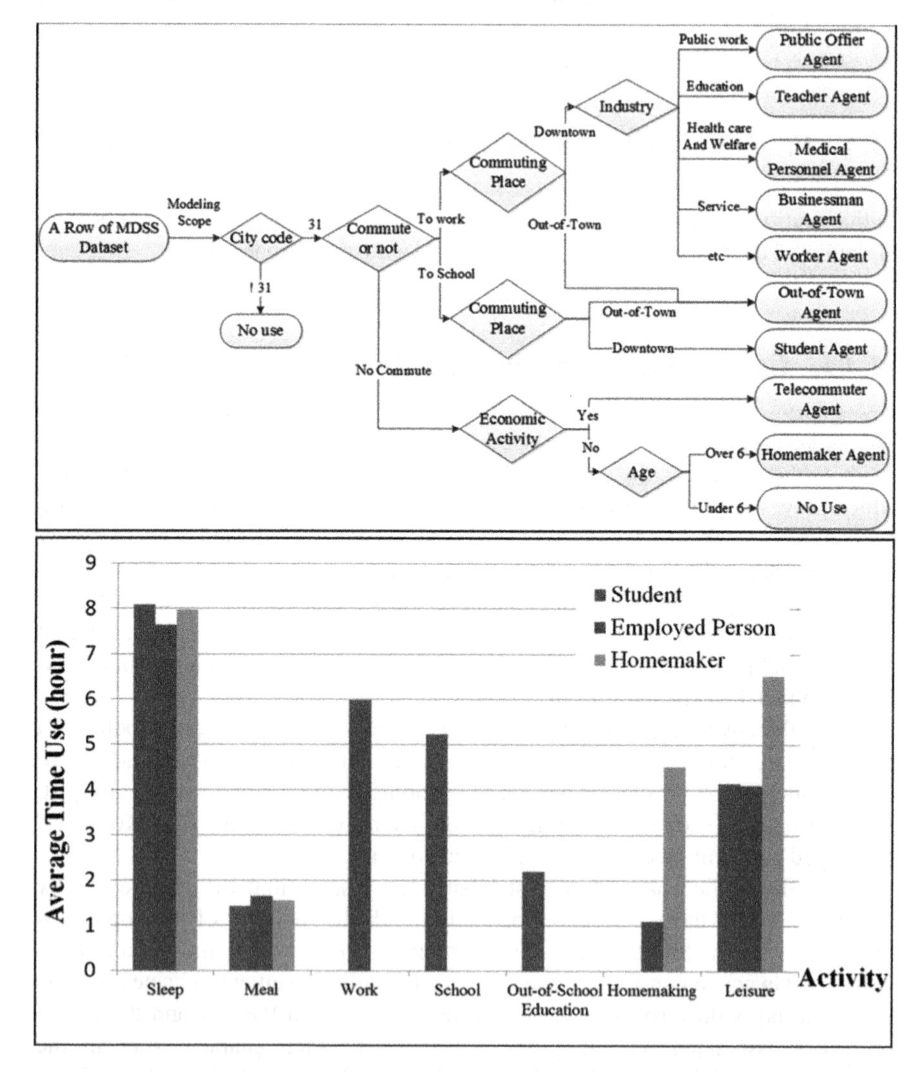

Fig. 11.2 (*Top*) Flowchart of agent-type categorization with MDSS dataset and (*bottom*) an example of time-use statistics of daily activity by student, employed persons, and homemaker

In many countries, time-use data is used to gauge productivity, daily life, and infrastructure efficiency of a population. In Korea, the Korean National Statistical Office provides time-use data that specify how much time an individual with a certain job spends performing a certain activity (see the bottom of Fig. 11.2). This data provides three types of modeling information. First, from the time-use data, we enumerate activity states of agent types and their transitions. Second, the transition time for states of an agent is also specified by the dataset. Lastly, we can develop commuting time of agents, which is important in analyzing and simulating traffic patterns, shopping-in behavior, and regional characteristics.

In the agent-based model, agents are corresponding to vehicles, and they move through the road network in Gwacheon city. Thus, the structure of the road network affects to the simulation objective which is to see the traffic flow in the target city. To reflect geospatial information to the agent-based model, we utilized GIS data of the target region. The data were downloaded from OpenStreetMap, and the data include the information of roads and buildings, such as coordinates, type, and identification. Figure 11.3 shows GIS data in Gwacheon city and replicated road network in the agent-based model using the GIS data. In particular, we selected the commercial buildings of interests to see the city commerce in the agent-based model (buildings with numbering in the right of Fig. 11.3).

On the other hand, some researchers raised a doubt about why agent-based model is applied to investigate the change of the city commerce. They stated that for the city commerce model, statistical model would be a better choice than agent-based model from the perspective of the accuracy in the model prediction.

For the comparison of accuracy from the two models, we develop several statistical models using a linear regression method. It is difficult to consider the above real data, such as MDSS, time-use data, and GIS data, in the development of statistical models, so we applied another data to the model development. In the development of statistical models, data that directly reflect the traffic flow would be

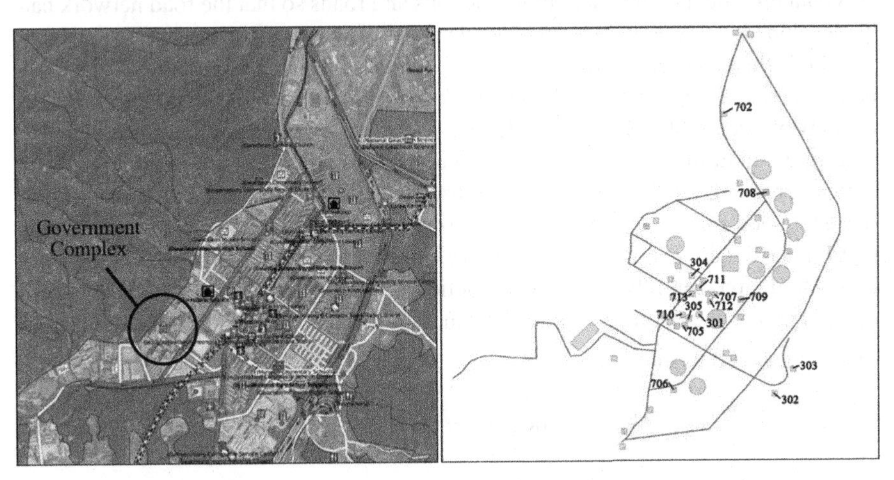

Fig. 11.3 (*Left*) GIS data of Gwacheon city from OpenStreetMap and (*right*) replicated road network and commercial buildings of ID in Gwacheon city using GIS data

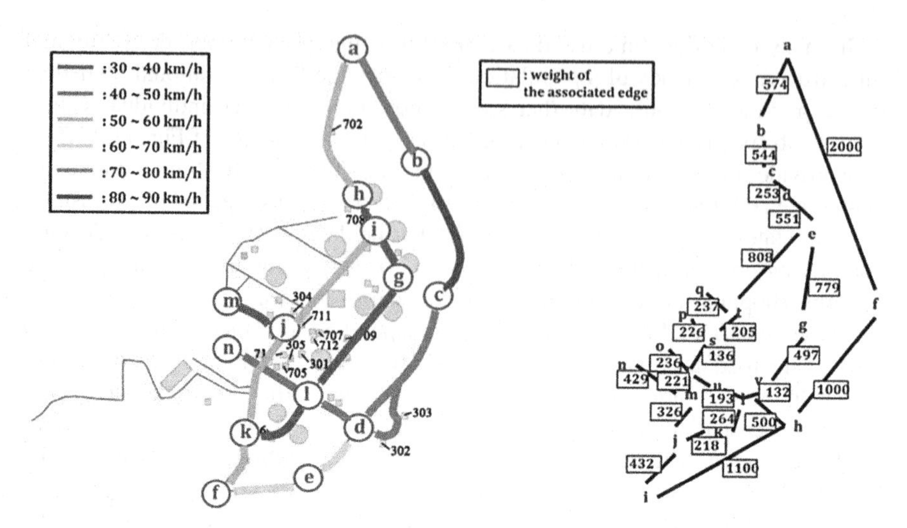

Fig. 11.4 (*Left*) Average traffic velocity of the roads and (*right*) road network with edge weight of road distance

more appropriate, so we collected two kinds of such data: traffic velocity and centrality of the road network.

The traffic velocity indicates average velocity of the roads in Gwacheon city, and it can be a direct barometer for the traffic flow. Traffic velocity of the roads in Gwacheon is opened in Intelligent Transport Systems (ITS) in Gwacheon. Based on the traffic velocity from ITS Gwacheon, we calculated the average velocity of the roads at each hour for a week during 6 months (Mar 2012–Jun 2012 and Mar 2013–Jun 2013). The left in Fig. 11.4 shows the average velocity of the roads in Gwacheon city.

The other is the road network structure derived from GIS data in Gwacheon city. GIS data provide the information of junctions and roads so that the road network can be developed by its vertices as the junctions and its edges as the roads. The right of Fig. 11.4 illustrated the road network of which edge weights represent the distances between two junctions. Using this network structure, we can calculate measures for the network centralities for the traffic flow: degree and betweenness centrality.

Degree centrality is defined as the number of edges that are incident upon a vertex (Freeman 1979). The degree centrality can be interpreted as the immediate risk of a node for catching whatever is flowing through the network, which means that higher degree centrality shows more central node in the network. In the road network, degree centrality (C_D) of a junction (j_i) is expressed by Eq. (11.1), where deg (j_i) means the degree centrality of junction j_i and $\sum_{i\gamma_{ij}}$ indicates the number of roads connected to junction j_i

$$C_D(j_i) = \deg(j_i) = \sum\nolimits_{i \in \text{all junctions}} \gamma_{ij} \tag{11.1}$$

Because the degree centrality is purely a local measure, we applied betweenness centrality, which is a useful measure for both the load and the importance of a node.

The load of a node describes a global effect in a network, whereas the importance of a node shows a local effect of the node. In the road network, betweenness centrality (C_B) of a junction (j_i) is expressed by Eq. (11.2). In Eq. (11.2), σ_{st} is the total number of shortest paths from node s to node t, and $\sigma_{st}(j_i)$ is the number of those paths that pass through j_i:

$$C_B = \sum_{s \neq t \neq j_i} \frac{\sigma_{st}(j_i)}{\sigma_{st}}, \text{ where } s, t \in \text{all junctions} \tag{11.2}$$

11.4.2 Comparison of Predictions from Agent-Based and Statistical Models

To evaluate our assumption, which is that the city commerce would have positive correlations with the mobility of the citizen, we developed agent-based model describing the daily movement in the target city and statistical model using real data that is directly related to the assumption. Now, we intend to compare the accuracy of their predictions by calculating their correlations to a value in the real world.

We collect an indicator of the real city commerce in the target region, which are the rent rates of the commercial buildings in the target region. The direct measure of commercial status of a city might be a sales amount of shops and malls in the target region, but such information is difficult to collect in the city-wide area. Therefore, we chose and collected the indirect measure, and this measure can be collected by personal visits to real estate agencies in the target region.

Before the comparison, we needed a value for calculating the correlations from the agent-based model related with the interested buildings. To do that, we counted the number of the passing-by agents which go through front roads around the interested areas. Similarly, for the statistical models, we mapped (1) traffic velocities to the buildings by calculating the average velocity of the roads around the building and (2) network centralities to the buildings by calculating the average centrality for the roads and the junctions around the building.

We started correlation analyses of the two from drawing scattering plots between the rent rates and the data from the agent-based model and the statistical model (see Fig. 11.5). Although it is difficult to see strong correlations between the two, they represented different trends: a scattering plot for the number of passing-by agents illustrates a positive correlation to the rent rates, yet the plot for traffic velocity shows little correlation, and the plot for network centralities shows, even, negative correlations to the rent rates. Also, we can find that several outliers in each scatter plot and some of them are included in all the scatter plots, such as building 711, 706, and 302.

Since we cannot confirm the correlations from Fig. 11.5 and say that the assumption is surely true, we calculated the correlations between the rent rate and the four data. In particular, we calculated three different correlations in each case: Pearson's value correlation, Spearman's rank correlation, and Kendall's tau rank

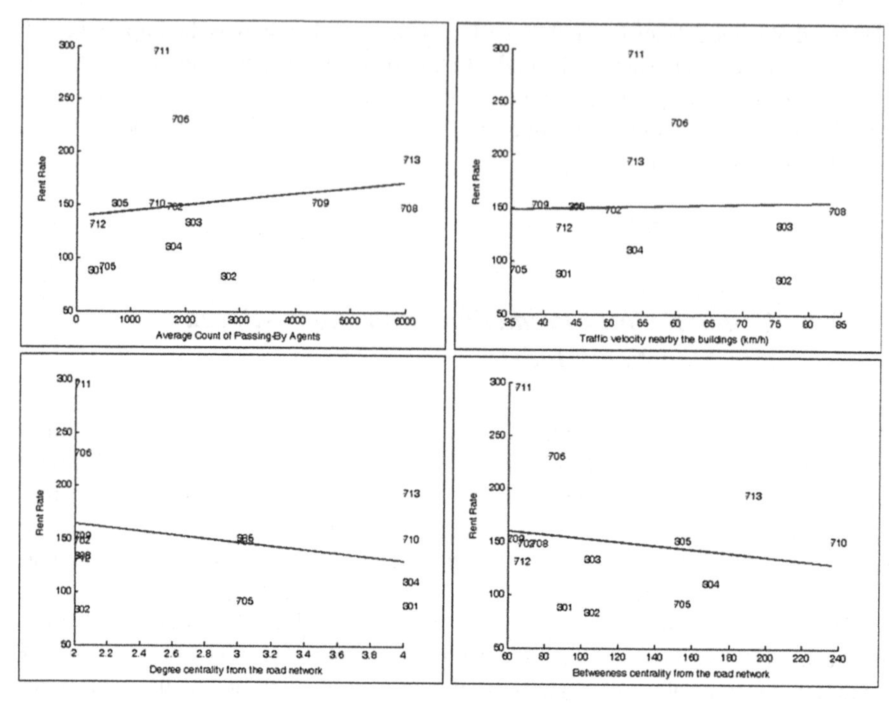

Fig. 11.5 Scattering plots and its linear fitted lines for the rent rates from the real world: (*left-top*) number of passing-by agents, (*right-top*) average of traffic velocity, (*left-bottom*) degree centrality, and (*right-bottom*) betweenness centrality

correlation. Table 11.1 shows the summary of this correlation analysis. When we include every building of interests, the number of passing-by agents and betweenness centrality is recorded the highest positive and negative correlation values, respectively. However, when we exclude the common outliers in Fig. 11.5, the correlation value of the number of passing-by agents becomes 68.3 and 68.5 % in value and rank correlations. This would be a good quality of validation considering the difficulty of the validation of social simulations. However, the correlation values of other data are much lower than the one of the number of passing-by agents.

Also, we built linear regression models for the rent rate as an independent variable and the four factors as the dependent variables (see Table 11.2). Model M1 including all the four variables shows the highest value of R-squared (0.680), which suggests that the prediction from this model is more reliable than any other models with confidence. However, some of the dependent variables showed p-value over 0.05. To find a significant regression model, we performed the backward elimination to M1 and, eventually, developed M6. M6 holds one dependent variable, as the number of the passing-by agents with p-value <0.05 and shows higher R-square value. On the other hand, other factors, such as traffic velocity, degree

Table 11.1 Correlation analysis between the rent rate and passing-by agent counts (agent), average traffic velocity (velocity), degree centrality (degree), and betweenness centrality (betweenness)

Correlation	14 buildings				11 buildings (excluding outliers: building 711, 706, and 302)			
	Agent	Velocity	Degree	Betweenness	Agent	Velocity	Degree	Betweenness
Pearson	0.184	0.036	−0.213	−0.266	0.683	0.239	0.115	−0.078
Spearman	0.337	0.110	0.062	−0.162	0.685	0.211	0.175	−0.068
Kendall	0.256	0.091	0.069	−0.122	0.537	0.133	0.133	−0.043

Table 11.2 Linear regression models for the rent rate with varying variables

Model	Variables	Std. coefficients	*p*-value	R^2
M1	Agent	0.878	0.019	0.680
	Velocity	−0.238	0.424	
	Degree	−0.481	0.204	
	Betweenness	0.519	0.177	
M2	Velocity	0.234	0.534	0.137
	Degree	−0.356	0.508	
	Betweenness	0.471	0.392	
M3	Agent	0.844	0.017	0.544
	Velocity	−0.266	0.373	
M4	Agent	0.706	0.023	0.499
	Degree	−0.072	0.780	
M5	Agent	0.713	0.018	0.536
	Betweenness	0.206	0.418	
M6	Agent	0.703	0.016	0.494
M7	Velocity	0.184	0.589	0.034
M8	Degree	−0.038	0.911	0.001
M9	Betweenness	0.170	0.616	0.029

Agent the number of passing-by agents, *Velocity* average of the traffic velocity, *Degree* degree centrality, and *Betweenness* betweenness centrality

centrality, and betweenness centrality, are not significant variables in all the regression models.

The above correlation analyses and linear regression models show that the number of passing-by agents from the agent-based model has strong relationship with the rent rates from the real world. These results indicate that the agent-based model is well validated and expected to generate reliable predictions in the assumption, yet traffic velocity and network centralities, which are directly related to the assumption, did not provide significant predictions. It is because both traffic velocity and network centralities are certainly related to the daily movements of the citizens, but each is not sufficient to express them. On the other hand, the agent-based model is capable of representing such dynamic behaviors, that's why the number of passing-by agents from the agent-based model shows the highest correlation.

11.5 Conclusion

Agent-based model has been applied to various problems, yet the method has been questioned for its prediction accuracy. It is because that contrary to statistical models, the prediction of the agent-based model is difficult to be evaluated, even if it is validated. However, as a generative model, agent-based model provides invaluable results which cannot be generated by statistical models. In certain cases, these invaluable results provide more accurate predictions than statistical models.

This chapter provides one example of the certain cases. To investigate relationships between the simulated and the real world data from the city commerce, we calculated the correlation between the two datasets, and the correlation was high. Although this example does not provide an answer for doubts about the accuracy of the agent-based model, it is sufficient to consider a trade-off for selecting a modeling method for different cases.

References

Barlas Y (1996) Formal aspects of model validity and validation in system dynamics. Syst Dyn Rev. http://www.ie.boun.edu.tr/labs/sesdyn/publications/articles/Barlas_1996.pdf

Bonabeau E (2002) Agent-based modeling: methods and techniques for simulating human systems. Proc Natl Acad Sci U S A 99(3):7280–7287

Brown DG, Page S, Riolo R (2005) Path dependence and the validation of agent-based spatial models of land use. Int J Geogr Inf Sci. http://www.tandfonline.com/doi/abs/10.1080/13658810410001713399

Chrissis MB, Konrad M, Shrum S (2003) CMMI guidlines for process integration and product improvement. http://dl.acm.org/citation.cfm?id=773274

Davis JP, Eisenhardt KM, Bingham CB (2007) Developing theory through simulation methods. Acad Manage Rev. http://amr.aom.org/content/32/2/480.short

Freeman LC (1979) Centrality in social networks conceptual clarification. Soc Netw 1(3):215–239

Grimm V, Revilla E, Berger U, Jeltsch F, Mooij WM, Railsback SF, Thulke H-H, Weiner J, Wiegand T, DeAngelis DL (2005) Pattern-oriented modeling of agent-based complex systems: lessons from ecology. Science (New York, NY) 310(5750):987–991. doi:10.1126/science.1116681. http://www.sciencemag.org/content/310/5750/987.short

Inman RL, Helson JC, Campbell JE (1981) An approach to sensitivity analysis of computer models: part II-ranking of input variables, response surface validation, distribution effect and technique synopsis. J Qual Technol. https://secure.asq.org/perl/msg.pl?prvurl=/data/subscriptions/jqt_open/1981/oct/jqtv13i4iman.pdf

Kleijnen JPC (1995) Verification and validation of simulation models. Eur J Oper Res. http://www.sciencedirect.com/science/article/pii/0377221794000166

Lee SH, Shin JS, Lee GH, Moon I-C (2013) Impact of population relocation to city commerce: micro-level estimation with agent-based model. In: Proceedings of the agent-directed simulation symposium, vol 11. Society for Computer Simulation International, Orlando

Lee G, Bae JW, Oh N, Hong JH, Moon I-C (2014) Simulation experiment of disaster response organizational structures with alternative optimization techniques. Soc Sci Comput Rev. doi:10.1177/0894439314544628

Marshall J (2007) Public sector relocation policies in the UK and Ireland. Eur Plan Stud 15(5):645–666

Moss S (2008) Alternative approaches to the empirical validation of agent-based models. J Artif Soc Soc Simul. http://jasss.soc.surrey.ac.uk/11/1/5.html

Sargent RG (2005) Verification and validation of simulation models. In: Proceedings of the 37th conference on Winter simulation. http://dl.acm.org/citation.cfm?id=1162736

Schelling TC (1971) Dynamic models of segregation. J Math Sociol 1(2):143–186. http://www.tandfonline.com/doi/abs/10.1080/0022250X.1971.9989794

Sheng G, Elzas MS, Ören TI Cronhjort BT (1993) Model validation: a systemic and systematic approach. Reliab Eng Syst Saf 42(2–3):247–259. doi:10.1016/0951-8320(93)90092-D. http://www.sciencedirect.com/science/article/pii/095183209390092D
Tesfatsion L (2003) Agent-based computational economics: modeling economies as complex adaptive systems. Inf Sci. http://www.sciencedirect.com/science/article/pii/S0020025502002803
Windrum P (2007) Empirical validation of agent-based models: alternatives and prospects. J Artif Soc Soc Simul. http://jasss.soc.surrey.ac.uk/10/2/8.html
Yilmaz L, Ören T (2009) Agent-directed simulation and systems engineering. Wiley, New York

Chapter 12
Generalized Discrete Events for Accurate Modeling and Simulation of Logic Gates

Maamar El Amine Hamri, Norbert Giambiasi, and Aziz Naamane

12.1 Introduction

We had the opportunity to welcome Dr. Ören in our laboratory LSIS (www.lsis.org) as invited professor in many occasions during 2006 until 2008. He participated actively in the life of our lab by having done several seminars, participated in the days of our VERSIM workgroup (toward a theory of the simulation) at Toulouse, and was member of jury of the Ph.D. thesis of Seck (2007) and of the Accreditation to Supervise Research of Santoni (2008). We are also grateful to him for having registered our workgroup on the label McLeod Modeling and Simulation Network (http://www.m-s-net.org/). Moreover, we had also published a French–English lexical dictionary (Ören et al. 2006) for the modeling and simulation readers.

I (Amine) remember having dinner with Dr. Ören at Vieux Port in Marseilles one night of 2008. We discussed for a long time on research on M&S (modeling and simulation) and the challenges of this discipline, which showed me the devotion of Dr. Ören for the sciences.

In the past, we have shown the advantages of the Generalized Discrete EVent system Specification (GDEVS) to build more accurate discrete-event models of dynamic systems. These theoretical concepts are applied to the field of logic gate design and analysis in order to get more accurate and fast simulations. In fact, states are represented with linear piecewise trajectories contrary to the classical Boolean logic models where states have constant piecewise trajectories (0 and 1). With GDEVS models, the transition from a low level to a high one and vice versa is a linear trajectory more realistic than the instantaneous transitions of classical logic

M.E.A. Hamri (✉) • N. Giambiasi • A. Naamane
Aix Marseille Université, CNRS, ENSAM, Université de Toulon, LSIS UMR 7296, Marseille 13397, France
e-mail: amine.hamri@lsis.org; aziz.naamane@lsis.org

L. Yilmaz (ed.), *Concepts and Methodologies for Modeling and Simulation*, Simulation Foundations, Methods and Applications,
DOI 10.1007/978-3-319-15096-3_12

gate models. Note that this more accurate representation does not imply any more computations than in Discrete EVent system Specification (DEVS).

This study is supplied with a software plug-in providing to the final user a framework to define new logic gate models by reusing those designed in an eXtensible Markup Language (XML) and to simulate the specified behavior based on object code generated from these specifications.

In this chapter, we propose an original approach to build discrete-event models of logic gates without the classical constraint of piecewise constant input—output trajectories. This allows building event-driven simulators using piecewise polynomial trajectories as in classical continuous simulators but with a continuous time representation and with an event-driven technique in place of a time-stepped one. A first work on this kind of approaches was proposed in (Giambiasi et al. 2000; Ghosh and Giambiasi 2001); here, we present a more general approach based on the GDEVS formalism for accurate modeling of dynamic systems.

GDEVS formalism has been introduced recently in the literature to enable the synthesis of accurate discrete-event models of highly dynamic continuous processes (Giambiasi et al. 2000). Its originality stems from the use of polynomials of arbitrary degree, as opposed to constant values, to represent the piecewise input–output trajectories. Thus, in essence, GDEVS constitutes a generalization of the concept of discrete event and of the classical discrete-event modeling approaches including DEVS (Zeigler 1989; Zeigler et al. 2000).

One of the concrete repercussions from the proposed generalization of the classical discrete-event abstraction process (to use any kind of trajectories instead of piecewise constant trajectories) is the possibility to simulate continuous or hybrid models with a good approximation by using event-driven techniques. The basic idea for the generalization of the concept of event was introduced for the first time in Giambiasi et al. (1994) for the purpose of modeling gate delays using fuzzy distributions.

Note that our work is a part of recent research works on normative views for M&S methodologies for which Ören and Zeigler (1979) have been the precursors. In addition, in the field of specification languages in order to keep user models, we propose through our framework to keep networks of logic gates. Ören (1984) has proposed GEST, a specification language, and proved thus the importance of specification languages which make the modeling task easier for the user.

In electronics, a logic gate is a physical device implementing a boolean function that performs a logical operation on one or more logical inputs and produces a single logical output (Cappochi et al. 2006; Paoli et al. 2004). Consequently, logic circuits are the basis for modern digital computer systems, and digital electronic circuits are usually made from large assemblies of logic gates, which are simple electronic representations of boolean logic functions. In addition, most hardware of these days is based on this boolean logic, where simple boolean operations like *And*, *Or*, and *Not* can be composed into basic building blocks used to construct more sophisticated logic circuits (calculators, modems, CPUs, etc.).

In this chapter, it is shown that with a discrete-event modeling formalism and event-driven simulation, logic gates can be modeled and simulated.

The remainder of this chapter is organized as follows. Section 12.2 recalls the DEVS formalism. Section 12.3 discusses functional abstraction for logic gates in the GDEVS paradigm. Section 12.4 presents the design and the implementation procedure, and Section 12.5 offers concluding remarks.

12.2 Overview

The key to building a discrete-event abstraction of a dynamic system consists in mapping the piecewise constant segments of the dynamic system, obtained perhaps through threshold sensor signals, into discrete events to generate the input, output, and state trajectories of the discrete-event model.

12.2.1 Discrete-Event Abstraction

Formally, a transformation from a piecewise constant segment, w, into discrete-event segments, w^*, may be expressed as follows:

Where A is a set of integers, reals, or symbols, Φ^e is the nonevent, and TE = $\{te_1, te_2, \ldots, te_n\}$ the set of occurrence times of events. A pictorial representation is shown in Fig. 12.1.

The reverse transformation, i.e., from discrete events, w^*, to a piecewise constant segment, may be expressed as follows:

$$\forall t \in [te_i, te_{i+1}], w_{[te_i,\ te_{i+1}]} = w^*(t)$$

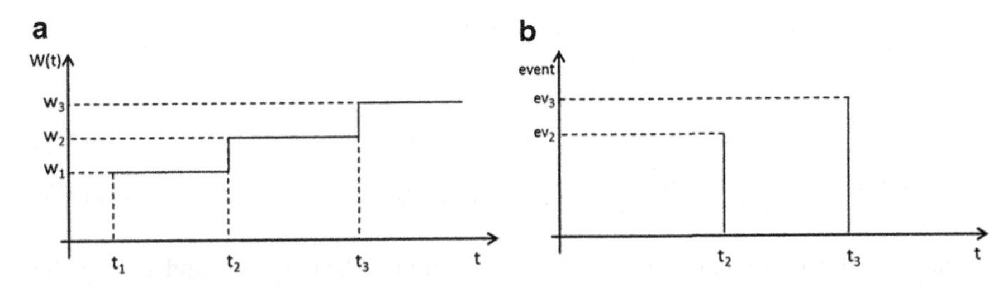

Fig. 12.1 Transforming a piecewise constant segment into a set of discrete events. (**a**) Piecewise constant segment. (**b**) Discrete event values

12.2.2 Generalized Discrete Events and GDEVS

For complex real-world systems that are highly dynamic, the use of piecewise constant input–output trajectories, for a given sampling time interval, may not succeed in accurately modeling the system behavior. Traditionally, under these circumstances, the sampling time interval is shortened to limit the representational error and achieve acceptable accuracy at the cost of increased simulation execution time. GDEVS adopts a radically new approach wherein it focuses on a system characteristic, namely, the function that represents the system behavior in the given time period, and increases its complexity from an identity function (classical Discrete-Event System models) to a higher-order function.

In this section, we recall the concept of generalized discrete events and of the GDEVS formalism. For more clarity, we consider first piecewise linear trajectories.

12.2.3 Specifying a Piecewise Polynomial Trajectory

A piecewise polynomial trajectory, expressed through the symbol w and shown in Fig. 12.2, is a collection of individual segments over a continuous time base.

The key characteristics include the following:

- There exists a finite number of time intervals $[te_i,\ te_{i+1}]$, with which tuples (a_0, a_1,. . ., a_n) are associated, where the a_i are constants.
- $\forall t \in [te_i,\ te_{i+1}]\ w(t) = a_0 + a_1 t + \cdots + a_n t^n$

  $w_{[te_0,\ te_n]} = w_{[te_0,\ te_1]} o\, w_{[te_1,\ te_2]} o \ldots o\, w_{[te_{n-1},\ te_n]}$ where o represents the left concatenation operator over the individual segments.

12.2.4 Example: Piecewise Linear Trajectory

In Fig. 12.3a, the piecewise linear trajectory, $w_{[te_0,\ te_3]}$, ranging between time instants t_0 through t_3, may be expressed as

$w_{[te_0,\ te_3]} = w_{[te_0,\ te_1]} o\, w_{[te_1,\ te_2]} o\, w_{[te_2,\ te_3]}$ where:

$$w_{[te_0,\ te_1]} = a_1 t + b_1 \tag{12.1}$$

$$w_{[te_1,\ te_2]} = a_2 t + b_2 \tag{12.2}$$

$$w_{[te_2,\ te_3]} = a_3 t + b_3 \tag{12.3}$$

In each of the individual Eqs. (12.1), (12.2), and (12.3), a_1, a_2, and a_3 are the gradients, and b_1, b_2 and b_3 are the intercepts of the individual segments. The origin of time, $t = 0$, is assumed to occur at the start of each segment.

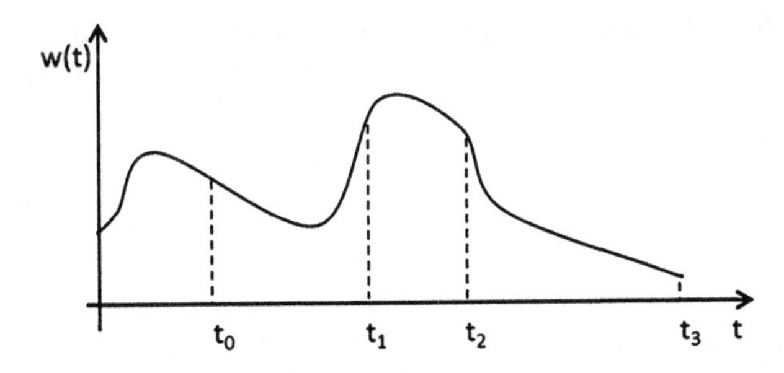

Fig. 12.2 Piecewise polynomial trajectory

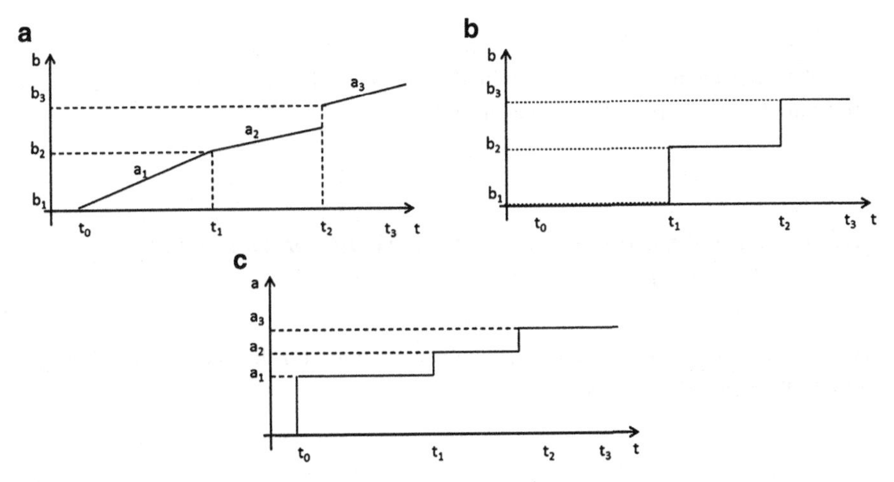

Fig. 12.3 Trajectories as a function of time. (**a**) Piecewise linear trajectory. (**b**) Intercept coefficients. (**c**) Gradient coefficients

12.2.5 *Coefficient Values of a Segment*

For an individual polynomial segment $w_{[ti,\ tj]}$, its coefficient values are defined by the tuple $(a_0, a_1, \ldots, a_n)$. Formally, the Coef function associates the coefficient values of a polynomial with all continuous polynomial segments $w_{[ti,\ tj]}$ over a time interval $[t_i, t_j]$. Thus, Coef : $\Psi \to A^n$ where Ψ represents a set of polynomials and A a subset of the real numbers. Also, the following function composition holds: for a given continuous polynomial segment $w_{[ti,\ tj]}$ over the time interval $[t_i, t_j]$, the components of the coefficients are $n+1$ constants.

Figures 12.3b and 12.3c describe the trajectories of the coefficients graphically over the time interval for a piecewise linear trajectory shown in Fig. 12.3a:

$$\text{In the interval } [t_0,\ t_1] : \text{Coef}\left(w_{[t0,\ t1]}\right) = (a_1,\ b_1)$$
$$\text{In the interval } [t_1,\ t_2] : \text{Coef}\left(w_{[t1,\ t2]}\right) = (a_2,\ b_2)$$
$$\text{In the interval } [t_2,\ t_3] : \text{Coef}\left(w_{[t2,\ t3]}\right) = (a_3,\ b_3)$$

In order to determine a polynomial trajectory on a time interval given the coefficients as a function of time, the inverse function Coef^{-1} is defined:

$$\mathrm{Coef}^{-1} : A^n \rightarrow L$$

$$\mathrm{Coef}^{-1}(a_0,\, a_1,\, \ldots,\, a_n) = a_0 + a_1 t + \cdots + a_n t^n \quad \text{and}$$

$$w_{[ti,\ tj]} o \mathrm{Coef}^{-1} : A^n \rightarrow A'$$

For example, in the time interval $[t_0, t_1]$, the coefficient values are (a_1, b_1), and the value of w at time t is through

$$w_{[ti,\ tj]} o \mathrm{Coef}^{-1}(a_1,\, b_1) = a_1 t + b_1$$

In the coefficient space, shown in Fig. 12.3, a piecewise linear trajectory is represented through piecewise constant values.

12.2.6 A New Concept: Coefficient Events or Generalized Discrete Events

As a generalization, under GDEVS, events are defined for the coefficients obtained from a linear piecewise trajectory.

12.2.6.1 Generalized Discrete Event

A coefficient event or a generalized discrete event, for a piecewise polynomial trajectory, is an instantaneous change of at least one of the elements of the tuple that defines the coefficient values.

In a time interval $[t_0, t_n]$ of a piecewise polynomial trajectory, there exists a generalized discrete event at time t_i if

$$\mathrm{Coef}\left(w_{[t_{i-1},\ t_i]}\right) \neq \mathrm{Coef}\left(w_{[t_i,\ t_{i+1}]}\right)$$

For example, consider the piecewise linear trajectory shown in Fig. 12.4a. In the coefficient space, shown in Fig. 12.4b, a coefficient event occurs at time $t = t_1$ that corresponds to a change in the gradient and the intercept of the trajectory, and at time $t = t_2$, there is an event corresponding to a change in the intercept value.

$$\forall te \in \{te_0,\, te_1,\, \ldots,\, te_n\} \wedge \forall t \in [te_i,\, te_{i+1}]$$

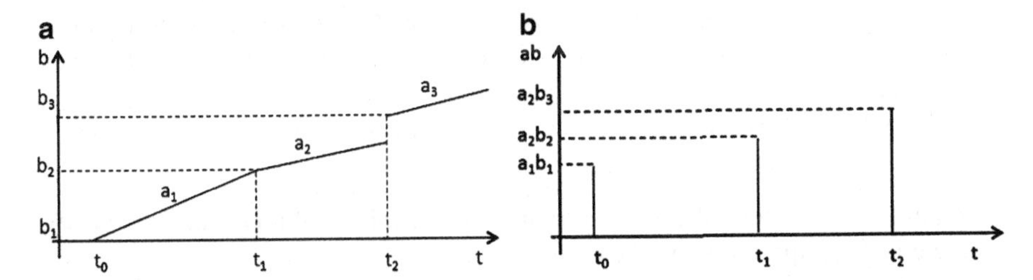

Fig. 12.4 Coefficient events for a piecewise linear trajectory. (**a**) Piecewise linear trajectory (**b**) GDEVS gradient and intercept coefficients

$$w^*[t_0,\ t_n] = \begin{bmatrix} x_i = (x_0,\ x_1,\ \ldots,\ x_n) & \text{if } t = te_i \\ x_i = \Phi^e \text{ is the non event} & \text{if } t \neq te_i \end{bmatrix} \tag{12.4}$$

12.2.6.2 Order of a Generalized Discrete Event

The order of an event is equal to the number of coefficients of the underlying polynomial minus one.

For piecewise linear trajectories, the order of the events is 1, while for classical piecewise constant trajectories, the order of the events is 0.

12.2.6.3 GDEVS Atomic Model

A GDEVS model M is a structure:

$$M = \ < X,\ Y,\ S, \delta_{\text{int}}, \delta_{\text{ext}}, \lambda,\ \text{lifetime} > \text{with}$$

- X: the set of input events, $X = A\text{x}A\text{x} \ldots \text{x}A$.
- Y: the set of output events, $Y = A' \times A' \times \ldots \ \times A'$.
- S: the set of states.
- lifetime: $S \rightarrow R^+$, the function defining the maximum length or lifetime of a state. Thus, for a given state, lifetime(s_i) represents the time during which the model will remain in the state s_i if no external event occurs.

A state s_i with an infinite lifetime is said to be a stable state. A state with a finite lifetime is a transitory state. Denoting S_S as the subset of steady states and S_T as the subset of transitory states, we have the following:

$$\sigma_i = \infty \Leftrightarrow s_i \in S_S \wedge \sigma_i < \infty \Leftrightarrow s_i \in S_T$$
$$S_S \cup S_T = S \wedge S_S \cap S_T = \varnothing$$

- $\delta_{\text{ext}} : Q \rightarrow S$, is the external transition function that specifies the state changes due to external events.

- $\delta_{\text{int}} : S_{\text{T}} \rightarrow S$, is the internal transition function that defines the state changes caused by internal events (autonomous evolution of the model). Where M arrives in state s_i at time t_i, it will transition to state $s_j = \delta_{\text{int}}(s_i)$, at time $t = t_i + \text{lifetime}(s_i)$, provided no external event occurs earlier than this time.
- $\lambda : S_{\text{T}} \rightarrow Y$, is the output function.

As in the DEVS formalism, we introduce the definition of total state $q_i = (s_i, e)$, where e is the elapsed time in the current state s_i.

$Q = \{(s_i, e) | s_i \in S \wedge e \in R^+ \cup \{\infty\}\}$. The elapsed time e is reset to 0 when a discrete transition takes place.

Evidently, the polynomial state provides knowledge of the value of the state at any time instant over the polynomial segment. Lastly, the polynomial segment degenerates to the case of the traditional piecewise constant segment (classical DES) where a_0 is the only non-null coefficient.

12.2.6.4 GDEVS Coupled Model

GDEVS promotes modular modeling to reduce the complexity of the system to describe as in DEVS. The GDEVS coupled structure MC allows formalizing the modeled system in a set of interconnected and reused components.

$$\text{MC} = \left(X_{\text{MC}}, Y_{\text{MC}}, D_{\text{MC}}, M_{d|d \in D}, \text{EIC, EOC, IC, Select}\right), \text{where}$$

X_{MC}: set of external events
Y_{MC}: set of output events
D_{MC}: set of component names
M_d: GDEVS model named d
EIC: External Input Coupling relations
EOC: External Output Coupling relations
IC: Internal Coupling relations
Select: defines a priority between simultaneous events intended for different components

This formalism is proved by the closure under coupling property, which shows that a GDEVS coupled model has an equivalent GDEVS atomic one (for more details, refer to Giambiasi et al. (2000)).

12.2.6.5 Abstract Simulator of GDEVS

Like in DEVS, the GDEVS abstract simulator (see Fig. 12.5b) consists of a root coordinator, which manages the simulation time, subcoordinators, which dispatch messages according to the specific couplings of the coupled model that attempt to

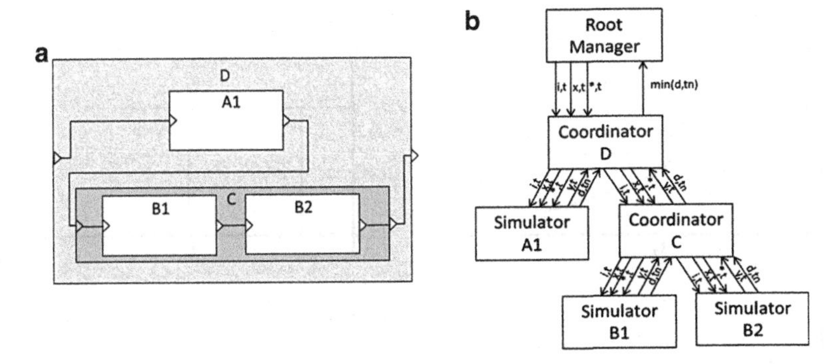

Fig. 12.5 GDEVS simulation structure. (**a**) GDEVS model (**b**) GDEVS abstract simulator

simulate, and basic simulators related to atomic models. Each process behaves according to the received messages from parent and child processes.

12.3 GDEVS Model of Logic Gates

GDEVS promises to permit the development of models of greater accuracy while preserving the computational advantages of discrete-event simulation. As an example, consider the process underlying a digital system design at different levels of abstraction. At the transistor level, signals may be represented through continuous graphs, as shown in Fig. 12.6a. At the higher logic gate level, the classical discrete-event abstraction employs a boolean variable, and the signal model utilizes piecewise constant values 0 and 1, as shown in Fig. 12.6b (classical DES model).

Figure 12.6d represents a piecewise linear approximation of continuous segments, which offers, clearly, higher accuracy in modeling the real-world behavior. The event trajectory is first-order discrete-event abstraction under GDEVS (see Fig. 12.6e), which is computationally faster than the transistor-level model and offers greater accuracy than the classical discrete-event modeling (see Fig. 12.6c).

12.3.1 Logic Gate Models

In order to model a logic gate, we use three basic components, as represented in Fig. 12.7:

- A boolean ideal function
- A delay block
- An amplifier block

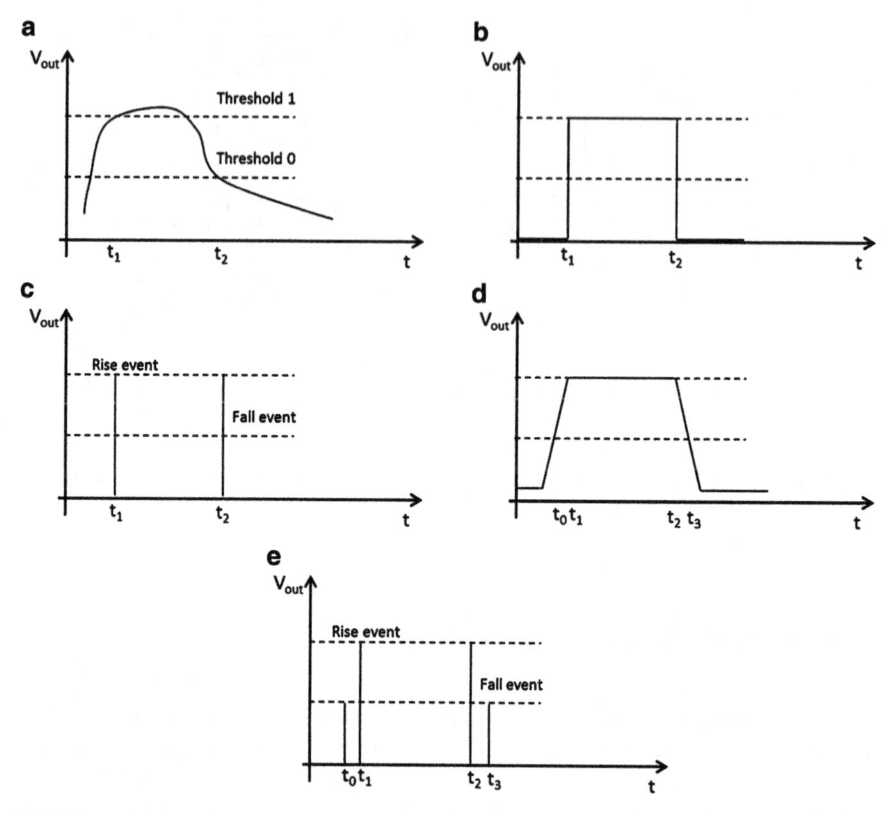

Fig. 12.6 Discrete abstraction of a signal. (**a**) Continuous signal. (**b**) Boolean signal. (**c**) Discrete events. (**d**) Piecewise linear. (**e**) Discrete events of order one

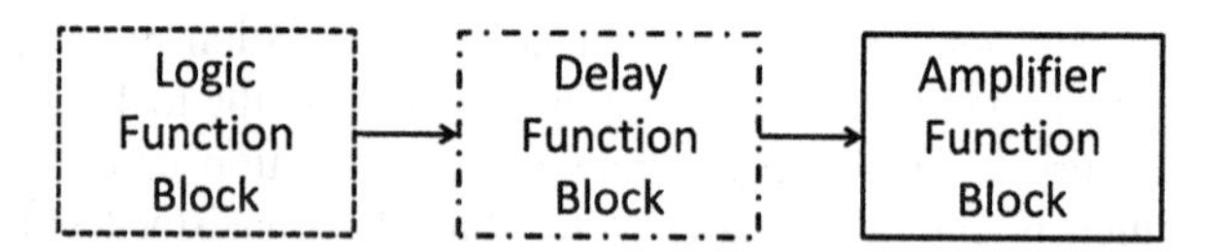

Fig. 12.7 Logic gate structure

The boolean ideal function block has the purpose to model the boolean function of the gate. This function block has a fixed number of inputs and computes one output signal. The output of the function block V_{out} is a function of the input signals V_i at time t, i.e.,

$$V_{out}(t) = f(V_1(t), V_2(t), \ldots) \tag{12.5}$$

12.3.1.1 Boolean Function Model

The basic boolean functions are defined as follows:

$$And: V_{\text{out}}(t) = \min V_i(t) \tag{12.6}$$

$$Or: V_{\text{out}}(t) = \max V_i(t) \tag{12.7}$$

$$Not: V_{\text{out}}(t) = V_{\text{offset}} - V_i(t) \tag{12.8}$$

We build a first-order discrete-event abstraction of these continuous functions under GDEVS considering piecewise linear input–output signals. More accurate representations can be defined using high-order polynomials. In our case, a first-order event will be a list of two values (a, b) representing the gradient a and the intercept b of the corresponding linear trajectory. The boolean operators apply the previous rules to the piecewise linear segments defined by the first-order events occurring on its inputs.

Let us notice that one input event can generate several output events. In general, an input event can create several transitory (active) states of the GDEVS model that implies several output events.

Note that at a conceptual level, the simulation consists in activating the external transition function when an external event occurs and in activating the output and the internal transition functions when the lifetime duration of the present state is elapsed.

The activation of the internal and external transitions and the output functions of the GDEVS gates *And* and *Or* is done as follows. These gates hold

- The two input ports input_1 and input_2 and the output port output
- The variables a_i, b_i, a_s, b_s, and a_n, b_n : the gradient and intercept of the input i, the gradient and intercept of the current output s, and the gradient and intercept of the output at crossing time t_n respectively.

We consider the two piecewise linear segments applied to input_1 at time t_1 and to input_2 at time t_2:

$$y_1 = a_1(\text{time} - t_1) + b_1 \tag{12.9}$$

$$y_2 = a_2(\text{time} - t_2) + b_2 \tag{12.10}$$

However, at crossing time t_n at which the two signal magnitudes are equal, we have

$$a_1(t_n - t_1) + b_1 = a_2(t_n - t_2) + b_2 \tag{12.11}$$

The gate *Not* has only one input port on which the gate receives the input signal (a, b) that will be limited between the two trajectories $V_{\min}(t)$ and $V_{\max}(t)$.

12.3.1.2 Delay Model

The delay model has to simulate the minimal time that an input signal spends to cross a digital component. The delay model sends out, through its unique output, the signal coefficients after a specified delay *d*.

$$(a_{t+d},\ b_{t+d}) = \text{input}(a_t,\ b_t)\big|d > 0 \tag{12.12}$$

12.3.1.3 GDEVS Amplifier Model

The purpose of the amplifier model is to model finite gain, supply rail clipping, and slew rate limiting. First, the input signal is multiplied by the gain $A{\in}R$ of the amplifier. This is easily accomplished by replacing input events of the form $ev = (t;\ (n;\ b;\ a))$ with events of the form $ev = (t;\ n;\ (Ab;\ Aa))$

The second function of the amplifier block is to model power supply rail clipping, i.e., the output of the block is constrained to be between some minimum output voltage V_{min}, and some maximum output voltage V_{max}. Given an input signal $V_{\text{in}}(t)$ at time *t*, the clipped voltage is given by

$$V_{\text{clip}}(t) = \begin{cases} V_{\text{min}} & \text{if}\, V_{\text{in}}(t) \leq V_{\text{min}} \\ V_{\text{in}}(t) & \text{if}\, V_{\text{min}} < V_{\text{in}}(t) < V_{\text{max}} \\ V_{\text{max}} & \text{if}\, V_{\text{in}}(t) \geq V_{\text{max}} \end{cases} \tag{12.13}$$

The third function of the amplifier block is to model limited slew rate (i.e., voltage time derivative) of the circuit component. The maximum slew rate of a component is a function of the nominal output capacitance of the component, the input capacitance of the devices driven by this component, and the capacitance of the interconnect. Slew rate limiting is characterized by two parameters δr and δf, which are the maximum time derivatives of rising and falling signals (respectively); i.e., in a slew rate limited event sequence, the component of each event satisfies $\delta f{<}a{<}\delta r$.

Slew rate limit would appear to only affect a component of signal events. However, simple clipping of a component of events results in signals that are not continuous; e.g., consider two consecutive events, $ev = (t,\ (n,\ b,\ a))$ and $ev_{i+1} = (t_{i+1},\ (n, b_{i+1},\ a_{i+1}))$. If a is positive and exceeds δr and if it is simply replaced by δr, then the signal is no longer continuous since $b_i + \delta r(t_{i+1} - t_i) \neq b_{i+1}$. A method for constructing a slew rate limited signal is given in the following explanation.

Given a sequence of events of the first order, $S(n) = \{ev_0, ev_1, ev_2, \ldots\}$, where $ev_i = (t_i,\ (n, b_i,\ a_i))$, the goal is to construct a new sequence of events, $S(n) =$

$\{f_0, f_1, f_2, \ldots, f_n\}$, where $f_j = (t_j, (n, w_j, w_j'))$, such that $S(n)$ is piecewise linear, continuous, and slew rate limited, i.e., $\delta f \le w'_j \le \delta r$. The following is a recursive method for computing $S(n)$.

First, the base case: Given $e_0 = (t_0, (n, b, a))$ as follows.

Let $t_0 = t$ and $w_0 = b_0$. Then, if $a_0 \le \delta f$, let $w_0' = \delta f$; if $a_0 \ge \delta r$, let $w_0' = \delta r$; otherwise, let $w_0' = a_0$.

Next, the recursive step: Given $e_0, e_1, \ldots, e_i$ and $f_0, f_1, \ldots, f_{j-1}$ such that $t_{i-1} \le t_{j-1} \le t_i$, compute f_j as follows. First, let $t_j = t_i$ and $w_j = w_{j-1} + w_{j-1}'(t_j - t_{j-1})$. To compute w_j', there are three cases to consider:

1. $w_j = a_i$

$$w_j' = \begin{cases} \delta f & \text{if } a_i < \delta f \\ \delta r & \text{if } a_i \ge \delta r \\ v_i & \text{otherwise} \end{cases} \tag{12.14}$$

2. $w_j < a_i$: in this case, the slew rate limited signal is below the input signal. Let $w_j' = \delta r$. Next, compute $t_x = t_i + (v_i - w_i)/(w_j' - v_j')$. At time t_x, the slew rate limited signal will meet the input signal. If the signals meet before the next input event, e_{i+1}, i.e., if $t_x < t_{i+1}$, then an extra output event, f_{j+1}, is computed as follows. Let $t_{j+1} = t_x$ and $w_{j+1} = w_j + w_j'(t_{j+1} - t_j)$
3. Finally, $a_i \le \delta f$:

$$w_{j+1}' = \begin{cases} \delta f & \text{if } a_i < \delta f \\ \delta r & \text{if } a_i \ge \delta r \\ v_i' & \text{otherwise} \end{cases} \tag{12.15}$$

The GDEVS models of the gates *And*, *Or*, *Not*, *Delay*, and *Amplifier* can be consulted in Hamri et al. (2014).

12.4 Design and Implementation

In order to permit the design of composite gates, we develop a software framework combining XML and java to develop simulations. Firstly, the user defines an XML file which corresponds to a coupled gate; this file has a specific grammar that respects the GDEVS coupled definition.

```
<?xml version="1.0" encoding="UTF-8"?>
<CoupledModel>
  <Name>Gen_Two_And</Name>
  <Ports>
                <OutPort>Out</OutPort>
  </Ports>
              <SubModels>
                <SubModel>
                  <Name>G1</Name>
      <InstanceOf>Generator</InstanceOf>
                                <FileName>scenarios/scenario_1</FileName>
                </SubModel>
                <SubModel>
                  <Name>G2</Name>
      <InstanceOf>Generator</InstanceOf>
                                <FileName>scenarios/scenario_2</FileName>
                </SubModel>
                <SubModel>
                  <Name>G3</Name>
      <InstanceOf>Generator</InstanceOf>
                                <FileName>scenarios/scenario_3</FileName>
                </SubModel>
                <SubModel>
                  <Name>And1</Name>
      <InstanceOf>And</InstanceOf>
                </SubModel>
                <SubModel>
                  <Name>And2</Name>
      <InstanceOf>And</InstanceOf>
                </SubModel>
              </SubModels>

....
```

Then, from a valid XML file, the framework generates a java class which corresponds to the designed coupled gate to simulate and with the name specified by the user in an XML file. Another java class called Main, which defines the simulation context is generated. The method main() instantiates the designed gate and starts the simulation. While the simulation advances, the input and output events of each gate are displayed to the user, who follows the simulation trace. In addition to this trace, for each gate (the whole gate and reused ones), a formatted file is generated containing the computed output events; using an appropriate software or toolkit (MATLAB©, EXCEL©, etc.), the user may load these files to plot the simulation and may make conclusions at critical times.

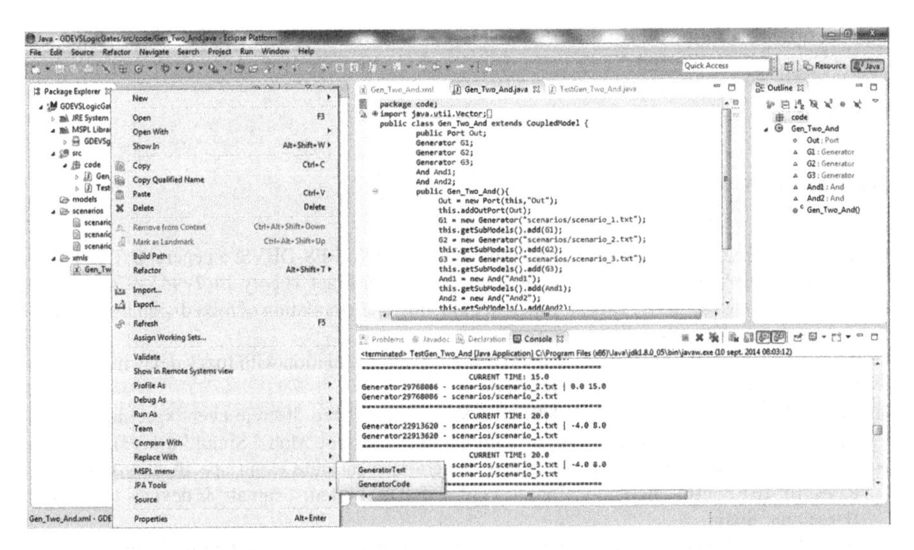

Fig. 12.8 Logic gate plug-in under eclipse©

To automatize the chain of M&S of logical gates, we develop an eclipse© plug-in to make easy the passage from a step to another one (generation of java files, start simulation, etc.). Figure 12.8 gives an overview about this plug-in.

12.5 Conclusion

In this research work, an original approach to build discrete-event models of logic gates without the classical constraint of piecewise constant input–output trajectories is proposed. This approach allows the use of GDEVS either of the first order or of higher order according to the requested accuracy. In such cases, a key advantage of GDEVS is that where higher accuracy is desired, higher-order polynomials may be used without any commensurate increase in the number of events. In contrast, in the classical continuous simulation approach, higher accuracies generally require a reduction in the simulation time step, which in turn implies more execution steps and higher execution time. Thus, the performance advantage of GDEVS over the classical continuous simulation approach is likely to be superior where the use of higher-order models is warranted.

One of the concrete outcomes from the proposed generalization of the classical discrete-event abstraction process, namely, the use of any kind of trajectories instead of piecewise constant trajectories, is the possibility to simulate continuous or hybrid models with a good approximation by using event-driven techniques.

In the near future, we will conduct practical comparisons with commercial simulation software of logic gates (PSpice©, MATLAB©, etc.) in order to confirm our best performances with GDEVS logic gates and the capacity of designing large-scale circuits.

References

Cappochi L, Bernardi F, Federici D, Bisgambiligia PB (2006) BFS-DEVS: a general DEVS-based formalism for behavioral fault simulation. Simul Model Pract Theory 14(7):945–970

Ghosh S, Giambiasi N (2001) Breakthrough in modeling and simulation of mixed-signal electronic designs in nVHDL. Simulation 76(5):279–281

Giambiasi N, Smaili M, Frydman C (1994) Discrete event simulation with fuzzy delay models. In: European simulation symposium 94, Istanbul

Giambiasi N, Escudé B, Ghosh S (2000) GDEVS: a generalized discrete event specification for accurate modeling of dynamic systems. Simulation Trans Soc Model Simul Int 17(3):120–134

Hamri M, Naamane A, Giambiasi N (2014) Generalized discrete event specifications of logic gates. In: IEEE 11th international multi-conference on systems, signals & devices (SSD'14), Barcelona, Spain

Ören TI (1984) GEST a modelling and simulation language based on system theoretic concepts. Simul Model Based Methodologies Integr View 10:281–335

Ören TI, Zeigler BP (1979) Concepts for advanced simulation methodologies. Simulation Trans Soc Model Simul Int 32(3):69–82

Ören TI, Torres L, Amblard F, Belaud J-P, Caussanel J, Dalle O, Duboz R, Ferrarini A, Frydmann C, Hamri M, Hill D, Naamane A, Siron P, Tranvouez E, Zacharewicz G (2006) Modeling and simulation dictionary - English-French-Turkish. In: GDR I3 – LSIS, Marseille, mai 2006 ISBN 2-9524747-0-2

Paoli C, Nivet M, Bernardi F, Capocchi L (2004) Simulation-based validation of VHDL descriptions using constraints logic programming. In: Proceedings of the 5th IEEE workshop on RTL and high level testing, Osaka

Santoni C (2008) Contribution Méthodologique à la Conception d'Interfaces Homme-Machine pour les Systèmes de Supervision, Habilitation à Diriger des Recherches. Université Paul Cézanne, Mars

Seck M (2007) Modélisation et simulation à événements discrets de comportements humains, Université Paul Cézanne (Aix-Marseille III), 2007 Sous la direction de: C. Frydman et N. Giambiasi

Zeigler BP (1989) DEVS representation of dynamical systems, academic press, Proc IEEE 77:72–80

Zeigler BP, Praehofer H, Kim TG (2000) Theory of modeling and simulation. Academic Press

Part IV
Cognitive, Emotive, and Social Simulation

Chapter 13
Specification and Implementation of Social Science Models

Paul K. Davis

13.1 Introduction

I was delighted with the opportunity to submit a chapter to the testimonial volume on Tuncer Ören's work, for all the reasons discussed in the introduction. Then, of course, I had to figure out what to write about—always a problem after agreeing to do something. In this case, the answer popped up immediately because of a paper that a colleague and I had done last year (Davis and O'Mahony 2013), a paper motivated in part by a talk I had heard from Tuncer no less than 30 years ago (Ören 1984). The ideas I got from the paper were significant abstractions that bore little resemblance to details of what Tuncer discussed, but that is sometimes the way we learn: an idea in one context generates an idea for another.

Tuncer's paper was presented in a conference seeking to integrate various simulation and modeling methodologies (Ören et al. 1984)—always a worthy effort and one that could just as well be attempted again in 2014, since success in such matters seems never to be complete. Tuncer's paper was about something called GEST, which I have never actually looked at in detail or used, but which I saw as reflecting a sense of fundamentals. GEST was actually a computer language, but not what people usually have in mind. Not being a computer scientist or programmer myself, I didn't know then how much traction the concept had gotten or would get over the years, but I knew that I liked some of the ideas. In particular, "a GEST program is highly descriptive and acts as documentation (for communication

P.K. Davis (✉)
Department of Engineering and Applied Sciences, RAND Corporation,
Santa Monica, CA, USA
e-mail: pdavis@rand.org

L. Yilmaz (ed.), *Concepts and Methodologies for Modeling and Simulation*,
Simulation Foundations, Methods and Applications,
DOI 10.1007/978-3-319-15096-3_13

among humans) as well as a specification (for man–machine communications)" (Ören 1984, p. 282).[1]

My favorable reaction reflected my graduate work in theoretical chemistry and physics, during which I learned how valuable it was to distinguish between the concepts of quantum statistical mechanics and the attempts to address particular problems by computational evaluation of differential equations with all the complexities and annoyances that so beset researchers in those days. Some of my fellow graduate students spent the majority of their dissertation period forcing unruly computers to do the right thing when they would rather have been focused on the science. Fortunately my research was more abstract, using elegant mathematical methods developed by physicist Eugene Wigner, John Kirkwood, and my dissertation advisor, Irwin Oppenheim. Wigner's beautiful work in the early 1930s expressed the then–new concepts of quantum mechanics in a Hamiltonian/Liouville formalism that extended classical physics to the quantum domain. Quantum mechanics is surely mysterious, but seeing its relationships to classical mechanics helps. Also, as with Paul Dirac's approach to quantum mechanics, operator-based mathematics allows one to think about the underlying physics rather than how to solve or compute solutions to differential equations.

Tuncer's paper about a specification model, then, fell on friendly ears, even though I was interpreting it a bit differently—i.e., that such a thing was useful because one should not confuse the concepts of a theory with the mathematical details, much less with computational procedures. My interest was even greater because, at the time, I was directing a project at RAND that involved building a large computer system for "analytic war gaming"—a system for studying paths from peacetime through crisis into conflict or, in favorable cases, back to peace (Davis 1985a, b). We had agents representing the leadership and military command levels of the Warsaw Pact and NATO and the leadership of third countries. Humans could substitute for agents. The system also included a large and complex simulation of military activities, including mobilization, deployment, and combat. This being the 1980s rather than the 2010s, everything was difficult. Our agents, for example, were uniquely constructed using UNIX coprocess mechanisms because the agent-based modeling and multi-model technologies that we now take for granted were only beginning to emerge. In addition, the military simulation was highly complicated because it was global in scope with multiple levels of detail (e.g., army brigades up to decisions by theater commanders or presidents). It was also variable-structure simulation in that top leaders might change in the course of the simulation. All of this required heavy-duty programming informed by serious computer science. However, even the deep-thinking computer scientists tended to focus on the programming challenges, not the real-world concepts that we were

[1] A scanned version is available at http://www.site.uottawa.ca/~oren/pubs-pres/1984/pub-1984-03_GEST_NATO_ASI.pdf. Tuncer and Bernie Zeigler have recently described the history of GEST and subsequent developments building on system theoretic foundations laid out by A. Wayne Wymore (Ören and Zeigler 2012).

attempting to model and simulate. It was a constant struggle for the team to rise above the machine-related difficulties to discuss the phenomena.

I found it even more difficult to encourage or demand model *design*. To a programmer, "design" can involve dealing with memory, processing speed, communication among elements of the system, and so on. Substantively, however, we needed to worry about how to represent various forms of combat and the decision processes generating escalation of conflict or, in more favorable cases, de-escalation. Design, in that context, means theory development.

The substantive modeling required knowledge from fields such as political science, psychology, organizational theory, and artificial intelligence. The link to artificial intelligence was due not to computational issues but rather to the fact that some computer scientist pioneers such as Herbert Simon were also deeply involved with political science, economics, decision-making, and organizational behavior. The kind of thinking that was most relevant to our modeling was to be found in artificial intelligence research rather than textbook political science. Simon, in particular, had "blown the whistle" on the notion that real-world decision-makers use or should use the rational-actor model on complicated problems. He pointed out that they lack the necessary information and, even if they had it, could not do the necessary computations because of uncertainty. He later won the Nobel Prize in economics for these observations (Simon 1978), although the field of economics made minimal use of them for decades. He also wrote inspirationally on nearly decomposable systems (Simon 1996), anticipating key elements of what we now associate with complex adaptive system research. So also we drew on the work of Daniel Kahneman and his colleagues, for which Kahneman won a Nobel Prize (Kahneman 2002). As it turns out, real people must draw on various heuristics to cope with complex situations. Various approaches existed for attempting to represent such heuristics. These included script-based methods as in the work of Roger Schank, methods used by chess masters to "read" a chessboard at a glance, and models in a style then referred to as rule-based systems or knowledge-based systems. Most or all of these methods were discussed to greater or lesser extent in the several conferences held in the 1980s in which Tuncer played a major role (Elzas et al. 1986, 1989).

Tuncer's 1984 paper, off-line discussions with him and Bernie Zeigler about modeling and artificial intelligence, a paper by Robert G. Sargent (1984), and the subsequent conferences relating simulation to artificial intelligence research all affected my thinking. One of Tuncer's papers was a forward-looking survey of the key challenges ahead, including some that were meaningful to my own work (e.g., goal-directed agents, anticipatory simulation, agent perceptions, and multi-faceted models) (Ören 1989). In any case, I became a believer in (1) distinguishing sharply between the conceptual model and its implementation in a particular programming language, (2) the value of describing the conceptual model separately and coherently, and (3) the Sisyphusian nature of trying to force programmers to design conceptually and document the results. Over the next half-dozen years, I strived mightily to apply those principles while running the model-building project mentioned above. Interestingly, there was always resistance. It seems to me that, especially in the United States rather than Europe, most programmers have resisted anything that gets between them and "cutting code." Bernie Zeigler's early text was a notable effort to encourage doing better (Zeigler 1984).

As an aside that I cannot resist, another memory from the time of the 1984 conference and its successors was that Tuncer *loves* taxonomies.[2] Perhaps because Tuncer believes in careful thinking and design, but has also worked as a computer scientist in engineering contexts, he likes nothing better than to lay out all the distinctions, niches, and crevices of a problem. Although I can't say that I emulated such work (it's too painful for this lazy author), I was at least admiring.

Many years have intervened, and I have often found Tuncer's papers to be provocative and insightful, especially his many contributions to agent-based modeling (Yilmaz and Ören 2009) and his recognition that agent modeling could help in understanding international disputes and crises in the context of peace studies. In the remainder of this paper, I sketch work that drew on the principles mentioned above that benefited from Tuncer's early influence.

13.2 A Modern Attempt at a Kind of Specification Language

13.2.1 Context of Work

Over the last decade, the United States and allies have been involved in conflicts and wars that could not be more different from the "industrial wars" of the twentieth century (WWI and WWII). Instead, they have involved terrorism, insurgency, and competitions for public support. Modeling the phenomena raises entirely different issues from those dominant in military modeling of the past. In the period 2002–2008, a good deal of computer modeling sought to represent those phenomena, often with agent-based simulations. By about 2008, however, Department of Defense officials responsible for modeling and analysis had concluded that they were quite unhappy with the results. In particular, they were seeing large computer programs that purported to simulate incredibly complicated social and military issues, but their content was inaccessible and/or incomprehensible. Further, the officials doubted that the programs actually represented the best social science available. They then took the remarkable step of backing up and saying to themselves "Whoa! We should insist on understanding the underlying science before proceeding further with modeling and programming." They asked for related studies, including a large RAND study that I led with co-author Kim Cragin—unlike myself, a bona fide social scientist (Davis and Cragin 2009). Our team

[2] As documentary evidence of this assertion, I note that Tuncer's personal website (https://www.site.uottawa.ca/~oren/) lists "Taxonomies" as one of the categories in which he lists his publications, right along with "Ethics" and, e.g., "Agent-Directed Simulation."

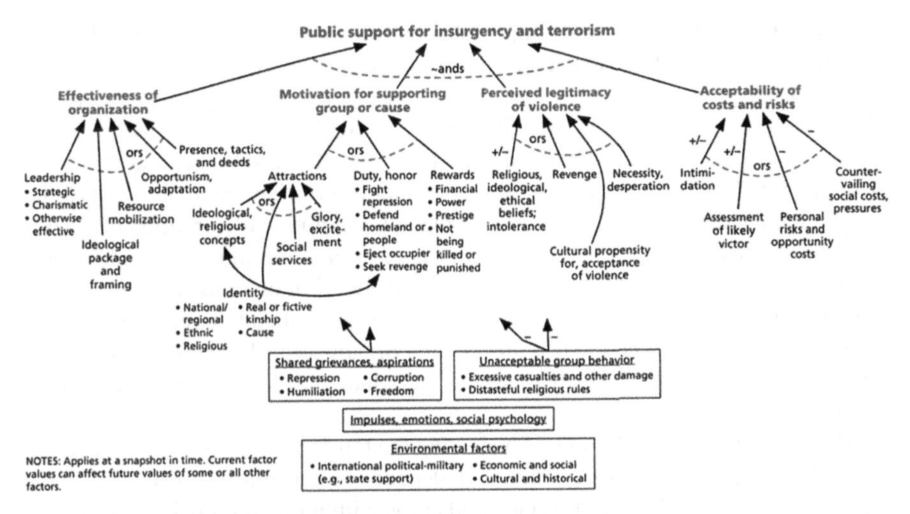

Fig. 13.1 Factor tree of public support for insurgency and terrorism

included people with backgrounds in psychology, political science, anthropology, sociology, and economics. We approached the study focused on the social science, not modeling. After a few months, however, it became clear that pulling the pieces together—i.e., going beyond just stapling essays together—required "modeling," even if it was not what people usually think of a modeling (there were no computers or equations in sight). We introduced what came to be called the *factor-tree* methodology that summarized our critical surveys of the relevant literature with diagrams that highlighted the factors at work causing developments such as individuals becoming terrorists, publics supporting terrorist movements, and so on. This approach focused on the content of social science that is most reliable: disciplinary experts, who have spent their professional lives working on subjects, really do know the factors at work, some of which are not at all obvious. In contrast, experts are notoriously unreliable in predicting consequences or even in judging the reliability of their own predictions (Tetlock 2005). Figure 13.1 shows the factor tree for public support of insurgency and terrorism stemming from the 2009 work and a subsequent study that validated the qualitative model with new case studies, which also motivated refinements (Davis et al. 2012). We have used factor trees successfully in conducting interdisciplinary meetings with both academic scholars and military officers.

Although factor trees such as Fig. 13.1 appear simple and can be shown in viewgraphs to audiences of varied backgrounds, they embody a great deal of nontrivial knowledge. It was common in the 2000s for people (even academics who should have known better) to argue that "the" cause of terrorism and its support were something specific such as the Islamic religion, poverty, or nationalism. Such "theories" (shown in quotes as a sign of derision) were and are simple minded. As indicated in Fig. 13.1, terrorism and its public support can arise from any of several causes. Further, all major religions have justified terrorism at one time or another, as have many nonreligious ideological causes. Moreover, the motivations are often more about identity than anything intellectual such as religious or ideological concepts (Davis et al. 2012). In addition, some who participate are driven by thrill-seeking on the one hand or fear of the consequences if they do *not* participate. The social science on these matters is strong. Terrorists and their supporters, then, are motivated by something. Beyond that, they typically need to feel that terrorism is "legitimate." Further, to participate in or support terrorism, motivations have to outweigh the risks and negatives that they recognize emotionally or rationally. And, finally, potential terrorists often never get around to anything significant unless they are part of an organization providing a mixture of motivational support and practical matters such as planning, logistics, leadership, and intelligence. Similarly, public support tends to be modest unless such organizational cohesion exists. The top-level factors of Fig. 13.1, then, are all important and, to a first approximation, necessary.

This is suggested by the "ands" connecting the top-level factors. As shown, each of the four top-level factors in Fig. 13.1 depends in turn on lower-level factors. Some of these can substitute for one another, as with motivation being due to any one or a combination of causes (as suggested by "ors").

The factor tree for public support includes tens of factors, but they are arranged in layers because most factors have their effect through higher-level factors. Technically, this is multiresolution modeling; it also reflects the way humans routinely reason (Davis and Bigelow 1998; Davis et al. 2001). It is not strictly hierarchical modeling. Some factors, shown at the bottom of the figure, are cross-cutting, and, over time, there are feedbacks and more cross-cutting. For example, if the United States were drawn into a new war in the Middle East, that would probably affect the subsequent values of numerous factors in the tree itself. Some of these cross-factors can be seen as environmental; others relate to culture and emotions.

Such conceptual modeling allows convergent, constructive discussion among people with different backgrounds and dispositions. It is relatively easy for people to accept that the causes that they have studied or believed in are not unique, so long as their perspective is included in the larger picture. That is, someone who sees nationalism behind terrorism can accept that another stream of research related to revolution against tyranny is also legitimate. A second value is that factor trees can get across a "system perspective," as in noting that people usually become terrorists only when the four factors mentioned above are *all* present.

My colleagues and I believe that social science knowledge on such matters is best approached qualitatively and that efforts to force research into quantitative form (as with statistical analysis of, say, terrorist incidents versus alleged determinants) are often counterproductive because the data analysis of historical events is intellectually flawed, with problems of hidden variables, uncontrolled variables, and poor proxy variables (Davis 2011). Qualitative causal models are more informative even if they cannot be used to make predictions as one would use a weather model or a model of whether precision-guided munitions would be able to destroy some particular target. They can be valuable for structuring issues, explaining what happens, and planning under uncertainty.

13.2.2 The Transition to a Computational Modeling

Despite our strategy of emphasizing qualitative models—something that changed the focus of DoD work in important respects—the intent from the outset was to come back to the question of whether computational modeling was possible and useful. In 2012–2013, I did so with colleague Angela O'Mahony (Davis and O'Mahony 2013). The question was whether we could do something more than the factor trees. We "should" be able to do so, since available social science knowledge goes beyond what is in such trees. That is, even if the factors of Fig. 13.1 are correct, what is the result of the combined influences? Social science has *some* things to say, even if predictions are at best highly contingent and, even then, uncertain.

The approach that we took was to treat the factor tree for public support of insurgency and terrorism as the qualitative "specification model" and to ask what was needed to move to a fully specified model with the same concepts but with the additional details needed to compute consequences. We would need "combining rules" consistent with social science and reasoning. Since a theme of our work was dealing with uncertainties, the model should be of a special variety: one that would assist analysts making assessments amidst great uncertainty. To enforce this philosophy, the interface should emphasize the uncertainty of factor values and certain structural aspects of the model itself (i.e., combining rules). It should be difficult for a model user to even generate a "point" conclusion, but—if all worked out—it would be possible to get a sense of situation-dependent propensities and possibilities. We would routinely show model results as a function of parameters, thereby making results contingent on contexts and assumptions. This is quite different from constructing a single model treated as though it were "correct," constructing an approved database for all the input assumptions, running the model, and reporting the results as a point prediction (perhaps with minor sensitivity testing).

Finally, a key element of the approach was to focus on static snapshot-in-time relationships. Dynamics are certainly important, but I felt that much of the insight could be gained with snapshot-in-time modeling, with dynamics treated as a next step. Moreover, while I have long concluded that the causal-loop or influence-diagram methods of system dynamics are powerful qualitatively, the uncertainties in dynamics are significantly worse even than those in a static situation. As a first step, then, we would focus on the static depiction and then discuss dynamics qualitatively as a next step.

13.2.3 The Research Challenges

Building a computational model based on a factor tree posed numerous challenges as summarized in Table 13.1. This list guided our research. The challenges were in four blocks, as indicated by shading in the table: (1) defining the factors and their

Table 13.1 Research challenges

Challenge	Issues
Define factors and factor values	How many values are sufficient? How can soft and fuzzy variables be reasonably defined?
Define "and" connections mathematically	How rigid should the relationship be? How can uncertainties be reflected?
Define "or" connections mathematically	How many alternative functional relationships are needed?
Define ambiguous and conflicting influences (± signs) mathematically	What does the ambiguity mean? How can it be represented?
Represent implications of line thickness in factor trees	How should relative importance of factors be understood and represented in the model?
Represent uncertainty of factor values	Should this be done by giving ranges of parameter values or by using probabilistic methods?
Represent structural uncertainty of combining relationships	Should this be done with alternative models, structural parameterization, or both?
Build model for exploratory analysis under uncertainty	How should far-reaching exploratory analysis be accomplished? When should probabilistic methods be used?
Assess "confidence" of nominal factor-value estimates and of model outputs	How should this best be accomplished?
Implement model in understandable high-level language	What language? How can the model be made transparent, comprehensible, and easy to re-implement (a form of reuse)?

values; (2) defining how to reflect cryptic factor-tree "and" and "or" relationships, ambiguous influences indicated by ± signs, and the varied significance of influences sometimes indicated by arrow thickness (illustrated in the text, but not in Fig. 13.1); (3) dealing with uncertainty about factor values and combining rules and showing results of exploratory analysis across uncertainties; and (4) implementing the model in a computer program in which substantive content is transparent, comprehensible, and as language independent as possible so as to facilitate model reuse, model composition, or rapid reprogramming.

How we dealt with the challenges is discussed in our study (Davis and O'Mahony 2013). Let me just mention a few highlights that are related to the issue of a specification language, i.e., of separating model content from program.

First, we characterized inputs in qualitative terms such as very low, low, medium, high, and very high, but—within the model (and in outputs, so that they could be graphed)—we mapped such values into numbers 1, 3, 5, 7, and 9. This required identifying reasonably recognizable factor levels that are "equally spaced," rather than merely ordinal.

Perhaps the most difficult challenge was developing building-block combining rules. Real-world behaviors are not consistently predicted by any one set of combining rules, but how many alternatives did we need? We concluded that a great deal could be done with methods that we called "thresholded linear sums" and a "primary factor" method. That is, sometimes, the effect of a number of factors would be a linear weighted sum of those factors exceeding threshold values. In other cases, the largest factor dominates: someone's overall motivation for an action is then driven by whatever *one* motivation is strong. The absence of other possible motivations is irrelevant. Which mathematics is the better approximation is context dependent. Thus, our modeling tool kit had to provide a set of possibilities, but we thought that it need not be large.

The tool kit of combining relationships that we settled on can be regarded as hypotheses about combining relationships to be tested in psychological research and case studies, both to determine the adequacy of the choices we provide and to better understand when one applies rather than another (e.g., When are threshold effects stronger and weaker? When does the strongest factor altogether dominate?).

Another crucial issue was dealing with uncertainty. We designed the model for exploratory analysis (i.e., analysis that shows results for all combinations of the relevant factor values). I prefer to do exploratory analysis deterministically—establishing a discrete range of plausible values for each input, running all the possible combination cases, and then looking for patterns indicating what combinations of factor values lead to results that are good, bad, or indifferent. The analyst and decision-maker can then make judgments about the relative plausibility of the different domains before reaching conclusions or taking actions. This approach has been used by RAND in many studies over the past two decades in connection with planning for adaptiveness and robust decision-making (Davis 2012).

We also used an alternative approach that characterizes each input with a probability distribution. This is sometimes valuable but can obfuscate causal relationships and be misleading if correlations are ignored. Given numerous uncertain factors, it is sometimes best to use the hybrid approach of treating the most important of them deterministically while treating the others stochastically.

We wanted to implement the model (as a program) so as to permit in-depth review, reuse, and composability. We used *Analytica®* which is maintained and sold by Lumina Corporation. The model's content can be largely comprehended without dealing with programming issues. The model is expressed visually in influence diagrams and, at the next level of detail, in a relatively simple syntax closely tied to mathematics rather than procedural programming. The result is intuitive for those with background in vectors, matrices, and arrays. This tie to mathematics also makes the model especially suitable as a specification model available to researchers generally. Figure 13.2 is a screenshot from the program itself, which is called PSOT, for Public Support of Terrorism. The bubbles are modules, functions, or primitive variables and constants. Important for the philosophical approach taken, *the structure of the program is identical to that of the qualitative factor tree*. That is valuable for review, communication, documentation, and reuse.

Figure 13.2 and various drill-downs indicate the model's *structure*, but the problem remains of how to represent the nitty-gritty of combining relationships. In documentation, we could do this with straightforward linear algebra as in Fig. 13.3. In this case, the output function is simply a linear weighted sum of

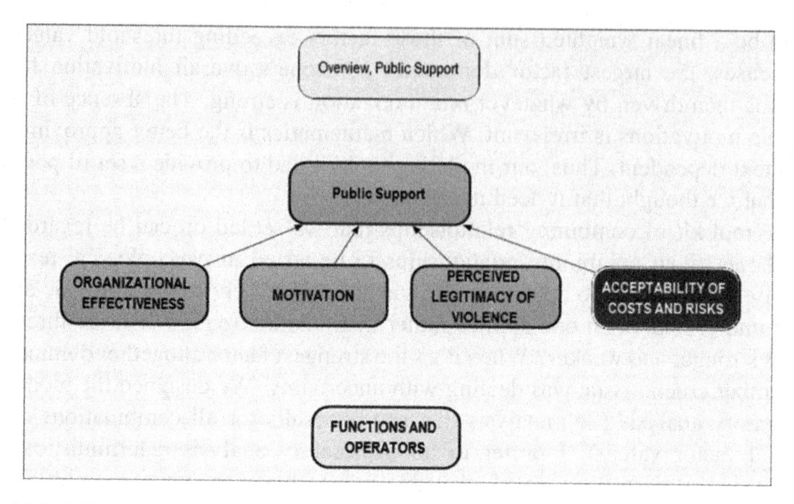

Fig. 13.2 Influence diagram in Analytica corresponding to factor-tree structure

Fig. 13.3 Specifying the algorithm with simple array mathematics

$$\mathbf{F}^0 = \{F_1^0, \ldots F_n^0\}; \mathbf{W}^0 = \{W_i^0\}$$

$$\mathbf{F} = \{F_1, \ldots F_n\}; \mathbf{W} = \{W_i\}$$

$$\mathbf{F} = Max(\mathbf{F}^0 - \mathbf{T}, 0)$$

$$\mathbf{W}^0 * \mathbf{F} \equiv \{W_1^0 F_1, \ldots W_n^0 F_n\}$$

$$\mathbf{W} = \frac{\mathbf{W}^0 * \mathbf{F}}{\sum_{j \in \{\text{set with } F \neq 0\}} W_j}$$

$$N = \mathbf{W} \bullet \mathbf{F} = \sum_i W_i F_i$$

factors remaining *after* all of them have been compared with threshold values and after those failing to reach threshold values are discarded.

In Analytica the scalar product of vectors *W* and *F* is represented by "sum (W, F, I)," where I is the name of the index over which the sum is to be performed, which is necessary because *W* and *B* may be n-dimensional arrays distinguishing between, say, region, year, tribe, or an uncertain parameter. In other languages, such scalar products may be accomplished with for loops or other such devices.

A next issue is what the outputs should look like. Figure 13.3 shows illustrative uncertainty-sensitive PSOT results for the extent to which "the public" will regard the costs and risks of supporting the insurgency and its terrorism as acceptable. That is, it indicates what factor combinations would create a net sense of unacceptable costs (such factors are colored green in Fig. 13.2, because they are good from the counterinsurgency perspective) or low costs (shown in red). Results are shown as a function of five factors: (1) intimidation by the insurgents, which raises acceptability of support by making it dangerous to not to support the insurgency; (2) intimidation by the government; (3) fear that the insurgents will win; (4) countervailing social pressures (e.g., the urgings of family respected leaders not to provide support); and (5) other personal costs of support. The "message" of Fig. 13.3 is that one can be in "bad" situations (the red areas, indicated also with 9s) as the result of different combinations of the contributing factors. Similarly, various combinations of those factors can lead to "good" situations (the green areas). Strategy involves trying to manipulate some of the variables to move in the right direction. Figure 13.3 reflects computations using a particular set of algorithms (the primary factor approach mentioned above). Other such figures would show how much the choice of algorithm affects the big picture.

Such a depiction can help in diagnosing the seriousness of a situation, discussing what factors must be changed to move to a better situation, and assessing the relative leverage of factors (which may or may not be subject to influence). The model, then, is not about prediction but about improved diagnosis and reasoning. It can be especially valuable in dampening enthusiasms when one factor is subject to influence but the net effect is unlikely to be significant or in suggesting approaches in which moderate influence on two or more factors may have synergistic favorable effects.

We completed our prototype work and published results. It remains to be seen how successful it will be judged to be by others, but initial feedback has been positive. Ideally, next steps would include using the model to improve knowledge elicitation, holding workshops to review and debate the scientific content without much programming overhead, and building analogous specification models for other aspects of terrorism, insurgency, and irregular warfare. I believe, however, that *many* social-policy problems could be modeled using similar techniques, whether in education, health, or other domains.

Social science is notoriously difficult—and, as the cliché goes, much "harder" than the hard sciences. Nonetheless, social science contains extraordinary amounts of knowledge. Our report was one step in the process of learning how to better represent that knowledge in increasingly rigorous, albeit often qualitative and uncertainty-sensitive, systemic models.

13.3 Concluding Remarks

The original concept of a "specification model" had much to recommend it. However, my interest has been less in distinguishing between model and simulation than between model and computer program (whether or not dynamic). In the old days, that meant something like specifying the model with rigorous mathematics and passing the specification over the transom to the programmers. Today that is impractical: most of us work at the computer from the outset. We "think" at the computer, simultaneously creating, designing, experimenting, and programming. The old-school ideal of mathematical specifications on paper now

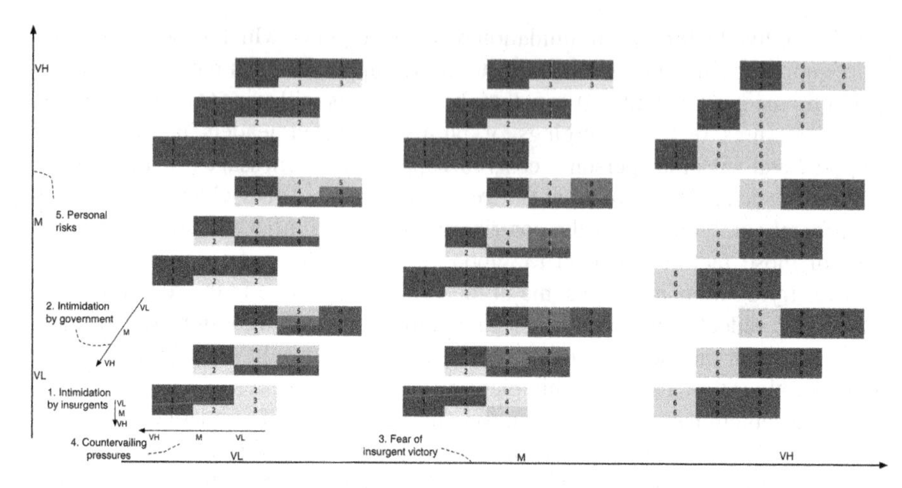

Fig. 13.4 Illustrative exploratory analysis. Note: The numbers shown are the values used internally in PSOT. Because low acceptability of costs and risks is good for the counterinsurgent side, the colors for *1*, *3*, *5*, *7*, and *9* are *green*, *light green*, *yellow*, *orange*, and *red*, respectively

seems absurd, although it is sometimes valuable to step back and resort to archaic tools such as pen and paper to clear our minds.

Because the man–machine linkage is so strong in today's world, it seems that a good analog to the old separation of mathematics and program is a combination of qualitative conceptual model (to include diagrams) and an implementation faithful to its structure in a very-high-level language. This is especially feasible with high-level visual languages, such as *Analytic*, the System Dynamic languages such as *iThink*, and certain other languages such as *Netica*. The result can be comprehensible even to people of varied disciplines with only modest programming skill. Further, if the model is expressed in mathematical concepts, reprogramming should be straightforward for the use in a different environment. To put things differently, my intent was to make model content reviewable by non-mathematician non-programming professors and easily adaptable by graduate students or others with reasonable computer skills. Whether we succeeded can be judged by others, but it was an interesting experiment and a good example of how Tuncer Ören's influence has manifested itself in unusual ways and places, even many years after his initial contributions on a subject. I can only hope that he will view the result with interest, even though he may well have ideas about how to do the same thing better. After all, he continues to be constructively pushy and forward-looking despite having contributed more than his fair share of good ideas over the years.

References

Davis PK (1985a) Applying artificial intelligence techniques to strategic level gaming and simulation. In: Elzas MS, Ören TI, Zeigler BP (eds) Knowledge based modelling and simulation methodologies. North Holland Publishing Company, Amsterdam, pp 315–338

Davis PK (1985b) Applying artificial intelligence techniques to strategic-level gaming and simulation. RAND Corp., Santa Monica, p 7120
Davis PK (ed) (2011) Dilemmas of intervention: social science for stabilization and reconstruction. RAND Corp., Santa Monica
Davis PK (2012) Lessons from RAND's work on planning under uncertainty for national security. RAND Corp., Santa Monica
Davis PK, Bigelow JH (1998) Experiments in multiresolution modeling (MRM). RAND Corp., MR-1004-DARPA, RAND Corp., Santa Monica
Davis PK, Cragin K (eds) (2009) Social science for counterterrorism: putting the pieces together. RAND Corp., Santa Monica
Davis PK, O'Mahony A (2013) A computational model of public support for insurgency and terrorism: a prototype for more general social-science modeling, TR-1220. RAND Corp., Santa Monica
Davis PK, Bigelow JH, McEver J (2001) Exploratory analysis and a case history of multiresolution, multiperspective modeling. RAND Corp., RP-925, Santa Monica
Davis PK, Larson E, Haldeman Z, Oguz M, Rana Y (2012) Understanding and influencing public support for insurgency and terrorism. RAND Corp., Santa Monica
Elzas MS, Oren TI, Zeigler BP (eds) (1986) Modeling and simulation methodology in the artificial intelligence era. North Holland Publishing Company, Amsterdam
Elzas MS, Ören TI, Zeigler BP (eds) (1989) Modelling and simulation methodology: knowledge systems' paradigms. North-Holland Publishing Company, Amsterdam
Kahneman D (2002), Maps of bounded rationality: a perspective on intuitive judgment and choice (Nobel Prize Lecture)
Ören TI (1984) GEST—a modeling and simulation language based on system theoretic concepts. In: Oren TI, Zeigler BP, Elzase MS (eds) Simulation and modelling methodologies: an integrative view. Springer, Heidelberg
Ören TI (1989) Bases for advanced simulation: paradigms for the future. In: Elzas MS, Ören TI, Zeigler BP (eds) Modeling and simulation methodology: knowledge systems' paradigms. North Holland Publishing Company, Amsterdam, pp 29–39
Ören TI, Zeigler BP (2012) System theoretic foundations of modeling and simulation: a historic perspective and the legacy of A. Wayne Wymore. Simulation 88(9):1033–1046
Ören TI, Zeigler BP, Elzas MS (eds) (1984) Simulation and modelling methodologies: an integrative view. Springer, Heidelberg
Sargent RG (1984) Verification and validation of simulation models. In: Ören TI, Zeigler BP, Elzas MS (eds) Simulation and model-based methodologies: an integrative view. Springer, Heidelberg, pp 537–555
Simon HA (1978) Nobel prize lecture: rational decision-making in business organizations. NobelPrize.org. As of 3 Feb 2014. http://www.nobelprize.org/nobel_prizes/economic-sciences/laureates/1978/simon-lecture.html
Simon HA (1996) The sciences of the artificial, 3rd edn. The MIT Press, Cambridge
Tetlock PE (2005) Expert political judgment: how good is it? How can we know? Princeton University Press, Princeton
Yilmaz L, Oren T (2009) Agent-directed simulation and systems engineering. Weinheim, Germany
Zeigler B (1984) Multifacetted modelling and discrete event simulation. Academic, Ontario
*All RAND publications can be downloaded from www.rand.org. Or, for the author's particular publications. From http://www.rand.org/about/people/d/davis_paul.html#publications

Chapter 14
Simulating Human Social Behaviors

Yu Zhang

14.1 Introduction

While computer simulations are a widely accepted method of research in the natural sciences, they have only begun to gain widespread acceptance among the social sciences. Initial apprehension of the social science community toward computer simulations grew out of a long-held belief that the experimental methodology employed by researchers in the natural sciences would not be a suitable mechanism for understanding social phenomena (Roehner 2007). With little quantitative knowledge on human social interaction, social scientists are eager to use computer code to transform their once textual-only social theories into virtual realities. Sophisticated computer simulations can serve as virtual laboratories to investigate feedback mechanisms, emergence, and the micro- and macrointeractions among agents in artificial societies.

The value of these simulations extends far beyond just proof and discovery (Axelrod 1997). Computer simulations of artificial societies can decompose complex inputs and generate predictions ranging all the way from the level of individual agents to the system as a whole. Simulations of human behavior can be carried out solely for performance reasons in order to mimic human behavior, which could lead to more accurate or optimal results, for example, medical diagnosis. Simulations could also serve as training mechanisms for helping children deal with bullying as well as military personnel or business management by providing dynamic, responsive, and reasonably accurate representations of their human colleagues (Aylett et al. 2004). Simulation of human social behavior can also serve a

Y. Zhang (✉)
Department of Computer Science, College of St. Benedict, St. John's University, Collegeville, MN 56321, USA
e-mail: yzhang@csbsju.edu

L. Yilmaz (ed.), *Concepts and Methodologies for Modeling and Simulation*,
Simulation Foundations, Methods and Applications,
DOI 10.1007/978-3-319-15096-3_14

purely entertainment purpose, as in the case of Will Wright's popular video game The Sims.[1]

While agent-based simulations have been a subject of a great deal of research in recent years, to date there is no framework for describing social agents that captures the uniqueness of human decision-making while remaining applicable across a wide variety of domains. The challenges facing any framework describing agents embedded in a social environment exist in two aspects.

First, such a framework should provide a computational model that neither singly takes the point of view of the individual agent nor the entire society by reconciling the needs, commitments, and goals of individual agents with the behavior of the system as a whole (Castlefranchi 2000). The problem of balancing the microlevel behaviors of the individual agents with the macrolevel behavior of the overall system results in one of two extremes: oversocialization or undersocialization (Castlefranchi 1997). Oversocialization occurs when a framework takes an entirely macro or organizational approach to constructing the social environment; the system is very static, predictability is stunted, and the resemblance to human social systems is tenuous. Likewise, undersocialization occurs when a framework focuses entirely on the microbehaviors of the individual agents by recursively modeling the nested beliefs of other agents leading to a potential explosion in computational complexity and very little resemblance to human social systems (Kim 1999).

Existing approaches to achieving a balance between under- and oversocialization embed agents with a notion of social awareness through two general approaches: external incentives and sanctions that favor group participation or endow agents with prosocial attitudes (Conte et al. 1997). Incentives and sanctions reward or punish an agent for respectively obliging to or deviating from institutionalized social norms and conventions (Hales and Edmonds 2003; Portes and Sensenbrenner 1993). Likewise, prosocial attitudes such as altruism and cooperation can either be acquired at runtime through learning or other socialization behaviors or encoded initially in the design of the model itself (Jiang and Ishida 2007; Parsons and Woolridge 2002). However, these solutions to the micro–macro problem have major disadvantages:

- When modeling a complex human-based social system, the incentives and sanctions that lead to the desired behavior may be difficult if not impossible to identify.
- Furthermore, even if identified for one domain, social norms are not universal across all simulation domains.
- The degree to which social norms are enforced can greatly affect the overall system behavior—too strong and the system is relatively predictable and nonaccidental, too weak and the system is chaotic and unruly.
- Learning prosocial attitudes can be computationally expensive for larger multiagent systems.

[1] http://www.thesims.com

- Prosocial attitudes are one way, representing the influence an agent has on social structures but not the influence those structures exert back onto the individual agents.

In the second aspect, such a framework should incorporate into the individual agent decision model recent developments in cognitive psychology that involve important modifications to the classical concept of a rational decision-maker. These developments, drawn from observed human decision-making patterns, shift decision theory away from a world view where the decision-maker chooses among a set of fixed and known alternatives with known consequences toward a conception of the world in which alternatives are not given and the consequences that will follow are unknown (Simon 1949). While creating heavier, more cognitive agents, this paradigm shift minimizes the work done by individual agents, likewise avoiding any potential performance penalty normally associated with other cognitive architectures such as COGENT and CODAGE (Das and Grecu 2000; Kant and Thiriot 2006). Aside from performance benefits, this new fuller description of decision-making has three distinct theoretical advantages over classical descriptions:

- The model allows for the perception of incomplete and imperfect information that is subject to biases, omissions, and distortions.
- Pseudointuitive inference can be carried out on key pieces of information (anchors) that constitute only a small fraction of available information (accessibility) if an agent is constrained by some external resource (Kahneman and Tversky 1979; Kahneman 2002).
- Deliberative inference utilizes information-gathering mechanisms such as communication to expand an agent's knowledge base, then adopts either a notion of satiation (Stirling 2003) or maximization to reach a decision.

In this chapter, we will introduce innovative mechanisms that allow agents to exhibit social behaviors by balancing their individual wants and needs with the concerns of the entire society while retaining a high level of cognition.

14.2 Dr. Tuncer Ören's Contributions to Human Behavior Simulation

Dr. Tuncer Ören is one of the first researchers who have philosophical thoughts on the mode, scope, and originality of bridging human decision processes and computer simulation. His research in human behavior simulation has pursued a decision theory–centric focus. Through Tuncer's whole career, he investigates a variety of decision-making techniques to meet a diverse set of needs. For example, advances in game theory have carried over as methods for selecting partners across a variety of agent-to-agent interaction patterns, while more traditional economic notions such as expected value and utility have translated into winning strategies for decision-making agents.

One of Tuncer's work done in early 2000 is multimodels and multisimulation (Yilmaz et al. 2006), which is an advanced simulation-based problem-solving environment for social and political scientists to improve their ability to conceive, perceive, and foresee conflicting situations for human behavior simulation. The multimodels and multisimulation theory is based on interpretation of emergent, potentially unforeseen conditions to facilitate dynamic runtime simulation composition and simultaneous experimentation with multiple plausible models. This method explores the problem state space using feasible sequences or stages of models. This enables experimentation with alternative realities, potentially at different levels of resolution. It can also detect relevant and significant situations in a problem domain and therefore lead to interpretation capabilities regarding emergent conditions and causes of observed effects. Finally, observed effects need to be attributed to certain causes within the domain theory of the problem at hand. Such causes need to be appraised against the problem-solving goals and preferences to make recommendations for further, potentially simultaneous exploration of different realities. While this scheme can be characterized as forward multisimulation, this work also nicely examines the possibility of backtracking and replaying situated simulation histories with altered conditions as well as futures generated before exploring alternative realities.

Perceptions—including anticipations— are subjective and are prone to biases and influences. Some biases may stem from lack of relevant knowledge; others may be induced by others by influencing decisions. Tuncer's group uses fuzzy logic to simulate them properly (Ören and Yilmaz 2004; Ghasem-Aghaee and Ören TI 2004). This problem is hard because there is a wide range of a base for persuasion such as reciprocation, consistency, social validation, liking, authority, and scarcity. But despite the inherent difficulty of the problem, several researchers have pursued a line of research that can be roughly grouped under the title Socially Rational Decision-Making. The primary goal of this research is to develop a fuzzy agent–based decision model that produces decisions that are inherently rational from the individual perspective yet retain that property of rationality on upward toward the level of the entire system. Traditionally, this has been achieved by making an individual agent's autonomy subordinate to the needs and desires of the overall system. This kind of the top-down approach fails to exploit the inherent bottom-up and emergent properties that characterize any multiagent system. To retain the autonomy of individual agents, this research advocates classical decision-theoretic approaches by encoding social considerations into the utility functions of individual agents.

Cognitive complexity is an important factor in decision-making in problem solving. Seck et al. (2005) study human cognitive abilities in order to understand and test the mechanisms of several aspects of cognition to be able to incorporate them in simulation studies. They foresee two types of use: (1) enhance simulation studies and contribute to the advancement of the methodology and technology of cognitive simulation and (2) use cognitive simulation to test hypotheses about human cognition. Ören elaborated on the importance of increasing cognitive complexity of an individual to increase his/her effectiveness in coping with complex situations. This paper aims at the cognitive ability under stress and fatigue. Stress and fatigue can interfere dynamically in behavior in performance and decision-making (both variables can change within the course of a task). They distinguish on certain tasks the performance difference between high–cognitive complexity people and low–cognitive complexity people. A first distinction might be done concerning the time necessary to finish successfully a cognitive task; a second one can be made concerning decision-making as high–cognitive complexity people are known to be more fluent in ideas and more creative and thus generally find the best solution. To do so, each task of the DEVS atomic behavioral model will contain a variable representing the task's cognitive complexity. Different individuals with different personalities, the openness trait in particular, will have different performances in terms of both time and decision-making.

The ability to understand the emotions of others is critical for successful interactions among humans. Kazemifard et al. (2011) presented a framework for emotion understanding to enable intelligent agents to improve their emotional intelligence when interacting with other agents. This framework builds on a paradigm of machine understanding. It includes (1) a metamodel, (2) an analyzer, (3) an evaluator, and (4) a memory modulator. The metamodel consists of episodic memory and three versions of semantic memory, semantic graphs, a general semantic graph, and a lookup table of general information about emotions. The analyzer is a perceptual categorization mechanism. The evaluator consists of an interpreter that provides an understanding of the perceived agent (analyzer output) with respect to the contents of the different kinds of memory (the metamodel). The memory modulator updates episodic memory and semantic graphs. This paper addresses one of the major themes in individual decision-making, bounded rationality (Tisdell 1996; Simon 1957; Kahneman 2003), in an expanded social setting. Agents are bound by the amount of time and resources they can commit toward resolving a balance between their wants and needs and those of the entire system. The emotional bound in this paper is able to dynamically change as more resources become available to an agent allowing them to devote an increased amount of time and effort toward social considerations. If resources are scarce, an agent may opt to make a socially nonoptimal yet computationally cheap decision over one that is more computationally expensive and more aligned with the prevailing social norms at the time.

The major novel contribution of Tuncer's research to human behavior simulation is the formalization to modeling and simulation from theory to practice. He built up the conceptual foundations of a new exploratory multisimulation methodology with dynamic models and simulation. This solution presents an advanced problem-solving environment for social and political scientists to observe and examine the implications and plausible outcomes of decisions in conflict. He also contributes to individual agents' decision-making by quantitatively measuring the effect of an agent's action (based on the agent's personality and emotion-understanding ability) on the needs of other agents relative to its own. Tuncer's model stands out in that it retains an individual's preference or indifference between two alternatives from not only its personal perspective but from its societal standpoint as well. This model falls short in providing a direct mechanism for agents to influence the decisions and subsequent actions of others; rather, agents are left to passively infer new beliefs and desires from their understanding of the needs of other agents.

14.3 CASE: Cognitive Agents for Social Environments

This section introduces CASE, a multiagent architecture that is efficient and scalable in simulating large-scale social systems.

14.3.1 System Overview

Social behaviors are behaviors that are solely oriented toward another agent. Such behaviors consider the intention behind another agent's expression, create expectations about another agent's actions, and aim to evoke a distinguishable response from another agent (Rummel 1976). Social interaction occurs when the social behaviors of two or more agents are mutually oriented toward one another.

Most social interactions can be differentiated according to Weber (1947) into the following three categories:

- Accidental. This class of interactions is often not planned by either party in advance and rarely repeated with the same members. However, in rare instances, this initial unplanned contact between agents has the potential to develop into one of the other three more temporally permanent classes of social interactions. Example: A waiter asking a table of customers for their order.
- Repeated. Similar to accidental interactions, these are not planned meetings between two agents but likely to occur on a frequent basis because of spatial proximity, shared interests, or similar habits. Example: Coworkers sharing small talk over the water cooler.
- Regulated. These interactions are planned and tightly controlled by the laws, customs, norms, or other enforcement mechanisms put in place by members of the society. Example: Attendance at an employee staff meeting or visiting a courthouse for jury duty.

In all its forms, social interaction carries with it some degree of influence on the behavior of the agents involved. While sociologists differentiate between several types of social influence, namely, peer pressure, charisma, connections, force, and reputation (Cialdini 2001), CASE agents only concern themselves with the social structure through which the interaction they are currently experiencing occurs.

These structures represent a relatively stable and enduring pattern of shared relationships among agents within the society. Each structure subdivides the entire society of agents into interrelated sets where member agents share a common function, meaning, and/or purpose (Porpora 1989).

The likelihood an agent will respond to social influence or social impact of an agent, however, is intimately tied to the following dimensions (Tanford and Penrod 1984) of the social structure through which the agents are interacting:

- Strength. How important are the other agents who are attempting to influence you?
- Immediacy. How close to you, in either geographic or social space, are these agents?
- Number. How many agents are exerting this influence upon you?

Agents always respond to the influence of another agent by altering their perception of their relationship to the influencer, other agents, or society in general. This alteration in perception ultimately affects future decisions and behaviors of that agent. Latane and Darley (1970) generalized these principles to state that the more agents that were interacting within a social structure, the more influence each individual agent will have. However, while the impact of individual agents may grow as new agents are added, the rate of growth actually shrinks inversely to the number of agents. In addition to the rate of growth, the amount of influence any individual agent can exert shrinks inversely proportional to the number of agents.

To achieve such ends, many researchers have attempted to grow in silico fundamental social structures and group behaviors. Their primary aim is to identify the local or microinteractions among agents that are sufficient to generate the desired macroscopic behaviors and collective patterns they desire (Epstein and Axtell 1996). However, while providing a good computational model that takes into consideration both the individual and social behaviors of autonomous agents, it is hardly efficient, scalable, or robust.

The difficulty exists in modeling the system by holding both the societal view and the individual agent view simultaneously. The societal view involves the careful design of agent-to-agent interactions so that an individual agent's choices influence and are influenced by the choices made by others within the society. A stark contrast to the agent view involves only modeling the individual decision-making processes. While the single societal view mainly concentrates on the centralist, static approach to organizational design and specification of social structures and hence limits system dynamics, on the other hand, the single-agent view focuses solely on modeling the nested beliefs of the other agents and suffers from an explosion in computational complexity as the number of agents in the system grows.

Motivated by these observations, the interactions among CASE agents are embedded CASE agents in three social structures: group, which represents social connections; neighborhood, which represents space connections; and a social network, which spans social and space categories. These three structures reproduce the way information and social strategy is passed and therefore the way people influence each other. In our view, social structures are external to an individual agent and independent from its goals. However, they constrain the individual's commitment to goals and choices and contribute to the stability, predictability, and manageability of the system as a whole.

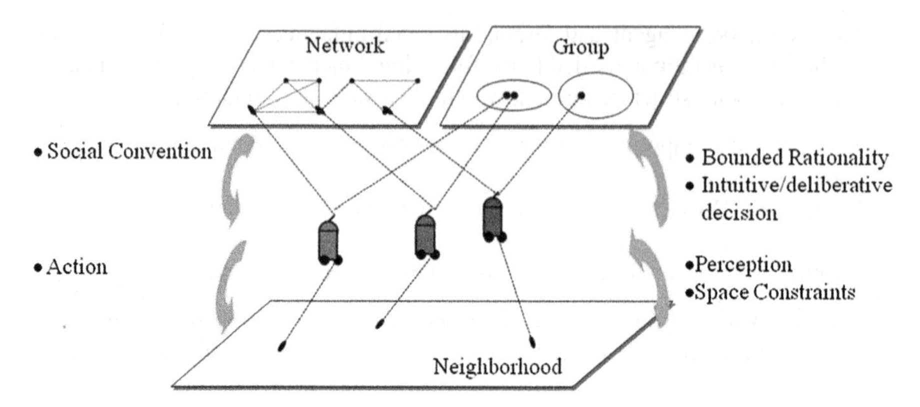

Fig. 14.1 Social realms for the CASE agent

We take up the classification proposed by Ferber (1999) that multiagent systems are an agent/society duality. There are two levels of organization in multiagent systems, which are illustrated in Fig. 14.1:

- The microagent level, which is in essence represented by the interactions between agents. There are three common types of interaction: cooperation, competition, and negotiation. Agents interact with each other through two ways: its sphere of influence in the environment and direct communication to other agents.
- The macrosociety level is represented by the dynamics of agents together with the general structure of the system and its evolution. Our work focuses on the mesolevel of the agent/society duality. Any society is the result of an interaction between agents, and the behavior of the agents is constrained by the assembly of societal structures. For this reason, a society is not necessarily a static structure, that is, an entity with predefined characteristics and actions.

14.3.2 Groups

A group is usually defined as a collection of agents who share certain characteristics, interact with one another, accept expectations and obligations as members of the group, and share a common identity (Sherif and Sherif 1948). Interactions within a group fall under Weber's regulated category as interactions within a group are tightly controlled by a communally established set of social enforcement mechanisms. A group differs from a mere aggregate of agents in that a group exhibits a sense of cohesiveness and stability through time. Groups may be formed on the basis of intimate relationships or more formal and institutional means. All agents maintain the concept of a reference group, i.e., if I am an A, then I am definitely not a B or a C. Indeed, it is by creating these disassociations with others in society that agents categorize, identify, and compare themselves with other agents by joining groups with whom they share commonalities.

CASE agents interact with other agents in their group with respect to the classical definition of the function and formation of a group as defined by Muzafer Sherif (1955):

- A common set of motives and goals
- An accepted division of labor, i.e., roles
- Established status (social rank) relationships
- An accepted set of social norms and values
- The development of accepted sanctions if and when social norms were respected or violated

Hence, CASE agents that share a similar preference for a class of decision problems form groups to reinforce their goals and objectives by diffusing their decision-making preferences to other agents. Each group maintains its own separate preference that is formulated based on a composite of its members' preferences as an analogue to that group's accepted set of social norms and values.

14.3.3 Neighborhood

An agent's neighborhood is a geographically localized community located within the environment and is comprised of all agents whose spatial location falls within some predefined distance of its own. Here, interactions are typically accidental in nature as an agent's neighborhood is subject to change as that agent moves through the environment. The size neighborhood of a CASE agent is directly related to the observation capabilities of the agent. The more an agent is able to observe, the larger its neighborhood will be. As an agent's neighborhood grows, so does the number of agents that are likely to influence it; however, in keeping with Latane and Darley's (1970) findings, the individual impact of each of its neighbors decreases relative to the size the entire neighborhood.

14.3.4 Social Network

A social network is a social structure made of nodes, here agents that are tied by one or more specific types of interdependency. The social network CASE agents utilize ties them together based on their communication patterns. This type of interaction is not frequently regulated like the interactions within a group are but typically are repeated on a regular basis between a small subset of agents. An agent's social network serves as a medium through which agents actively disseminate information and influence to other agents through explicit communicative acts.

14.3.5 *Varying the Sociability of Individual Decision-Making*

Let a given agent in the population be denoted as a, where A denotes the set of all agents and $a \in A$. Each agent has a social strategy. This social strategy can be either ordinal or cardinal. We denote the social strategy for agent a by S_a.

Let a given group in the population be denoted as g, where G is the set of all groups and $g \in G$. Groups are formulated on the basis of a common preference. Each agent identifies itself with any group such that the agent's preference falls within some threshold of the group's preference.

$$\forall a \in A \text{ and } g \in G,\ a \in g \text{ if } \operatorname{diff}\left(S_a,\ S_g\right) < d \tag{14.1}$$

where diff(S_a, S_g) is the difference between the agent's strategy S_a and the group's strategy S_g and d is the threshold. It can be seen that agent a can belong to more than one group at a time and can belong to different groups over time.

When an agent joins a group, it is given a rank in that group. An agent will have one rank for every group it belongs to. The agent's rank can be evaluated based on the agent's importance, credibility, popularity, etc. It defines how much the agent will influence the group as well as how much the group will influence the agent. A high-ranking agent influences the group and therefore its members more than a low-ranking agent and at the same time is influenced more than a low-ranking agent. An agent's rank is specific to the domain and may change over time. At each time step, every group will update its strategy. The update is determined by its members' strategy and the percentage of the total group rank they hold. At each time step, every group will update their strategy.

$$S_g = \sum_{a \in g} S_a \times \frac{R_a^g}{\sum_{b \in g} R_b^g} \tag{14.2}$$

where R_a^g denotes agent a's group rank. This allows for groups to be completely dynamic because both their members and their strategy can change at each time step. Just like the rank an agent holds in groups, an agent also has a rank in its neighborhood and network. Each agent keeps track of the agents in its neighborhood and the agents it communicates with. Every time an agent observes another agent in its neighborhood, that agent's neighborhood rank will increase. Also, each time an agent communicates with another agent, that agent's communication rank increases. Therefore, every agent will have a rank value for every agent it interacts with and a separate rank for every agent it communicates with. When an agent updates its strategy, it will take into account these ranks. Agents with a high rank relative to the other agents will have a stronger influence. Therefore, the longer two agents are near each other, the more they will influence each other. The same is true for communications. Below is the update function for the neighborhoods strategy and the networks strategy:

$$S_n = \sum_{a \in n} S_a \times \frac{R_a^n}{\sum_{b \in n} R_b^n} \tag{14.3}$$

$$S_w = \sum_{a \in w} S_a \times \frac{R_a^w}{\sum_{b \in w} R_b^w} \tag{14.4}$$

where S_n is the strategy for neighborhood n, S_w is the strategy for network w, R_a^n is agent a's neighborhood rank, and R_a^w is agent a's network rank.

At each time step, every agent also updates their strategy. An agent's update function is defined as

$$S_a^{'} = \alpha \times S_a + \beta \times S_g + \gamma \times S_n + \lambda \times S_w \tag{14.5}$$

where α, β, γ, and $\lambda \in [0, 1]$ and $\alpha + \beta + \gamma + \lambda = 1$. These values represent what percentage of influence the agent takes from itself, its group, its neighborhood, and its network. They allow for multiple agent types. For example, (1, 0, 0, 0) represents a selfish agent because it cares nothing about the whole society, and (0, 0.33, 0.33, 0.34) represents a selfless agent who cares about the three social structures equally.

14.3.6 The Psychophysics of Individual-Agent Decision-Making

Traditionally, the design of intelligent agents has centered around the common abstract notion of an agent execution cycle. This structure serves as a high-level map for the internal components of any agent-based system. This relates not only the data structures that comprise an agent's knowledge about the environment but the algorithms that act on and control that flows between these structures. In a vast majority of cases, agent architectures differ only by the data structures and algorithms they choose to utilize. Figure 14.2 illustrates this cycle graphically, with details about each of the five major steps listed as follows:

- Observation. This step collects information on current environmental conditions and maps those conditions to precepts. It is important to note that this step is absolutely domain dependent and limited in its scope by its implementation. For example, if this model were to be implemented within some sort of robotic system that utilizes a video camera for input, then the agent's observation step would be limited in the amount and types of information it could take in as sensory input.
- Updating KB (Knowledge Base). An agent's knowledge base will be updated under two cases: (1) when the agent observes the environment, it will assert new

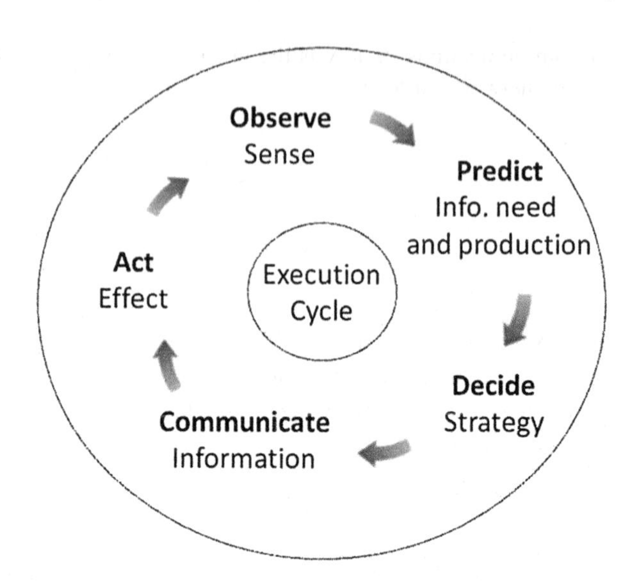

Fig. 14.2 Traditional agent execution cycle

percepts to the knowledge base; (2) when the agent performs an action, it will assert the effects of the action to the knowledge base. For both cases, the function update must check the entire knowledge base for inconsistencies.

- Decision. Here, agents make two separate decisions: (1) what act to perform and (2) what message to communicate and to whom.
- Communication. In general, intelligent agents working within a multiagent environment cannot force other agents to perform a specific action or directly alter their internal state. However, they can exert influence over other agents through communicative actions. Multiagent researchers have built upon John Searle's speech-act theory (Sherif and Sherif 1948) to develop a number of formal languages and ontologies such as FIPA-ACL and KQML (Labrou et al. 1999) so intelligent agents can understand one another.
- Action. The functional nature of an agent's action step is rather intuitive and simple; its purpose is to ensure a successful, coherent, and fault-proof execution of the optimal action that was recommended by the agent's decision-making mechanism. No real further explanation of act is necessary as this function is highly dependent on the implementation.

14.3.6.1 CASE Agent Execution Cycle

Kahneman and Tversky (1979) suggest a two-phase decision model for descriptive decision-making (see Fig. 14.3): an early phase of editing and a subsequent phase of evaluation. In the editing phase, the decision-maker constructs a representation of the acts, contingencies, and outcomes that are relevant to the decision. In the

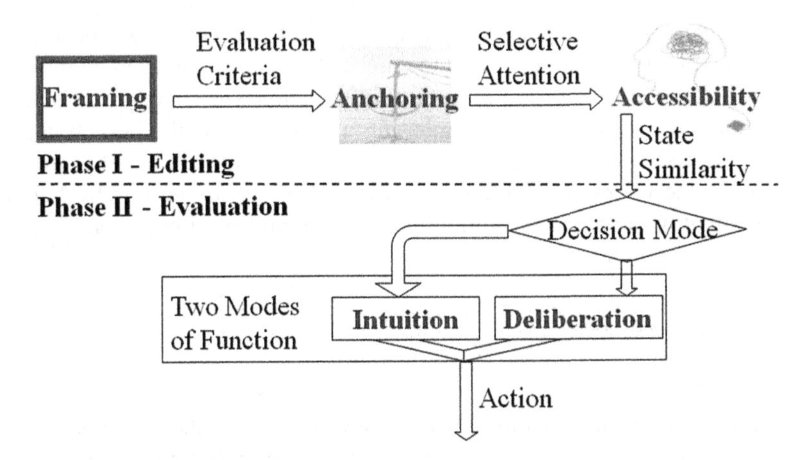

Fig. 14.3 Two phase decision-making process

evaluation phase, the agent assesses the value of each alternative and chooses the alternative of highest value. Our decision model incorporates their idea and specifies it by the following five mechanisms:

14.3.6.2 Editing

- Framing: the agent frames an outcome or transaction in its mind and the utility it expects to receive.
- Anchoring: the agent's tendency to overly or heavily rely on one trait or piece of information when making decisions.
- Accessibility: the importance of a fact within an agent's selective attention.

14.3.6.3 Evaluation

- Two modes of cognitive function: intuition and deliberation.
- Satisfying theory: the goal is no longer optimality, and decisions are accepted when they are good enough.

14.3.6.4 Editing Phase

One important feature of the descriptive model is that it is reference based. This notion grew out of another central notion called framing where agents subjectively frame an outcome or transaction in their minds and the utility they expect to receive is thus affected. This closely patterns the manner in which humans make rational decisions under conditions of uncertainty. CASE agents frame their current situational context by forming an attitude or weight, w, toward one class of decisions or outcomes.

Framing can lead to another phenomenon referred to as anchoring. Anchoring or focalism is a psychological term used to describe the human tendency to overly or heavily rely (anchor) on one trait or piece of information when making decisions. A classic example would be a man purchasing a used automobile; he may tend to anchor his decision on the odometer reading and year of the car rather than the condition of the engine or make of the car. CASE agents anchor by building selective attention on relevant information. The salience of information i is determined by

$$\Delta_i = \frac{\sum_c i \text{ is used}}{\mathrm{Card}(I \text{ is used})}, \quad i \in I \tag{14.6}$$

where Δ_i is the frequency that information i was used under the context c.

If the salience of i is higher than the threshold, i becomes the anchored information:

$$I^* = \{ i \mid \Delta_i > \text{threshold} \} \tag{14.7}$$

Accessibility is the ease with which particular aspects and elements of a situation, the different objects in a scene, and the different attributes of an object come to mind. As it is used here, the concept of accessibility subsumes the notions of stimulus salience, selective attention, and response activation or priming. CASE agents determine the similarity between states only with I^* establishing the relation

$$S_t \sim S_m \text{ if } d_{c,I^*}(S_t, S_m) < D \tag{14.8}$$

where S_t is the current state, S_m is a state in the agent's memory, and $d_{c,I^*}(S_t, S_m)$ is the distance between S_t and S_m regarding all anchored information I^* under the context c. Those states that are most similar to the current one are said to be more accessible than others.

14.3.6.5 Evaluation Phase

In the evaluation phase, there exist two modes of cognitive function: an intuitive mode, in which decisions are made automatically and rapidly, and a deliberative mode, which is effortful and slower. The operations of the intuition function are fast, effortless, associative, and difficult to control or modify, while the operations of the deliberation function are slower, serial, and controlled; they are also relatively flexible and potentially rule governed. Intuitive decisions occupy a position between the automatic operations of perception and the deliberate operations of reasoning. Intuitions are thoughts and preferences that come to mind quickly and without much reflection. In psychology, intuition can encompass the ability to know valid solutions to problems and decision-making.

Our technical solution to achieve this behavior is that if S_t, the current state, is close to a state in memory, S_m, then the optimal policy $\pi^*(S_t)$ and $\pi^*(S_m)$ should be close as well. Hence, the agent uses an optimal policy that it has employed before in a similar state and updates its state memory by adding the current state. If the policy the agent employed was successful, then the reward associated with that policy and its accessibility will be increased. The slower, serial, and controlled process of deliberation determines the state similarity across all information available to the agent, not just that which is anchored, I^*. Traversing its memory, an agent attempts to reoptimize a previously used policy stored in memory:

$$\pi^*(S_m) = \text{argmax}_x E\left[\sum_{i=0}^{\infty} \gamma^i w R(S_i) | \pi\right], \ 0 < \gamma < 1 \qquad (14.9)$$

where r is the time discount factor and $R(S_i)$ is the reward an agent receives when it arrives at state S_i.

In keeping with the notions of satisficing theory under their intuitive mode, CASE agents do not compute an optimal policy to use in the current S_t if there is a state in the agent's memory S_m that is similar and the policy utilized under that state can be used once again.

14.3.7 *Experiments and Results*

We tested the CASE architecture and its new decision-making mechanism within a number of domains ranging from the classic prisoner's dilemma to an artificial stock market as well as initial work on such real-world applications as the subprime lending crisis.

14.3.7.1 An Extended Prisoner's Dilemma: Investigating Intuitive Attitudes Toward Risk

We choose an extension of the classical prisoner's dilemma as the domain for our initial experiment. In the classical prisoner's dilemma, a game comprises two agents: A and B. Each agent is given the option to either cooperate with or defect from its opponent with various outcomes for each choice.

We extend this classical prisoner's dilemma in two distinct ways. First, outcomes are cumulated in our domain. In the classical dilemma, the outcomes of each game are not cumulative. Even in an iterated prisoner's dilemma scenario, the outcomes of a previous game have no effect on an agent's decision in subsequent games. The only outside factor that influences an agent's decision in an iterated prisoner's dilemma is an agent's knowledge of what action(s) his opponent has taken in the past. The change to cumulative outcomes allows agents to assign value to gains and losses rather than final assets. Since the current asset (prisoner sentence) of an agent serves as a reference point for subsequent decisions, the cumulative value of an agent's assets can have a tremendous effect on that agent's later performance. Second, while in the classical prisoner's dilemma, the four outcomes are fixed, here we allow them to be uncertain.

Our experiment involved a total number of 2,000 agents within either one or two societies. Each agent plays over 500 iterations. At each iteration, we randomly paired agents to play a prisoner's dilemma game. After each game, the assets of each agent will be changed to reflect the outcome of the game (gains or losses). This outcome was then used by the agents to the next iteration. At the start of each experiment, each agent was assigned a small positive number to represent its beginning asset position. In some of the experiments, we arbitrarily chose this number to force groups of agents into either initially risk-seeking or risk-averse attitudes. At other times, we allowed this number to be randomly generated to create a heterogeneous distribution of both risk-seeking and risk-averse agents.

Figure 14.4 shows the average asset position of all the agents in the experiment. The two upwardly curving lines reflect the two possible rewards each agent could receive for either cooperating (left line) or defecting (right line). There is an evident shift in the concentration of agents from one decision choice to another as time and assets progress. This is reflective of the fact that as the agent's overall assets increase, its individual behavior becomes increasingly risk averse. The increased density of points toward the upper end of the line reflects the congregation of agents

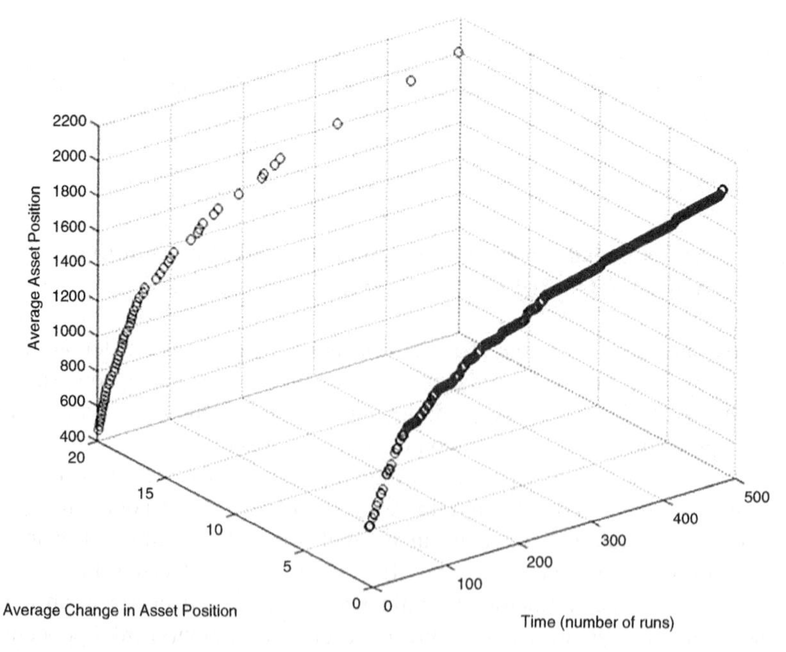

Fig. 14.4 The performance of agents assets

around a single risk averse decision. This result demonstrates that a few minor and conscientious alterations to individual agent decision processes are more than sufficient to create the attitudes toward risk that characterize observed human decision patterns.

14.3.7.2 An Artificial Stock Market: Evaluating the Performance of Intuitive and Deliberative Decisions

Our initial experiments involving the prisoner's dilemma only investigated a small portion of the entire CASE agent functionality and explored only a distinct subset of prescribed human behavior. Here, we aim to examine in detail the effectiveness and role of the mechanisms underlying an individual agent's two-phase decision process, most notably the two cognitive modes of intuition and deliberation. Twenty thousand agents were selected from among 30 unique stock indices for a time frame of 25 rounds. Every agent began the simulation with initial 10,000 cash, and no limitations were set on the amount of stock it could purchase each round as long as they had cash available to make a desired purchase. Stock prices changed each round based on traditional microeconomic supply and demand curves that accounted for the volume of buying and selling that occurred in the previous round. The more shares of a stock were purchased, indicative of a higher demand for that stock and a dwindling supply, the higher the price was driven up and vice versa. Agents bought and sold stock only to the market and did not engage in interagent purchases, sales, or trades for simplicity purposes.

Agents used the two-phase decision-making process, first editing the decision space by selecting only 10 stock indices from among the available 30 to serve as anchors each round. These anchor stocks could change from round to round and are selected as basis for predicting the overall market behavior. Anchors that do not seem to reflect observed market behavior are discarded at the end of each round, and new ones are added. However, each agent only keeps exactly 10 anchor stock indices at each time step. The second phase of the decision process utilizes the two modes, intuition and deliberation. In the intuitive mode, the 10 anchors are utilized to predict, by way of a simple polynomial fit, the expected behavior of each anchor stock index and likewise the predicted behavior of the overall market in the next round. A downturn in the overall market would signal the CASE agents to begin selling off their low-performing stocks, while an upturn would signal the need to purchase stocks on the rise. If half or more of the chosen anchors are the same stocks that the agent is holding, stock holdings that match current anchors are bought and sold, and no action occurs to holdings that do not match a current anchor. Otherwise, more information must be gathered and the deliberation process started to determine either buying new stocks or selling an agent's current holdings. This is done by computing the distance between several random points on the anchor's price function and a selected holding's price function. The anchor with the smallest distance was chosen to be representative of that particular holding.

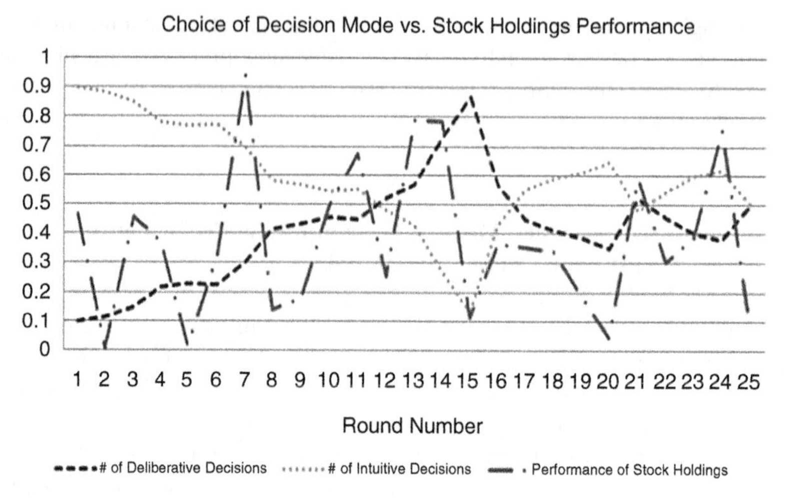

Fig. 14.5 Intuitive or deliberative decision vs. stock holding performance

Figure 14.5 illustrates the CASE agent's choice of cognitive function (i.e., intuitiveor deliberative) relative to their stock-holding performance. Here, we see a clear visual correlation between the number of agents utilizing the intuitive mode and positive performance in the CASE agent's stock holdings. This reflects the crucial role information plays in the simulation. The better the information the CASE agents have about their environment which is reflected in their choice of anchors, the better they are able to predict both positive and negative fluctuations in stock price and likewise react to those anticipated changes. A rise in the number of deliberative agents and slump in stock-holding performance can be explained in terms of information as well. Here, the reactive agents have either bought or sold a large point of stock changing the behavior of the overall system. Hence, the CASE agent's anchors no longer serve as a good predictor of overall market performance. This loss of good information on the part of the CASE agents results in a temporary downturn in their performance until the next round when new anchors can be chosen that better reflect the newly altered reality.

Investigating the performance of the CASE agents, alone is certainly not enough to validate the superior performance of the two-phase decision process. Likewise, Figs. 14.6 and 14.7 draw from a separate experiment run over 50 time steps (twice the length of the original) in which the performance of CASE (indicated by the lighter pink line) and a set of agents employing a classical decision-theoretic approach (indicated by the darker blue line) were compared.

In Fig. 14.6, when no limits are placed on how many shares of each stock are available for purchase, agents are absolutely guaranteed that they can purchase shares of any stock bearing in mind that they have sufficient funds to do so. This environmental characteristic essentially devalues the major competitive advantage of decision-making speed that CASE agents hold. Even in these shallow decision problems where complexity and available information are low as indicated by

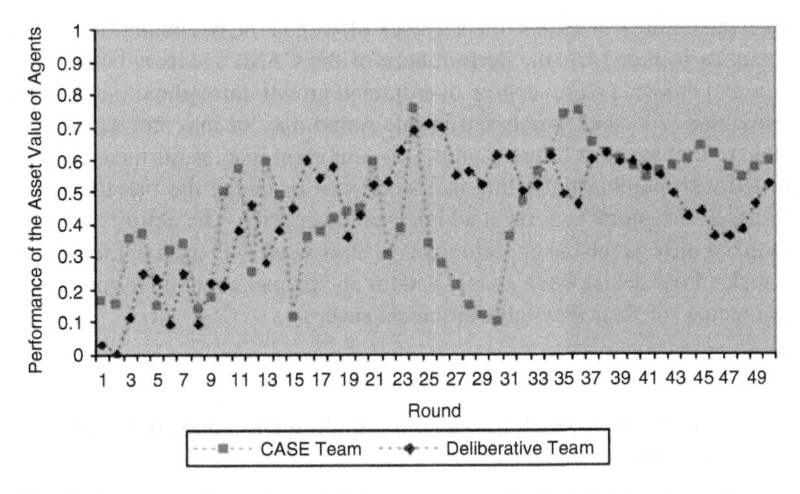

Fig. 14.6 Performance of two-phase decision process vs. classical decision-theoretic approach: no limits placed on stock volume

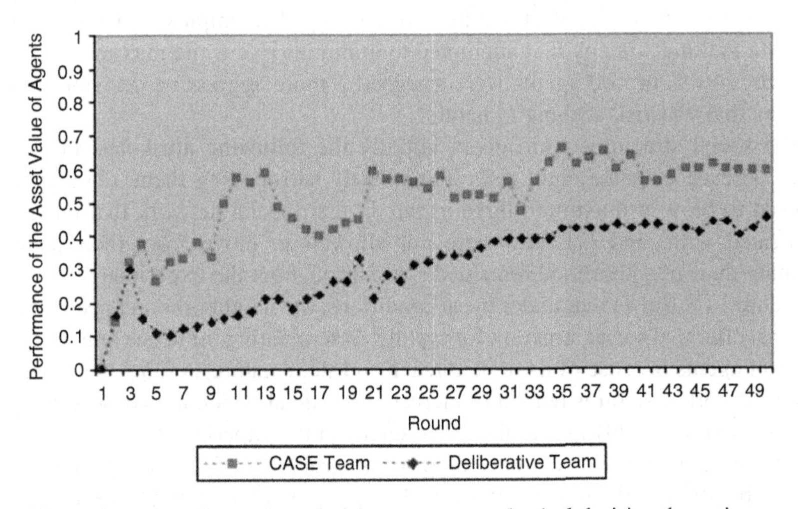

Fig. 14.7 Performance of two-phase decision process vs. classical decision-theoretic approach: limits placed on stock volume

Fig. 14.6, our CASE agents remain competitive with the classical decision-theoretic frameworks traditionally employed by agent-based researchers. What is most apparent from Fig. 14.6 is that even in areas when the CASE agent's performance drops below that of the classical decision-theoretic agents, their ability to return to a decision strategy that yields more optimal results is remarkably fast and usually within approximately five rounds of the simulation.

When the number of shares of each stock offered at the beginning of each round is limited, as in Fig. 14.7, the performance of the CASE agents is both markedly superior and enjoys a slight degree of sustained growth throughout the duration of the simulation. As stock purchased in one round may or may not necessarily be available to that agent in future rounds, it is important that agents measure the cost associated with purchasing/selling that stock now or taking the risk to potentially purchase/sell that stock later on at a higher or lower price. The ability of the CASE agents to not only gauge the opportunity cost associated with each of their decisions but to make those decisions in a rapid and timely manner using their intuition is the integral recipe for their inevitable sustained success.

14.3.7.3 An Artificial Stock Market: Evaluating Diffusion by Social Structures

To measure the influence of the three social structures we developed on the individual agent decision-making process, a 100×100 grid-based environment was created, and 1,000 agents were randomly dispersed across it. Sixty percent or 600 agents were assigned at random to a group that employed a conservative decision-making strategy that attempted to minimize risk while maximizing profit. Likewise, 40 % or 400 agents were assigned a more aggressive decision-making strategy that was risk seeking in nature.

The social structures were given initially the following attributes: (1) agents could observe only the eight cells immediately surrounding them, (2) they were allowed to have at maximum three agents in their social network that they communicated with, and (3) they were not allowed to move from their location, meaning their neighborhood remained static throughout the experiment

Figure 14.8 shows that under these conditions, the neighborhood appeared to be the least effective social structure for rapidly disseminating influence among a large group of agents because of its limited reach and static nature. Adding the group and social network structures tended to increase the rate at which the conservative and successful strategy diffused to the other agents in the experiment.

This pattern continues until all three social structures are in use, at which point the combination of the neighborhood and social network significantly outperforms the combination of all three. While initially puzzling, this result is indicative of the very nature of the group social structure. To maintain consistency with conventional sociological conceptions of a group, an agents group serves as a composite of influence its members receive through their neighborhood and network. This allows the group to serve as an important medium for widely broadcasting influence nondiscriminately to a number of agents not bound by any social or spatial context. The group also serves as a mechanism for resisting or smoothing rapid and sharp changes occurring in the underlying social structures. In a very limited sense, we can say that CASE agents through their groups not only maintain a sense of identity or commitment to a certain ideology (here taken to be aggressive or conservative)

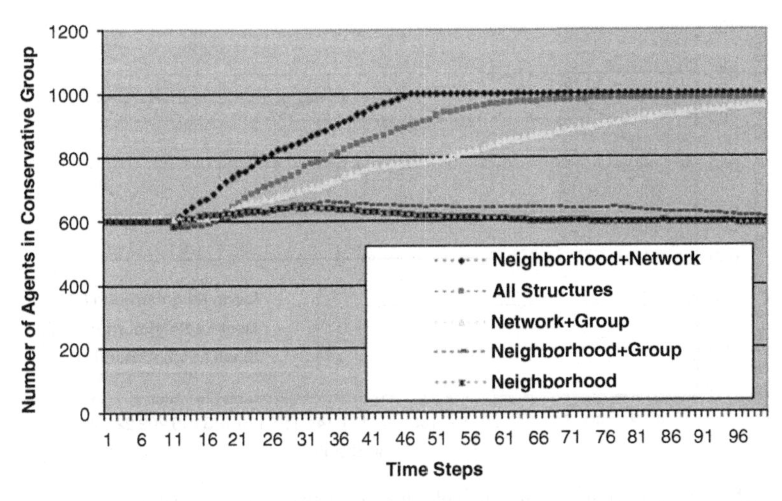

Fig. 14.8 Social structure diffusion rate: small neighborhood + network size: 3 + no walk

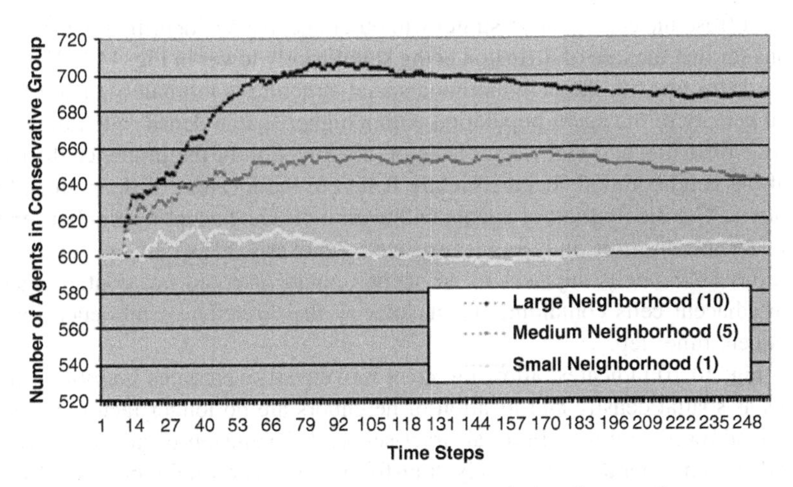

Fig. 14.9 Small, medium and large neighborhoods + network size: 3 + no walk

but actively try to maintain and propagate that sense of connection to other agents in a fashion that mimics observed human social behaviors.

A more thorough examination of the relationship between an agent's social network and neighborhood was carried out by extending the previous experiment along the following lines: (1) the duration was increased to 300 time steps to ensure adequate time for the diffusion rate to stabilize, (2) the size of both structures was varied along with the ability of the agents to move.

Figures 14.9 and 14.10 indicate a strong relationship between the movement of agents in the environment (no walk/walk) and the rate at which the neighborhood is

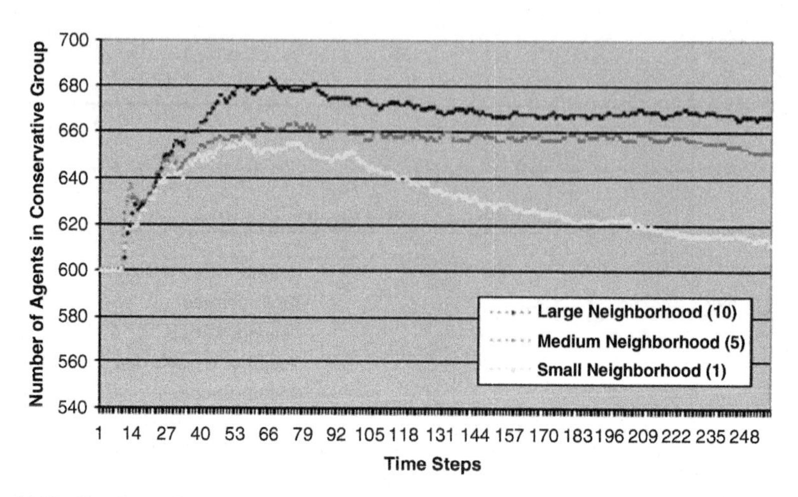

Fig. 14.10 Small, medium and large neighborhoods + network size: 3 + walk

able to diffuse the conservative strategy to other agents. We identified two primary reasons for that the rate of diffusion being significantly lower in Fig. 14.9. First, the rate of diffusion with the neighborhood social structure is intimately linked to the spatial density of the agent population with a higher spatial density yielding rapid, effective diffusion and vice versa. Second, the direction of the influence diffusing out of the neighborhood social structure is tied to the location of their immediate neighbors. The distribution of agents within an individual agent's neighborhood is by no means uniform and could very well be overly heavy in one or several directions as the cells adjacent to an agent's could or could not contain agents. Those adjacent cells containing agents specify the direction of influence for the subsequent time step.

As Fig. 14.10 indicates, allowing agent movement overcomes both these limitations as spatial density and location of neighbors are no longer factors when an agent is allowed to move. In an abstract sense, the inclusion of agent movement around the environment effectively transforms an agent's neighborhood from a static to dynamic entity. As Figs. 14.9 and 14.10 illustrate, this move to dynamism is also a dramatic move toward an increased rate of diffusion.

In contrast to the neighborhood, as Figs. 14.11 and 14.12 demonstrate, an agent's social network is seemingly unaffected by the spatial density and movement of the agent population as it exists outside the boundaries of physical space. However, a direct correlation does exist between the number of agents within an individual agent's network and the rate at which and/or degree of influence it can exert on those agents.

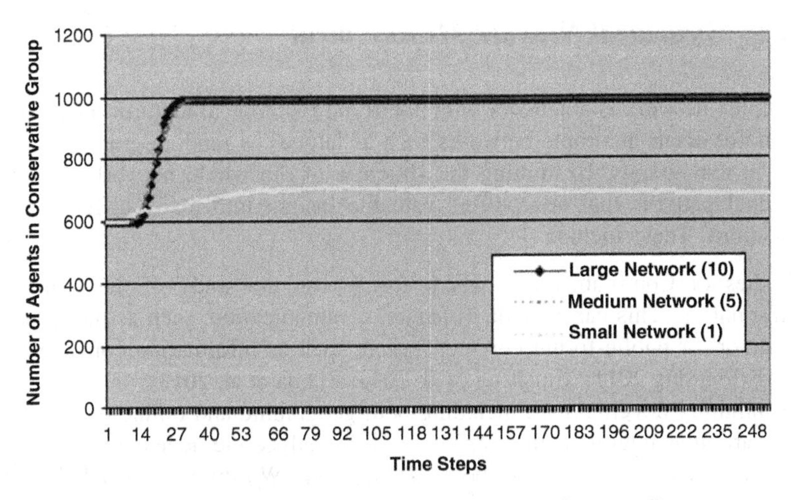

Fig. 14.11 Small neighborhood + small, medium and large network + no walk

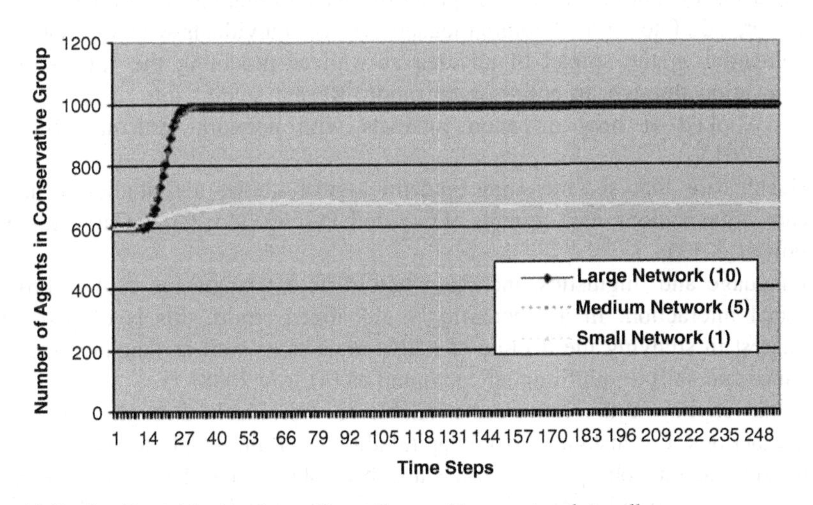

Fig. 14.12 Small neighborhood + small, medium and large network + walk

14.4 Grand Challenges on Simulating Human Social Behaviors

Human social behaviors are directed toward society. Therefore, these behaviors are influenced by the interactions with other people in the society. At the same time, human behaviors are also influenced by culture, attitudes, emotions, values, ethics, authority, rapport, hypnosis, persuasion, coercion, etc. Due to the paper length, we focus the grand challenges on how people interact with each other in social networks to maintain their relationships. In the following sections, we will discuss the challenges in complex social networks, temporal patterns, and network randomness.

14.4.1 Structural Network Measurement

A complex network is a network with nontrivial topological features, i.e., features that do not occur in simple networks such as lattices or random graphs but often occur in real graphs. Examining the structure of the whole network as well as individual patterns that arise offers valuable insight into many different social applications. These include

- Studies of Communication, which focuses on the study of the transfer of information. This can include in-person communication, such as the spread of a rumor, or public-forum communication, such as information conveyed on a blog (Fleming 2011; Minsheng et al. 2013; Zhoua et al. 2013)
- Community development, including both geographic and online communities. Of particular interest is developing tools to analyze the development of social media networks such as Facebook, Twitter, and Wordpress (Lapachelle 2011; Zhoua et al. 2013)
- Diffusion of innovations or the spread of ideas throughout a community. This can include finding the "opinion leaders" or the individuals who are especially influential in the spread of an idea as well as modeling the spread of an innovation through an entire organization. Recent studies into diffusion have also looked at how diffusion interacts with network structure (Stattner et al. 2013)
- Health care analysis, including epidemiological studies and studies of health care organizations and systems (Levy and Pescosolido 2002; Christakis and Fowler 2013)
- Language and linguistics, including how different languages evolve through social interaction. In an increasingly globalized world, this is of particular interest in studying the decline of native dialects as well as language maintenance and shift in multilingual communities (Milroy 2008)
- Social capital or the resources available to individuals through their social interactions. For instance, social capital allows certain people to access opportunities such as job openings. It has also been shown that there is a correlation between measured social capital and reported quality of life (Valenzuela et al. 2009).

As complex social networks can be used to analyze many real-world interaction types from social networking websites to interactions between animals, being able to effectively study their structures has become increasingly important in recent years (Pinter-Wollman et al. 2013)

Rumors, opinions, behaviors, and diseases spread to the population via social interactions. A blocker is an individual in the network that can most effectively slow down the spread of a process through the population. For example, to slow the spread of disease, it would be most efficient and effective to vaccinate one of the key blockers in the network. This paper attempts to find structural network measures that indicate the best blockers in dynamic and social networks. A dynamic network is a series of static networks that show the interactions of an individual at a certain time. An aggregate network shows a group of individuals and their interactions over a period of time. If two nodes have interaction during the observed period of time, it is represented by an edge; multiple interactions between a pair of individuals might be represented as a single edge, multiple edges, or possibly a weighted edge between the two nodes. A dynamic network is generally more useful because it shows time and keeps intact the order of interactions.

Structural network measures are like social properties within a network, for example, betweenness. This method used several of these measures to look at the entire network and other more localized measures to look at individual nodes. The global structural properties observed were density, the proportion of edges in a network to possible edges, dynamic density, the average density at one time, path, a distinct sequence of nodes, temporal path, a time-respecting path in a dynamic network, and diameter, the length of the longest shortest path. The localized properties used were degree, or a node's number of neighbors, dynamic degree, dynamic average degree, nodes in the neighborhood, edges in the neighborhood, betweenness (previously discussed), dynamic betweenness, closeness, or the average distance between one individual and other individuals in the network, dynamic closeness, clustering coefficient and dynamic clustering coefficient, the fraction of a node's neighbors that are neighbors to each other in previous time steps. These measures were all compared to determine the blocking ability of individuals.

The paper by Habiba et al. (2010) finds that the dynamic clustering coefficient, which basically measures how many of your friends are friends with each other, was a good indicator of the node's blocking ability. The other structural methods that best predicted blocking ability were node degree, number of edges in a node's neighborhood, and dynamic average degree. These methods must still be tested on larger and more complex models to determine whether they will be truly useful for realistic disease spread models. Another problem with this method is that it focused on practical applications and the theoretical structure of the problem is still not well known. A large problem with this method is that it cannot identify a set of top blockers because it goes through the data and tests nodes one at a time by removing them and measuring the spread of data. Finding the top set of blockers is computationally hard, and an exhaustive search is infeasible. Another interesting thing they find is that in networks where blocking spread was difficult, nodes were all ranked about the same; however, in networks where spread could be blocked by just removing a few individuals, the nodes had a wider range of rankings.

14.4.2 Temporal Patterns

A temporal network is a network in which the connection between the nodes is not continuous. The most important part of a temporal network is time. This is also the most difficult part of a temporal network to visualize and analyze because relationships in networks are changing and modeling the change in relationships while keeping the order and time aspect of the network is difficult to do in just one image. It almost takes a whole string of images or snapshots of the aggregate network at different points in time to demonstrate a temporal network. In our research, we examined the various methods that have been used to analyze different aspects of temporal networks. After becoming thoroughly acquainted with the various methods of examining properties of temporal networks, we compared each method and summarized the pros and cons of each method.

We examine several models each of which focuses on a specific problem in temporal patterns of social networks. The first method we examined was the betweenness preference (Pfitzner et al. 2013). This method focuses on the structural properties of a temporal network and looks at how likely certain nodes are to mediate interactions between any two nodes. Betweenness preference is based on the idea that certain nodes contact other nodes based on previous contact. The problem with using betweenness preference to analyze a temporal network is that it is not present in the time-aggregated network. For example, if given two temporal networks, when these networks are aggregated, we cannot determine between which time steps a node mediated an interaction between two other nodes or if these two nodes are able to interact through their previous contact. This becomes a problem because of the order that edges are made. In spite of this, betweenness preference is an important aspect of a temporal network. If we can keep betweenness preference intact when aggregating our network, this will help us see the flow of information throughout the network, which is typically lost when the network is collapsed.

Another problem with a network of time-stamped pairs defining who spoke to who or who tweeted at who is that a vital part of the flow of information is lost. For example, if A speaks to B in the morning, then B speaks to C in the afternoon, information might flow from A to C but not from C to A. The paper (Gindrod et al. 2011) suggests using a more natural definition of a walk on an evolving network. Specifically, a node's ability to both broadcast and receive information is calculated through a series of basic operations in linear algebra, and the lapse in time can be accounted for in the flow of data from node to node by utilizing the noncommutativity of matrix–matrix multiplication. A walk is a path that goes from node to node. A path is closed if it starts and ends at the same place and open if it starts and ends at different places. The Katz centrality of a person measures a person's centrality by taking into account the number of walks between a pair of people.

The third issue of temporal patterns is that in some networks, links between nodes are either positive, like a friendship, or negative, like an opposition. For a link in a social network, its sign is either positive or negative depending on the attitude of the creator of the link to the recipient. If we are given a network of nodes with edges of either positive or negative value and one of the links has a missing value, we want to find a way to determine, based on the other connections of the network, whether this missing link is positive or negative. The paper by Leskovec et al. (2010) uses the logic that the enemy of my friend is my enemy and the friend of my enemy is my enemy. This method looks at how the sign of a link interacts with the pattern of the signs of the links within a certain distance of the given link. This paper uses edge sign prediction to determine the state of a link given an almost complete network of either positive or negative links.

While many of these graphs are directed, it is sometimes useful to examine the graph and its links regardless of the direction of the connection between two nodes. However, when predicting the sign of an edge, they use directed links. For example, if we are trying to determine the sign of the edge from node u to node v, we look at the signs of the outgoing edges from u and the incoming edges to v. They also keep track of the number of outgoing positive and negative edges from u and the total number of common neighbors u and v have. When predicting the sign of a link, it is also helpful to form triads of nodes. For example, they consider the triad containing the edge (u, v) with a node w such that w has an edge either to or from u and an edge either to or from v. This theory is called structural balance theory. This theory is based on the idea stated above that the friend of my friend is my friend and the enemy of my friend is my enemy. Using the structural balance theory and the idea of triads, if w forms a triad with the edge (u, v), then (u, v) should have the sign that causes the triangle formed by w, u, v to have an odd number of positive signs. By using a balanced data set, we know that by guessing the status of a link, it can have a 50 % correct prediction rate. If we use a full data set, accuracy is improved to 80 %.

14.4.3 *Network Randomness*

Most social networks are dynamic networks where connections are being continuously made and broken. Being able to accurately predict a dynamic network in the future has many consequences, many of which are found within the business world. Large organizations can benefit by being able to suggest new collaborations and interactions within the organization. Security companies also benefit from the fact that we could use this information to analyze terrorist networks.

Our first problem is to predict which interactions among existing members are likely to occur in the near future. A supervised random walk method (Backstrom and Leskovec 2011) looks at how we can use only the given data of the existing relationships to accurately predict the future of the network. This is how social networks like Facebook send users recommendations on who they should befriend in the future. Supervised random walk combines the information from the network with node and edge attributes. These attributes are then used to guide a random walk on the graph. When looking at link prediction, we look at a network at a given time t and try to predict the edges that will be added to the network at a future time t.

There are a few problems with link prediction methods. One major problem is that social networks are sparse. For instance, Facebook users connect to an average of 100 nodes out of a 500 million-node network. For this reason, a good way to predict edges is to predict no new edges, allowing this method to be extremely accurate, albeit useless. Thus, supervised random walk helps with these problems. Using this method, Backstrom and Leskovec (2011) take a supervised random walk on the network which visits given nodes more often than other nodes. They use given node and edge features to determine the strength of the edges so the random walk will be more likely to visit positive nodes more than negative nodes. Positive nodes are nodes to which new edges will be created in the future, while negative nodes are all of the other nodes. A new function must be created then to assign strength to each edge so when we compute the random walk in a weighted network, nodes which a node will connect to in the future will have higher scores than those nodes which a node will not visit in the future.

Another way to analyze the randomness is to use Markov Chain Monte Carlo (Clauset et al. 2007). This analyzes the probability of nodes making connections by looking at the whole network. We can then use these probabilities to look at how the network will look at any given time. Markov Chain Monte Carlo uses Markov chains along with the law of large numbers to estimate the state of the network at any given time in the future. Markov Chain Monte Carlo uses stationary distribution, making the network easier to work with. The network has a given probability distribution, and Markov Chain Monte Carlo generates random elements with the same distribution. Markov Chain Monte Carlo then uses this information to predict the state of the given temporal network at any given time. This means that it will predict which connections will be made between which nodes at a certain time.

Markov Chain Monte Carlo (Tjelmeland 2007) is closely related to supervised random walk. A random walk is performed on the network, and each step has a probability associated with it. Markov Chain Monte Carlo creates a Markov chain with the same distribution of the network and uses a random walk to simulate the chain. The problem with the random walk method is the aspect of random walk that must be calculated and performed on each and every node in the network. These calculations in a large and expansive network such as Facebook or twitter can become exhaustingly long and tedious. Another problem with Markov Chain Monte Carlo is the rate of convergence. It is not very well known how to determine how long the chain must be used on a network to generate a suitable convergence.

14.5 Conclusion

One of the fundamental questions of human social simulation has traditionally been how should agents make decisions given they inhabit an environment where their actions may have unforeseen or unpredictable effects on others? This question often raises interesting points about the extent to which the individual autonomy of agents should be sacrificed for global needs and desires of society. As the economic and mathematical sciences have transitioned toward a more socially conscious decision-theoretic framework, they have discovered that human beings operating in real-world environments often rely not only on their own cognitive capabilities but of those of others around them as well through a network of complex social structures and institutions.

If multiagent systems are to provide a proper computational model of both human decision-making and social interaction, then these structures and institutions and the cognitive capabilities of the agents that comprise them must be modeled to a level where computational complexity is not sacrificed on behalf of realism. Our CASE model represents an important work that processes by combining recent advances in behavioral economics that point to a more bounded-rationality human mindset with the time-honored theories of socialism that cross disciplinary boundaries between both sociology and psychology.

References

Axelrod R (1997) Advancing the art of simulation in the social sciences, Simulating Social Phenomena. Springer, Berlin, pp 21–40

Aylett R, Louchart S, Pickering J (2004) A mechanism for acting and speaking for empathic agents. In: Proceedings of the 2004 autonomous agents and multi-agent systems workshop (AAMAS'04)

Backstrom L, Leskovec J (2011) Supervised random walks: predicting and recommending links in social networks. In: The 4th ACM international conference on web search and data mining, pp 634–636

Castlefranchi C (1997) Challenges for agent-based social simulation: the theory of social functions. In: Invited talk, SimSoc, Cortona, Sept 1997

Castlefranchi C (2000) Engineering social order, Lecture notes in computer science. Springer, Berlin, pp 1–18

Christakis N, Fowler J (2013) Social contagion theory: examining dynamic social networks and human behavior. Stat Med 32:556–577

Cialdini RB (2001) Influence: science and practice. Allyn and Bacon, Boston

Clauset A, Moore C, Newman M (2007) Structural inference of hierarchies in networks. In: Statistical network analysis: models, issues, and new directions, Lecture notes in computer science. Springer, Berlin, vol 4503. pp 1–13

Conte R, Hegselmann R, Terna P (1997) Studies in social simulation. Springer, Berlin

Das S, Grecu D (2000) Cogent: cognitive agent to amplify human perception and cognition. In: Proceedings of the fourth international conference on autonomous agents (Agents'00), pp. 443–450

Epstein JM, Axtell R (1996) Growing artificial societies: social science from the bottom up. Brookings Institute Press, Washington, DC

Ferber J (1999) Multi-agent system: an introduction to distributed artificial intelligence, intelligence design. Addison-Wesley, Harlow

Fleming C (2011) What is communication? The definition of communication. Communication, pp. 58–70

Ghasem-Aghaee N, Ören TI (2004) Effects of cognitive complexity in agent simulation: basics. In: Proceedings of SCSC 2004 – summer computer simulation conference, San Jose, 25–29 July 2004, pp 15–19

Gindrod P, Parsons MC, Higham DJ, Estrada E (2011) Communicability across evolving networks. Phys Rev E 83:046120

Habiba Y, Yu T, Berger-Wolf Y, Saia J (2010) Finding spread blockers in dynamic networks. Advances in social network mining and analysis, Lecture notes in computer science. Springer, Berlin, vol 5498, pp 55–76

Hales D, Edmonds B (2003) Evolving social rationality for mass using tags. In: Proceedings of the 2003 autonomous agents and multi-agent systems workshop (AAMAS'03), pp 497–503

Jiang Y, Ishida T (2007) A model for collective strategy diffusion in agent social law evolution. In: Proceedings of the 2007 international joint conference on artificial intelligence (IJCAI'07)
Kahneman D (2002) Maps of bounded rationality: a perspective on intuitive judgment and choice. In: Invited talk, Les Prix Nobel
Kahneman D (2003) Maps of bounded rationality: psychology for behavioral economics. Am Econ Rev 93(5):1449
Kahneman D, Tversky A (1979) Prospect theory: an analysis of decision under risk. Econometrica 47(2):263–292
Kant JD, Thiriot S (2006) Modeling one human decision maker with a multi-agent system: the Codage approach. In: Proceedings of the 2006 autonomous agents and multi-agent systems workshop (AAMAS'06), pp 50–57
Kazemifard M, Ghasem-Aghaeea N, Koenig B, Ören TI (2011) Toward effective emotional intelligence simulation: modeling understanding ability for emotive agents. Proceedings of the Fourth International Conference of Cognitive Science (ICCS), pp 154–160
Kim S (1999) Intelligent software agents and business-oriented application scenarios. In: Invited talk, AOIS, Hidelberg
Labrou Y, Finin T, Peng Y (1999) Agent communication languages: the current landscape. IEEE Int Syst 14:45
Lapachelle P (2011) The use of social networking in community development. CD Practice 17:1–8
Latane B, Darley JM (1970) The unresponsive bystander: why doesn't he help? Appleton-Century-Crofts, New York
Leskovec J, Huttenlocher D, Kleinberg J (2010) Predicting positive and negative links in online social networks. In: The nineteenth international conference on world wide web, pp 641–650
Levy J, Pescosolido B (2002) Social networks and health. JAI Press, Amsterdam
Milroy L (2008) Social networks. In: The handbook of language variation and change. Blackwell Publishing Ltd, Oxford
Minsheng T, Xinjun M, Guessoum Z, Huiping Z (2013) Rumor diffusion in an interests-based dynamic social network. Sci World J 2013:824505
Ören TI, Yilmaz L (2004) Behavioral anticipation in agent simulation. In: Proceedings of WSC 2004 – winter simulation conference, Washington, DC, 5–8 Dec 2004, pp 801–806
Parsons S, Woolridge M (2002) Game theory and decision theory in agent-based systems. Kluwer Academic Publishers, Boston
Pfitzner R, Scholtes I, Garas A, Tessone CJ, Schweitzer F (2013) Betweenness preference: quantifying correlations in the topological dynamics of temporal networks. Phys Rev Lett 110:198701
Pinter-Wollman N, Hobson E, Smith J et al (2013) The dynamics of animal social networks: analytical, conceptual, and theoretical advances. Behav Ecol 25:242–255, EprintarXiv
Porpora DV (1989) Four concepts of social structure. J Theory Soc Behav 19(2):195
Portes A, Sensenbrenner J (1993) Embeddedness and immigration: notes on the social determinants of economic action. Am J Sociol 98(6):1320–1350
Roehner BM (2007) Driving forces in physical, biological and socio-economic phenomena: a network science investigation of social bonds and interactions. Cambridge University Press, Cambridge, 2007
Rummel RJ (1976) The conflict helix. Sage, Beverly Hills
Seck M, Frydman C, Giambiasi N, Ören TI, Yilmaz L (2005) Use of a dynamic personality filter in discrete event simulation of human behavior under stress and fatigue. In: First international conference on augmented cognition, Las Vegas, 22–27 July 2005
Sherif M, Sherif CW (1948) An outline of social psychology. Harper Brothers, New York
Sherif M, White BJ, Harvey OJ (1955) Status in experimentally produced groups. Am J Sociol 60:370
Simon H (1949) Readings in applied microeconomic theory: market forces and solutions. Blackwell Publishing, New York
Simon H (1957) A behavioral model of rational choice. Wiley, New York

Stattner E, Collard M, Vidot N (2013) D2snet: dynamics of diffusion and dynamic human behavior in social networks. Comp Human Behav 29:496–509

Stirling W (2003) Satisficing games and decision making: with applications to engineering and computer science. Cambridge University Press, Cambridge/New York

Tanford S, Penrod S (1984) Social influence model: a formal integration of research on majority and minority influence processes. Psychol Bull 95:189

Tisdell C (1996) Bounded rationality and economic evolution: a contribution to decision making, economics, and management. Edward Elgar, Brookfield/Cheltenham

Tjelmeland H (2007) Introduction to Markov chain Monte Carlo — with examples from Bayesian statistics. Wiley, Norway

Valenzuela S, Park N, Kee K (2009) Is there social capital in a social network site? Facebook use and college students' life satisfaction, trust, and participation. J Comp Mediated Commun 14:875–901

Weber M (1947) The theory of social and economic organization. Free Press, New York

Yilmaz L, Ören TI, Ghasem-Aghaee N (2006) Simulation-based problem solving environments for conflict studies. Simul Gaming J 37(4):534–556

Zhoua L, Dingb L, Fininc T (2013) How is the semantic web evolving? A dynamic social network perspective. Comp Hum Behav, pp. 1–9

Part V
Body of Knowledge of Modeling & Simulation

Chapter 15
A Review of Extant M&S Literature Through Journal Profiling and Co-citation Analysis

Navonil Mustafee, Korina Katsaliaki, and Paul Fishwick

15.1 Introduction

It has only been about 60 years since the founding of one of the most important professional societies for M&S, the Society for Modeling and Simulation International (SCS) in 1952; however, the field already celebrates official recognition with the passage of the US House of Representatives *House Resolution 487* in 2007 identifying M&S as a national critical technology (Ören 2009). The high number of industry groups related to M&S, the growth of M&S academic programs, and the existence of multiple professional certification programs (Bair and Jackson 2013) are indicators of the importance of the field. The M&S knowledge base benefits from advancements in Computer Science, Systems Engineering, Software Engineering, Artificial Intelligence, and more (Fishwick 1992). On the other hand, M&S contributes to the development of these disciplines by being an enabler. Let us take the example of M&S and Computer Science to illustrate this symbiotic interdependence among research disciplines. Computers with multicore CPUs/GPGPUs enable faster execution of simulations and may bring about methodological advancements in M&S (e.g., how do we execute multiagent simulations over GPGPUs?).

N. Mustafee (✉)
Centre for Innovation and Service Research, University of Exeter Business School, Exeter, UK
e-mail: N.Mustafee@exeter.ac.uk

K. Katsaliaki
School of Economics, Business Administration & Legal Studies, International Hellenic University, Thessaloniki, Greece

P. Fishwick
Department of Computer Science, The University of Texas at Dallas, Richardson, TX, USA

L. Yilmaz (ed.), *Concepts and Methodologies for Modeling and Simulation*,
Simulation Foundations, Methods and Applications,
DOI 10.1007/978-3-319-15096-3_15

Similarly, developments in M&S methodology and applications may act as an enabler of complex systems modeling in CS (e.g., executing network protocol simulations over GPGPUs). However, it can sometimes be a challenge defining the boundaries, and by extension vocabularies, that align between M&S and the other disciplines, and Tolk argues that "a comprehensive and concise representation of concepts, terms, and activities is needed that make up a professional Body of Knowledge for the M&S discipline" (Tolk 2010). Fishwick (2014) recently posited that M&S is at the heart of computer science by virtue of the role of modeling in capturing fundamental concepts in computing. Due to the broad variety of contributors, this process is still ongoing, and according to Ören (2009), researchers ought to continue advancing it for consolidating and disseminating pertinent knowledge about M&S; assuring professional standards; and advancing M&S science, methodology, and technology to continue solving problems in hundreds of traditional application areas and new areas.

This research aims to contribute to the development of such a body of knowledge by presenting a profiling and cocitation analysis study of literature from a scholarly outlet published by SCS. SCS is a technical society that is devoted to furthering the field of M&S. The Society has effectively engaged the community it serves and has played a significant role in advancing research in simulation and allied computer arts, in applying research for solving real-world problems, in fostering networking among professionals, in organizing and sponsoring leading conferences in this area, in providing outlets for scholarly research (through Society publications), and in recognizing the achievements and contributions of both Society members and the M&S community at large (SCS 2014). With the objective of providing a comprehensive and integrative view of M&S body of knowledge, we have reviewed literature published in the Society's journal—*Simulation*: *Transactions of the Society for Modeling and Simulation International*. So as to eliminate the ambiguity between the name of the journal and the discipline that it caters to (both being "Simulation"), the journal will henceforth be referred to in uppercase italics, i.e., as *SIMULATION*. For the purpose of this research, we have considered scholarly articles published in the journal between 2000 and 2011; we use this underlying data set to conduct two forms of analyses, namely, a profiling study and a cocitation study, to realize the following seven objectives in pursuance of furthering knowledge on the scope and the breadth of our scholarly field:

- To determine the most commonly used M&S approaches and techniques
- To identify the broad areas/sectors associated with the application of M&S
- To identify the specific fields (within the aforementioned areas/sectors) where the application of M&S is widespread
- To identify the institutional departments associated with the majority of publications
- To determine the geographic location associated with the majority of publications

- To identify the underlying research clusters that have high article cocitation counts associated with them (using cocitation analysis)
- To identify journals which are frequently cocited (using cocitation analysis)

The limitation of this work is its sole reliance on one particular journal, and we would like to emphasize that the findings of this study should be regarded as indicative only of the journal's activity and its contribution in shaping the M&S knowledge base. However, it is also true that advances in research usually percolate through various sources of scholarly dissemination, and it is arguable that important topics, applications, and advancements in M&S would have found their way in all leading journals in M&S including the journal *SIMULATION*. It further follows that a methodological review of the M&S knowledge base using papers published in one journal would be akin to sampling a subset of literature in our field. Thus, the findings of this research can reasonably lay claim to being representative of the wider body of literature in M&S.

The remainder of this book chapter is organized as follows. In Section 15.2, we discuss the work of Tuncer Ören in developing an integrated view of the M&S knowledge base. In Section 15.3, we present the research methodology which includes (a) an overview of the journal *SIMULATION;* (b) the journal profiling methodology, which is used for the analyses of application sectors, specific fields, M&S techniques, institutional departments, and geographic location of authors; and (c) cocitation analysis, which is used for the identification of research clusters and journals that are frequently cited by authors publishing in the journal. Following this, we present the findings of our study in Section 15.4 and conclude by reinforcing the contribution of this research and pointers to future work (Section 15.5).

15.2 Tribute to Tuncer Ören for His Contribution in Developing the M&S Body of Knowledge (BOK)

Ören states in several of his publications that "Modeling and Simulation (M&S) –as a field, discipline, and profession– is progressing, maturing, and continues to be used in many conventional and unconventional application areas as a powerful infrastructure... and urgently needs to have its Body of Knowledge" (Ören 2011a; Ören and Waite 2010). It is arguable that Ören is the founding father of the M&S BoK, and this is aptly established in his numerous publications on this subject (over 35 papers since 2000) and his work toward compiling the M&S BoK Index. The Index is an online resource consisting of four parts (Ören 2013). We elaborate on the Index for the benefit of the readers to fully appreciate the corpus of M&S knowledge that has been archived by Ören.

- The *first part* provides a background to the project and to the M&S discipline and profession and its developments, incorporating also terminology and

relevant dictionaries and a comprehensive view of the area. One facet of Ören's work is dedicated to the terminology of M&S with the purpose of identifying a good definition which covers all aspects of M&S and allows a top-down decomposition of the field for elaboration of all relevant aspects (Ören 2011a). Relevant to this, he has published a compilation of definitions of simulation (Ören 2011b, c), collections of special terms, and M&S multilingual dictionaries. His comprehensive and consolidated view of the area is also identified in many of his publications (Ören 2009, 2010).
- The *second part* of the Index is dedicated to 11 core areas of M&S—Science/Methodology, Types of simulation, Life cycles of M&S, Technology, Infrastructure, Reliability, Ethics, History, Trends, Challenges and Desirable Features, Enterprise, and Maturity. These have multiple subcategories and may include specific terms and literature related to the overarching core theme.
- The *third part* concerns the M&S supporting domains that are independent of its application areas; these are categorized into Computers and Computation, Science Areas, Engineering Areas, Management Areas, and Education. The exploration of synergies among these supporting domains and M&S is also another well-researched area in Ören's work (Ören 2002, 2005).
- The final part of the Index provides links to important M&S portals, blogs, and references (by authors, application areas, and topics).

The updated M&S Index will also incorporate BoKs of other disciplines for adopting best practices and creating synergies among them. Subsequent to the M&S BoK Index being finalized, it will be hosted by the SimSummit collaborative environment (sim-summit.org/BoK07/). The finalization of the Index requires the views of professional simulation practitioners for updating information held in the website, and a special call toward this has been made by Ören and Waite (2010). A review committee is already formed for this purpose, which receives the recommendations from colleagues. This collaborative work through an open forum is an excellent mechanism for keeping the discipline active with a fast pace of growth. Ören's concluding remarks in one of his presentations includes the following: "M&S offers many opportunities and challenges to solve problems of unprecedented complexities. Simulationists can contribute by sharpening the tools used, abiding by ethics and offering their services" (Ören 2012). Therefore, the M&S BoK has a clear direction for the future.

Ören's work was an inspiration to this research, which shares some common ground with the M&S BoK in general terms. The aim of this study is in line with the profiling of M&S discipline and profession and its developments as in the first part of the BoK through the documentation of M&S publications by geographic locations, university departments, and publishing outlets. Moreover, our study classifies M&S publications in terms of techniques, application areas, and their context in a relevant way with the second and third parts of the BoK, which define the M&S core areas and supporting domains. Lastly, this paper provides a network analysis of the M&S publication relationships to other subdisciplines in line with the purpose of the updated M&S Index to identify synergies among M&S and other disciplines.

15.3 Methodology

The research presented in this chapter has employed two forms of content analyses, namely, *journal profiling* and *cocitation analysis*. These are described under separate subheadings and provide further information pertaining to the technique, the specific data sets, and, in the case of cocitation analysis, the tool used. As has been mentioned earlier, our data set comprises of articles published in *SIMULATION*, and we therefore consider it prudent to begin this section by presenting an overview of the journal.

15.3.1 A Brief Overview to the SIMULATION *Journal*

SIMULATION encourages submissions on methodology and applications and has a strong interdisciplinary focus. Presently in its 90th volume, the journal is indexed in numerous scholarly databases (including the Web of Science™) and has an impact factor of 0.656 according to the 2013 Journal Citation Report® (JCR Science edition) that is published by Thomson Reuters; it has a 5-year impact factor of 0.737. JCR classifies the journal under two categories, Computer Science: Interdisciplinary Applications and Computer Science: Software Engineering. The reputation of the journal has meant that it continues to attract a large number of submissions, which are then subjected to peer review (each submission is usually allocated three reviewers), and this constant throughput of original research and review articles have ensured that the journal has continued to offer a monthly publication frequency. For example, the number of research papers that were published in the time span 2000–2010 varied from a minimum of 39 in 2001 to a maximum of 56 articles in 2002, with a yearly average of around 48 papers. Yet another indicator of the journal's reputation is the number of special issues (SI) that have been published over the years.

Academics and practitioners acted as SI guest editors realizing the dissemination potential of the journal and its standing in the international M&S community. This is best demonstrated by the fact that the total number of special issue papers that were published between 2000 and 2010 was 267—this represented approximately half of all articles published. The special issue topics also demonstrate the focus of the journal on methodology and theoretical papers as well as application-oriented papers (refer to Mustafee et al. 2012 for the list of SI topics).

15.3.2 Journal Profiling Methodology

The profiling exercise required us having to undertake an exhaustive review of papers that were published in the journal from 2000 to 2010. *SIMULATION* is the monthly publication of the Society; thus, every volume (from 2002 onward) usually has 12 issues. The publication frequency is largely consistent during the period of analysis, the exception being the double issues that were published within this time frame.

The papers published in the journal generally belong to one of two categories: regular articles or special issue articles. However, between 2000 and 2004, articles were published under several other categories including introduction to special issues, columns on AI and simulation, the art of modeling, simulation in the service of society, spotlight on M&S activities, and special issue call for papers. Most of the articles under these supplementary categories cannot be considered as having undergone a peer review. Hence, in the analyses presented in this paper, we have only considered regular articles (258 papers) and special issue articles (267 papers). Thus, the total number of papers selected for the analyses is 525.

For every paper included in the analysis, we captured data on variables pertaining to, for example, the year of publication, the department and the geographic location pertaining to authors' affiliation, the simulation technique that was applied, the application domain/sector, and the context of its application. Extracting detailed information for the last three variables required reading the article and then adopting a peer review approach to limit any bias. For further discussion pertaining to the individual variables and for the complete journal profiling study, the reader is referred to Mustafee et al. (2012).

15.3.3 Cocitation Analysis

Cocitation analysis identifies clusters of "cocited" references by creating a link between two or more references when they co-occur in the reference lists of citing articles (Raghuram et al. 2009). Let us take an example where there are three articles (A1, A2, and A3), each of which cites two articles (B1, B2). Even though B1 and B2 may not directly cite each other, B1 and B2 form a kind of semantic cluster since A1, A2, and A3 all cite B1 and B2. B1 and B2 are, therefore, related by cocitation. The resultant cocitation networks provide important insights into knowledge domains by identifying frequently cocited papers, authors, and journals related to the domains in question, and this would have been overlooked if only conventional citation analysis techniques were used.

For this part of the analysis, citation data pertaining to the journal was downloaded from Web of Science™ (WOS). A search for papers associated with our target journal revealed that the journal was indexed in WOS starting from September 2001—the search criterion used was as follows: Publication Name = (*SIMULATION-TRANSACTIONS OF THE SOCIETY FOR MODELING AND SIMULATION INTERNATIONAL*); Databases = *SCI-EXPANDED*, *SSCI*, *A&HCI*, *CPCI-S*, *CPCI-SSH*. For this study, we considered a time span of 10 years starting from January 2002 (Volume 78, issue 1) to December 2011 (Volume 87, issue 12). In total, we downloaded citation data pertaining to 564 papers that were published during this period. For cocitation analysis, we used this downloaded data set and the knowledge domain visualization software called CiteSpace (Chen 2004). CiteSpace identifies turning points associated with articles from citation data. The use of CiteSpace requires careful selection of a multitude of options, and an acceptable options' combination frequently requires learning through "trial and error" as well as knowledge of the underlying research domain. For the specific CiteSpace option values that were selected and for the complete cocitation study, please refer to Mustafee et al. (2014).

15.4 Findings

Our journal profiling exercise and cocitation analysis concluded in a series of findings. These findings are described next under separate headings; each heading is associated with a particular objective which was defined in the introduction section. More specifically, findings that present a comprehensive and integrative view of M&S include the identification and categorization of M&S techniques (Section 15.4.1), identification of the broad areas/sectors associated with the application of M&S (Section 15.4.2), and the context of its application (Section 15.4.3); findings derived from authors' affiliation which show the interdisciplinary nature of M&S and its widespread prevalence include institutional departments (Section 15.4.4) and geographic locations (Section 15.4.5); findings from the cocitation analysis include identification of research clusters and important journals that are frequently cocited (Sections 15.4.6 and 15.4.7).

15.4.1 Analysis Based on M&S Technique

Two authors independently and critically reviewed the papers by reading their abstracts and, if in doubt, reading the whole article. Furthermore, the authors scrutinized papers that had coding discrepancies; the objective was to reconcile the differences pertaining to classification and to agree on a decision. Indeed, this exercise often necessitated revisiting previously classified papers for the sake of consistency. The authors then grouped the M&S technique–related data under specific headings. Since this required subjective decision-making, regrettably, the tables presenting this analysis cannot be recreated. The authors also admit that the inclusion of a third reviewer could have changed the groupings to an extent; however, it is arguable that the important M&S categories identified and their corresponding frequencies would still have remained largely consistent with the present findings.

Table 15.1 lists the M&S techniques that were reported in the papers published in the journal, grouped under 12 categories (including the category *Not Known*) and their corresponding frequency of use. We have assigned one M&S technique for

Table 15.1 M&S techniques

A. Simulation Techniques	**196**
Network Modeling and Simulation	76
Discrete Event Simulation	55
Monte Carlo Simulation; Numerical Simulation	9 each
Finite Element Method-based Modeling and Simulation; Real Time Simulation	7 each
Discrete-Event Simulation and Visualization; System Dynamics; Trace-based Simulation	4 each
Continuous Simulation/Flow Simulation; Statistical Simulation (including Regression and Poisson Simulation)	3 each
Rare Events Simulation; Software-In-The-Loop Simulation; Stochastic Simulation; Virtual Reality Simulation; Web-based Simulation	2 each
Chaos-based Simulation; Interval-based Microscopic Simulation; Qualitative Simulation and Prediction; Simulation Visualization; Spreadsheet Simulation	1 each
B. Parallel and Distributed Simulation	**69**
Parallel and Distributed Simulation	32
Distributed Simulation	22
Agent-Based Distributed Simulation	6
Parallel Simulation	4
Distributed Interactive Simulation	3
Grid-Based Simulation; Web-based Distributed simulation	1 each
C. Systems Modeling	**67**
Mathematical and Equation-based Modeling	25
Bond Graph Modeling; Petri Nets	9 each
Markov-chain Modeling	6
Multi-Paradigm Modeling	4
Statistical Modeling; Stochastic Modeling	3 each
Visual Interactive Modeling	2
Bayesian Networks; Discrete-Time Modeling; GERT -Graphical Evaluation and Review Technique; Meta-Modeling; Model Verification and Validation; Semi-Markov Model	1 each
D. Agent Based Modeling and Simulation	**44**
Agent-Based Modeling and Simulation	34
Multi-Agent Systems	9
Agent-Based Geo-Simulation	1
E. Discrete Event System Specification (DEVS) and other Formalisms	**37**
Devs	26
Devs – Cell-Devs	2
Composable Cellular Automata Formalism; Devs – Devs/soa; Devs – Dsdev; Devs – eUdevs; Devs – Gdevs; Devs – Rtdevs; Devs – Cell space approach (note: this is different from Cell-Devs); Formal Specification and Analysis (Maude); Heterogeneous Flow System Specification Formalism	1 each
F. Application-Specific Modeling and Simulation	**31**
Analysis of Algorithms (including Simulation of Algorithm)	8
Physics-based Modeling and Simulation (including N-body and VOXEL-based simulation)	3
Biological Pathway Modeling; Logic Simulation; Sound simulation	2

(continued)

Table 15.1 (continued)

Architecture Simulation; Chemical Simulation; Circuit Simulation; Computerized Tomography Simulation; Constructive Military Simulations; Drift Path Simulation; Embedded Simulation; Engineering Simulation; Job Shop Simulation; Landslide Simulation; Load Flow Modeling; Simulation and Gaming; Simulation of Flight Mechanics; Thermodynamic Simulation	1 each
G. Programming/Specification Languages/Frameworks/Methodology	**24**
Object Oriented Simulation	6
Programming (including, Fuzzy Linear Programming, Genetic Programming, Integer Programming, Integer Linear Programming)	4
Component-based Modeling and Simulation	2
Architecture Description Languages; Cellular Automata Programming Environment; Data Exchange Model; Extensible Battle Management Language; Finite State Machines Modeling Language; Formal Co-design Framework; GESAS II Methodology; Object-Oriented Modeling Language; Parallel Object-Oriented Specification Language; Programming Environment for Simulator; Programming Language; Service-Oriented Architecture (SOA) Simulation	1 each
H. Operations Research Techniques (including Optimization and AI-based approaches)	**22**
Optimization (including Genetic Algorithm Optimization, Metaheuristic-based Optimization, Particle Swan Optimization, Simulation-based Optimization)	10
Artificial intelligence (including Fuzzy Inductive Reasoning and Neural Networks)	6
Heuristics	3
Multiobjective Decision Analysis; Scheduling; Uncertainty Modeling	1 each
I. Multiple Techniques	**13**
Various	7
(Discrete-Event Simulation + Hardware-in-the-loop simulation); (Genetic algorithm-based optimisation + Finite-Element Method + Grid-enabled Parallel Simulation); (Kinematic Vehicle Modeling + VR Modeling); (Monte-Carlo Simulation + Petri Net Modeling); (Policy Specification Language + Policy Development Framework + Distributed Simulation); (Very High Speed Integrated Circuits Hardware Description Language [VHDL] + Artificial Neural Network + Fuzzy Logic)	1 each
J. Hybrid Methods	**8**
Intelligent Agents with queuing network model; Mesoscopic simulation (microscopic and macroscopic simulation)	2 each
Discrete-continuous combined simulation; Hybrid symbolic-numerical simulation method; Hybrid system examples; Monte Carlo–based Discrete Event Simulation	1 each
K. Not known	**8**
L. Uncategorised	**6**
Knowledge-based systems and expert Systems	3
Model-based information-processing systems; Performance evaluation of simulated systems; Reliability simulation	1 each
Total	**525**

each article. Articles that deal with multiple M&S techniques have been clustered either under *Multiple Techniques* (where there is equal emphasis on each technique and the techniques are applied independently) or *Hybrid Methods* (where the techniques are applied symbiotically). The data is presented in descending order, sorted on the number of occurrences identified for each of the 12 broad categories.

As can be seen from the table, category *Simulation Technique* has 196 occurrences; the different M&S techniques that make up this figure include *Network M&S* (76 occurrences), *Discrete-Event Simulation* (55), *Monte Carlo* and *Numerical Simulation* (9 each), etc. Owing to the large number of papers that relate to agents (44 occurrences), we have not included this under the *Simulation Technique* category but have created a separate category called *Agent-Based Modeling and Simulation*. The other prominent categories in Table 15.1 include *Parallel and Distributed Simulation* (69 occurrences), *System Modeling* with 67 occurrences (this includes mathematical and equation-based modeling, statistical modeling, Petri nets, Markov chains, Bayesian networks, etc.), *DEVS and Other Formalisms* with 37 occurrences, and *Operations Research Techniques* (22 occurrences).

15.4.2 Analysis Based on M&S Application Areas/Sectors

For the purposes of this analysis, we adopted a peer review approach similar to the one used above. The findings are presented in Table 15.2. We have identified a total of 28 application areas with *Telecommunications* reporting the highest number of studies. *Health care* and *Military/Defense* are placed fifth and the sixth

Table 15.2 Application areas/sectors

Application areas/sectors	Count	%
Telecommunications	98	18.82
Engineering	50	9.51
Distributed Computing	40	7.60
Manufacturing	30	5.70
Health care	26	4.94
Military/Defence	23	4.37
Computers	19	3.61
Environment	18	3.42
Air Transport	13	2.47
Automotive; Education	12 each	2.28 each
Road Transport; Urban studies	11 each	2.09 each
Systems Biology	9	1.71
Marine/Water Transport	6	1.14
Logistics; Supply chain	5 each	0.95 each
Rail Transport	4	0.76
Astronomy; Construction; Mobile Computing; Retailing and Wholesaling; Space	3 each	0.57 each
Mining/Metals	2	0.38
E-Business; Economics; Public Administration; Sports	1 each	0.19 each
Methodology papers (*not specific to any application area*)	112	21.29
Total	**525**	**100**

respectively. The majority of the papers (approx. 21 %) are not specific to an application area but are related to methodology. These papers are included at the end of the table (for obvious reasons, our count of 28 application areas excludes this category). The predominance of *Methodology* implies that majority of papers analyze and develop specific techniques and focus more on the method rather than on testing their application on a specific sector.

15.4.3 Analysis Pertaining to the Field (Within an Area/Sector)

Similar to the methodology adopted in the earlier sections, in this analysis we adopted a peer review approach with the objective of eliminating any unintended bias. Table 15.3 presents the context of the application of M&S within an area/ sector. We started with the 29 application categories that were identified in the previous analysis; the papers reporting on the use of M&S techniques and their application areas provided specific information on the application context.

As can be seen from the table, the category *Methodology* was applied in several contexts, for example, framework (10 occurrences), time management—related to parallel and distributed simulation (9), component-based M&S (3), etc. Similarly, M&S techniques were applied to the *Telecommunication* sector in contexts such as analysis of networks (12 occurrences), quality of service (6), analysis of protocols, e.g., routing protocol, flow control, physical layer, access/admission control (numerous occurrences), and network power management (4 occurrences).

15.4.4 Analysis Based on Authors' Departmental Affiliation

The purpose of this analysis is to investigate the interdisciplinary nature of M&S as evidenced by the breadth of the subject area of the authors' departmental affiliation (i.e., the home departments in which the authors are based). Unfortunately, for this variable we had a lot of missing data. From a total of 1,250 academic authors and coauthors, we could gather information for approximately 88 % (1,100 authors to be precise). Moreover, in order to present readable results, we had to cluster the names of the authors' departments/schools under more general and distinct headings. For example, the category *Computer Science, Information and Communication Technologies (ICT) and Electronics Engineering* consists of schools and departments related to Computer Science (including Applied CS), Computer Engineering, Computing and Mathematical Sciences, Electronics, Communications Engineering, Telecommunications, Information Sciences, M&S, etc.; all the specific Engineering departments (other than those in the aforementioned category) are classified under the *Engineering* category—e.g., Aerospace Engineering,

Table 15.3 Analysis pertaining to context of application (within an area/sector)

A. Methodology	**112**
Simulation Environment/Platform/Language	13
Framework	10
Time Management	9
Rare Event Simulation	6
Hybrid M&S	5
Performance Evaluation; Verification & Validation	4 each
Complex Systems; Component-based M&S; Optimization Algorithm; Simulation Experimentation/Experimentation Design; Simulation Output Analysis; VR Modeling/Virtual Environments	3 each
Collaborative Simulation Environment/Tool; Data Distribution Management; Hybrid Systems; Model Integration/Model ComposAbility; Poisson Simulation/Poisson Process; Real Time Systems; Visualization	2 each
Artificial Intelligence; Automatic Model Completion; Business Process Simulation; Chaos-based Simulation; Construction of Models; Continuous Systems; Derivative Estimation; Event List; Fault Tolerance; Graphical Models; Grid-based Simulation; Input Data Analysis; Large-Scale Simulation; Model Extraction; Model Selection; Model Transformation; Network Traffic; Proportion Estimation; Quantization-Based Simulation; Queuing Systems; Simulation Cloning; Simulation Interoperability; Simulation Model Reuse; Simulation Practice; State Management; Time-Parallel Simulation; Time-Series Forecasting; Training Simulator; Uncertainty Modeling	1 each
B. Telecommunications	**98**
Analysis of Networks	12
Network Security; Programming/Network Simulation Environment; Protocol M&S (Routing)	8 each
Design of Integrated Architectures	7
Network Quality of Service	5
Multimedia Services; Power Management; Protocol M&S (Congestion Control)	4 each
Protocol M&S (Flow Control)	3
Distributed Network Simulation/Parallel Network Simulation; Optimal Configuration of Networks; Protocol M&S (Access/Admission Control); Protocol M&S (Communication); Protocol M&S (Physical Layer); Protocol M&S (Scheduling); Reusability; Scalability of Networks; Speed of Simulation Execution	2 each
Empirical Models; End-User Studies; Execution Time; Intelligent Networks; Load Balancing; Network Emulation; Network Management; Network Mobility; Network Reconfiguration; Pricing; Protocol M&S (Deadlock Recovery); Protocol M&S (TDMA); Protocol M&S (Access/Admission Control); Protocol M&S (Wireless); Review; Voice Quality; Workload Modeling	1 each
C. Engineering	**50**
Power System Design/Power Transmission	12
M&S of Physical Systems	8
Design of Systems; Fault Diagnosis/Fault Detection and Isolation	6 each
Movement of Fluids/Flow Simulation	4
Control Systems/Factory Automation Systems/Expert Systems	3
M&S of Physical Processes; Modeling Framework; Training Simulator	2 each
Audio Signal Processing; Flood Management; Logic Simulation; Model Driven Engineering; Review	1 each

(continued)

Table 15.3 (continued)

D. Distributed Computing	**40**
Scheduling; WWW/SOA/Web Services	8 each
Design of Distributed Systems	5
Load Balancing/Resource Management	4
Communication; Execution/Programming Environment; Simulation of HPC systems	3 each
Data Replication; P2P Networks; Peer-to-Peer (P2P) Gaming; Scalability; Transaction Management; Virtual Environments	1 each
E. Manufacturing	**30**
Factory/Production Line/Job Shop Simulation; Simulation of Physical Systems/Process	6 each
Fault Diagnosis/Fault Detection and Isolation	4
Web-Based Simulation	2
Complex Manufacturing Systems; Execution Speed; Enterprise Decision-making Support; Grid-based Simulation; Inventory Management; Lean Manufacturing; Quality Improvement; Repair and Maintenance; Shop-Floor Control Systems; Simulation Interoperability; Simulation-based Order Acceptance; System Reconfiguration	1 each
F. Healthcare	**26**
Epidemic M&S; Modeling of Physical Systems/Computed Tomography	4 each
Hospital/Clinic Management; Scheduling	3 each
Healthcare Informatics; Operating theatres; Review	2 each
A&E; Lean/JIT; Simulation of Disorders; Supply Chain Simulation; Training; Viewpoint	1 each
G. Military/Defence	**23**
Simulation Interoperability; Training	4 each
Military Communications	3
Behaviour Representation	2
Airborne Operations; Availability of Weapon Platforms; Casualty Evacuations; Dynamic Behaviour of Simulation; Embedded simulation; Live–Virtual–Constructive (LVC) Simulation; Missile Threat Simulation; Radar Interference; Simulation State Updates; System Decomposition	1 each
H. Computers	**19**
Computer Architecture	6
Microprocessor Architecture	5
Emulation; Execution/Programming Environment; Formal Design Methods; GPU; Human-Computer Interface; Real Time Computers; Software Architecture; Ubiquitous Computing	1 each
I. Environment	**18**
Ecology Modeling	7
Spread of Fire	4
Modeling Forest Landscapes	3
Methodology for Environment Modeling; Terrain Modeling/Landslide Modeling	2 each
J. Air Transport	**13**
Aviation Safety	4
Air and Ground Traffic Control; Air Network Simulation; Evolution of the Airline Industry; Flight Control System; Future of Air Transportation; M&S Infrastructure for Airports; Risk Management; Training; Visualisation of Airport Operations	1 each

(continued)

Table 15.3 (continued)

K. Automotive	**12**
Design of Automobiles	5
Automobile Production Line	4
Automobile Safety; Driving Simulator; Sound modeling	1 each
L. Education	**12**
Simulation Pedagogy; Simulation-based Training and Teaching	4 each
Visual Interactive and Multimedia Simulations	3
Design of Simulation Course	1
M. Road Transport	**11**
Traffic Light Control/Traffic Signal Timings	3
Intelligent Transportation System	2
Driving Behaviour; Hybrid Modeling; Incident Management; Operation of a Toll Plaza; Surface Transportation System; Training Simulator	1 each
N. Urban studies	**11**
Behavioural M&S; Water Management	4 each
Crowd M&S	2
Organisational Adaption	1
O. Systems Biology	**9**
Biological Modeling	3
Experimental Design; Modeling Environment/Modeling Description Language	2 each
Functional Genomics; Model Decomposition	1
P. Marine/Water Transport	**6**
Analysis of Physical Systems; Control Systems; Design of Systems; Investment Decisions; Maritime Transport System; Training Simulator	1 each
Q. Logistics	**5**
Optimization	3
Planning; Quality Improvement	1 each
R. Supply chain	**5**
Distributed Supply Chain Simulation	3
Hybrid Supply Chain Simulation; Supply Chain Simulation	1 each
S. Rail Transport	**4**
Control Systems; Intermodal Transport Planning; Safety; Simulation of Physical Systems/Process	1 each
T. Astronomy	**3**
Astronomic telescope data processing; Galactic Simulation; Radiometer simulation	1 each
U. Construction	**3**
Construction Management; Highway Maintenance and Reconstruction; Stress Analysis of Materials	1 each
V. Mobile Computing	**3**
Location-Based Service; Mobile Network Performance; Mobility Prediction	1 each
W. Retailing and Wholesaling	**3**
Customer Experience; Inventory Control; Store Management	1 each

(continued)

Table 15.3 (continued)

X. Space	**3**
Design of Satellite Cluster System; Satellite Communication; Simulation of Physical System/Process	1 each
Y. Mining/Metals	**2**
Investment Decisions; Surface Mine Design	1 each
Z. E-Business	**1**
Business Process Reengineering	1
AA. Economics	**1**
Fiscal Modeling	1
AB. Public Administration	**1**
Institutional Reorganisation	1
AC. Sports	**1**
Agent Behaviour	1
Total	**525**

Table 15.4 Classification of the authors' departmental affiliation under eight broad categories

Academic departments	Total	Total (%)
Computer Science, Information & Communication Technologies (ICT) & Electronics Engineering	682	62.0
Engineering (Mechanical, Civil, Electrical, etc.)	197	17.9
Economics and Management	44	4.0
Mathematics, Statistics and Physics	39	3.5
Basic Sciences and Research	29	2.6
Medical-Health	21	1.9
Social Sciences	13	1.2
Others	75	6.8
Total	**1,100**	**100.0**

Bioengineering, Chemical and Materials Engineering, Civil Engineering, Electrical Engineering, General Engineering, Hydraulic Engineering, Industrial and Operations Engineering, Mechanical and Control Engineering, and Production Engineering; *Economics and Management* category consists of Administration, Business, Economics, Econometrics, Decision Sciences, Management Science, Organizational Science, Supply Chain Management, and other similar departments. In total, we formed eight such categories (shown in Table 15.4).

Our analysis of the department/school-specific affiliation information showed that the largest number of contributors were from departments/schools under the umbrella category of *Computer Science*, *Information* and *Communication Technologies (ICT) and Electronics Engineering* (62 %). This category is followed by *Engineering* (17.9 %), *Economics and Management* (4.0 %), and *Maths, Stats, and Physics* (3.5 %). Research labs have been classified under the category *Basic Sciences and Research*, and considering that this category only has a handful of research labs (e.g., IBM Austin Research, IBM T. J. Watson Research, IBM Zurich Research, Domaine Scientifique de la Doua—INSA Lyon, Google Taiwan R&D, Ford Scientific Research, and C&C Research Laboratories), 2.6 % of contribution is noteworthy.

15.4.5 Analysis Based on Authors' Geographic Location

The purpose of this analysis was to bring to attention the widespread prevalence of M&S as evidenced by the number of authors around the world. The geographic location of the authors' affiliations was the underlying data used for this analysis. This analysis has taken into consideration the double affiliations reported by seven authors. Our analysis revealed that contributors came from 58 different countries, with the USA (38.7 %) clearly dominating. The second (5.6 %) and third (5.3 %) largest categories were formed by authors affiliated to either Spanish or Canadian institutions respectively. France, South Korea, UK, and the Netherlands were next in the list. Table 15.5 shows the top 20 countries in terms of (a) the geographic

Table 15.5 List of the top twenty geographical locations based on (a) authors' affiliation (b) and total number of author contributions

Country (a)	Unique authors (a)	Total (%) (a)	Country (b)	Author contributions (b)	Total (%) (b)
US	484	38.7	US	581	38.7
Spain	70	5.6	Spain	78	5.2
Canada	66	5.3	Canada	76	5.1
France	57	4.6	South Korea (*incl. Korea = 21*)	68	4.5
South Korea (*incl. Korea = 20*)	53	4.2	France	65	4.3
UK	52	4.2	UK	62	4.1
Netherlands	50	4.0	Netherlands	59	3.9
China; Germany	47 each	3.8 each	Germany	51	3.4
Italy	44	3.5	China	50	3.3
Greece	26	2.1	Italy	48	3.2
Taiwan	25	2.0	Singapore	44	2.9
India	24	1.9	India	40	2.7
Singapore	20	1.6	Greece	35	2.3
Turkey	17	1.4	Taiwan	34	2.3
Iran	16	1.3	Iran	23	1.5
Australia; Brazil	13 each	1.0 each	Turkey	18	1.2
Sweden	12	1.0	Sweden	15	1.0
Hungary	9	0.7	Brazil	14	0.9
New Zealand	8	0.6	Australia	13	0.9
Slovenia	7	0.6	Hungary; New Zealand	9 each	0.6 each

location of the authors' affiliations (columns 1–3) and (b) the total region-specific contributions of the authors taking into consideration the fact that authors could have contributed to more than one paper (columns 4–6). The actual number of contributions is 1,494, but 7 of the authors appear in the database with double affiliation, and thus, the total contributions are considered to be 1,501.

It is perhaps not surprising that the largest contribution is from the USA. This is because the journal was created and established in the USA with US editors. However, the large representation of other countries indicates the journal's international audience and reputation. However, the findings must be carefully considered since the results are meaningful but not normalized by country population size. For example, Singapore has just over five million people, whereas the USA has 307 million. Table 15.5 shows 484 unique authors from the USA and 20 from Singapore. When normalized using per capita figures, Singapore shows 4 authors per million people and the USA 1.57 authors per million. One also needs to keep in mind relative densities: Singapore is highly concentrated in space with significantly high technology, whereas the spatial variations differ in other countries.

15.4.6 List of Research Clusters (Cocitation Analysis)

Nodes and links are the building blocks of a cocitation network. CiteSpace supports a total of 11 node types (NTs). In this and the subsequent finding (Section 4.7), we are interested in NTs "cited references" and "cited journals" and the resultant *Document Cocitation Network* (*DCN*) and *Journal Cocitation Network* (*JCN*) respectively. The following discussion applies equally to DCN and JCN. Each node in the DCN/JCN refers to an article/journal. The different time-sliced DCN/JCN cocitation networks are distinguished by their color. The colors indicate time, and through the use of the VIBGYOR spectrum, they represent the complete time interval of the analysis. The links can visually represent various characteristics of the underlying network; for example, the color of the link represents the year in which a connection between two nodes was first established (e.g., with regard to the DCN, it is the year in which two articles were first cocited), and the strength of connection between any two nodes is represented by the thickness of the link (e.g., in the context of JCN, the thicker the connection between two nodes, the greater the frequency of journal cocitation).

The DCN visualization allows us to identify underlying relationships among the cited articles. For example, a thick link between two nodes (denoting high cocitation count among the articles), both of which also have a relatively large diameter (denoting high citation count) and have been consistently cited over the years, would generally identify two papers that are equally important to a subject matter. But the question is: what is the subject matter? Is it possible to infer this from reading the abstracts of the papers with high citation count? However, this process is time consuming, and the interpretation is subjective as it is based on a researcher's domain knowledge. An alternative way to achieve this is to

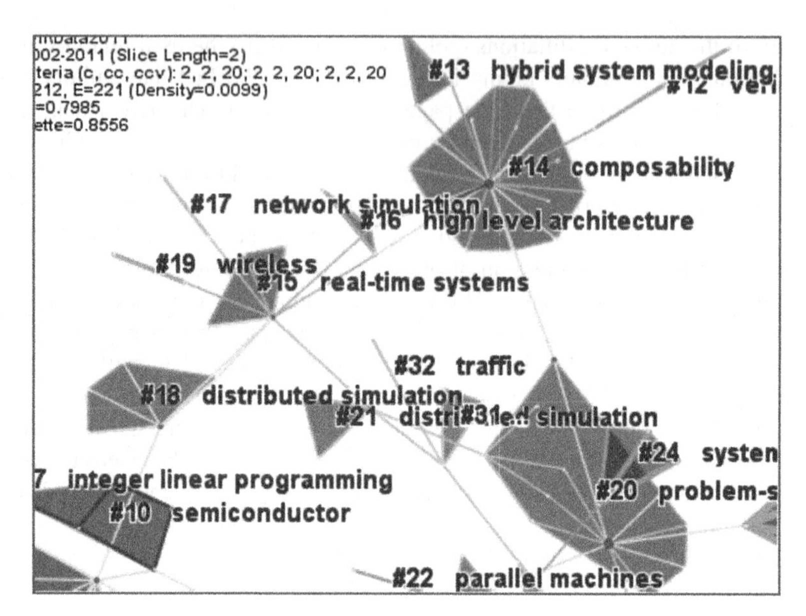

Fig. 15.1 Clusters identified in the DCN solution space and named using candidate labels

automatically assign meaningful labels to the cocitation clusters that are identified in a cocitation network; CiteSpace "characterizes clusters with candidate labels selected by multiple ranking algorithms from the citers of these clusters and reveals the nature of a cluster in terms of how it has been cited" (Chen et al. 2010). CiteSpace presently supports nine different ways of labeling the clusters—allowing selection of candidate terms from three sources (titles, abstracts, and index terms), all of which belong to the citing articles, and three ranking algorithms (tf*idf weighting, LLR, MI) (Chen et al. 2010). In our study, we have selected index terms and have used the tf*idf weighting algorithm. The output is shown in Fig. 15.1. The list of clusters is presented in Table 15.6. It shows a total of 35 unique clusters. These clusters were identified from among 212 nodes and a total of 221 links. Figure 15.2 focuses on one such cluster (*#40 scalability*), and it shows five papers with high cocitation count (De Boer 2006; Dupuis et al. 2007; Glasserman and Kou 1995; Kroese and Nicola 2002; Parekh 1989). All these papers are on queuing networks (tandem queues, Jackson network).

15.4.7 List of Frequently Cited Journals (Cocitation Analysis)

Table 15.7 presents a list of scholarly literature sources that are frequently cocited by authors of *SIMULATION*. The majority of the items identified are journals, with the exception of three books (authored by Zeigler BP; Law AM & Kelton WD;

Table 15.6 List of research clusters identified using DCN

Cluster	Cluster	Cluster
Policies	Composability	Health-care
Mathematical-theory	Real-time systems	Systems biology
Virus	High level architecture	Continuous system simulation
Traffic management	Network simulation	Traffic
Explicit window adaptation	Distributed simulation	Architecture
OMNET plus	Wireless	Combat simulation
Tomography	Problem-solving environment	System dynamics modelling
Dynamic structure discrete event system specification	Parallel machines	Telemedicine
Semiconductor	Support	Integer linear programming
Bifurcation	System dynamics	Demand
Verification and validation	Web	Scalability
Hybrid system modelling	Coherence protocol	

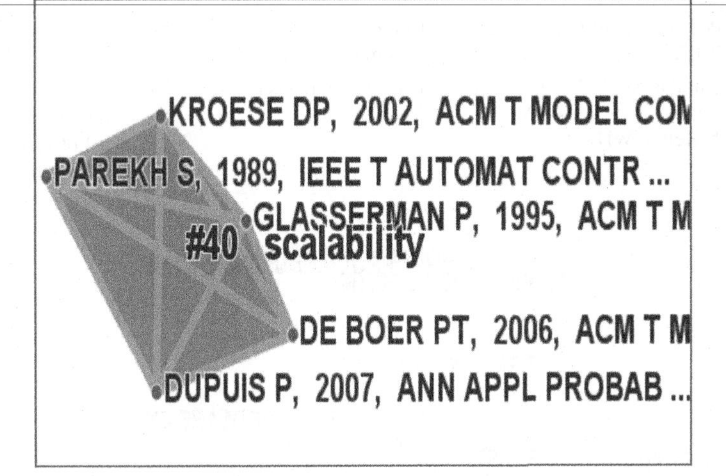

Fig. 15.2 Articles with high co-citation counts for cluster on scalability (using DCN)

Fujimoto RM), an edited book series by *Springer* (*Lecture Notes in Computer Science*), and five conference proceedings (*Proceedings of the Winter Simulation Conference—WSC, 1998, 2000, 2002, and 2005*; *Proceedings of the Annual Simulation Symposium—ASimS*). If we combine the *WSC* citations, then WSC is the third most popular citation source, with the number of citations being higher than *ACM Transactions on Modeling and Computer Simulation* (*TOMACS*).

The JCN has identified that *ACM TOMACS* is most frequently cocited with *Communications of the ACM*, *Journal of Parallel and Distributed Computing*, *IEEE Transactions on Software Engineering*, *Proc. of WSC* (*1997 and 2000*), and *Proc. from the Workshop on Principles of Advanced and Distributed Simulation* (*PADS*). This is shown in Fig. 15.3.

Table 15.7 List of frequently cited journals, books, conferences (freq ≥ 20) identified using JCN

Freq	Journal	Freq	Journal	Freq	Journal
193	Simulation: Transactions of the SCS	33	Proc. Annual Simulation Symposium (ASimS)	25	European Journal of Operational Research
102	Lecture Notes in Computer Science (LNCS)	31	Simulation Modelling Practice and Theory	25	Journal of Parallel and Distributed Computing
67	ACM TOMACS	30	Parallel and Distributed Simulation Systems (Fujimoto RM)	24	IEEE Transactions on Software Engineering
57	Theory of Modeling and Simulation (Zeigler BP)	30	Operations Research	22	Proc. 2000 Winter Simulation Conference (WSC)
55	Communications of the ACM	29	IEEE Transactions on Parallel and Distributed Systems	22	Proc. 1998 WSC
48	IEEE/ACM Transactions on Networking	29	Management Science	22	Artificial Intelligence
43	IEEE Journal on Selected Areas in Communications	28	IEEE Transactions on Computers	21	Journal of the Operational Research Society
38	Proceedings of the IEEE	26	Computer Communications	20	Proc. 2005 WSC
38	Simulation Modeling and Analysis (Law AM & Kelton WD)	26	IEEE Communications Magazine	20	Proc. 2002 WSC
				20	International Journal of Production Research

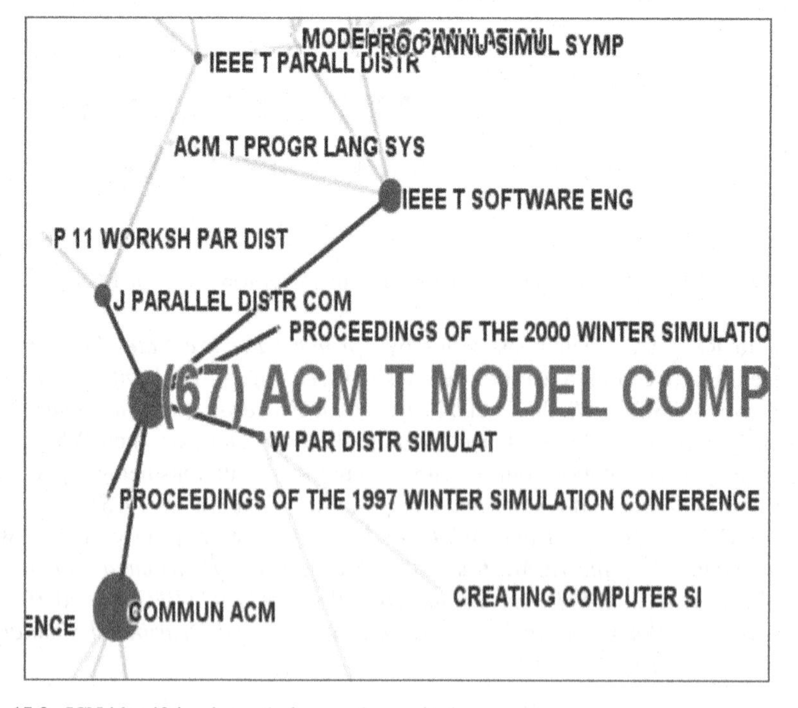

Fig. 15.3 JCN identifying journals frequently co-cited

15.5 Conclusions

The purpose of this research was to extract both qualitatively and quantitatively derived observations about a corpus of information centered on the publications of a society (The Society for Modeling and Simulation International). Two methods were used for this purpose. The first method relied on a *profiling study* by reading *SIMULATION* articles and then capturing variables in a methodological manner to present the findings. The resultant data from this analysis provide a reference point for discussions pertaining to the discipline of M&S. As the readers would note, the peer review approach was adopted for capturing variable values pertaining to the M&S technique used (Section 4.1), M&S application area/sector (Section 4.2), and the context of the application of M&S in particular areas/sectors (Section 4.3). The objective of this was to eliminate any unintended bias that could have been a result of our biased decision-making. However, as the peer review approach was being conducted, it became evident that the majority of the discrepancies arose from the differing categorization granularity we adopted. For example, whether a paper on "agent-based distributed simulation" is codified under a new category with the same name or under an existing category (e.g., "Agent-Based M&S" or "Parallel and Distributed Simulation") would be dependent on how specific we wanted the categorization to be (keeping in mind that the number of categories should be manageable) and, in instances where we independently decided against creating a new category, whether we felt the paper was better represented by one or the other of the available umbrella categories. In cases where there was no consensus with regard to codification, we created a new subcategory and assigned it to an over-arching category with the best-fit (this was unusually achieved subsequent to reading the full text). Taking the previous example, a subcategory called "Agent-Based Distributed Simulation" was created, and it was placed under the existing category of "Parallel and Distributed Simulation." In summary, the tables that we have collated have a wealth of information in them, and although we do not claim that our categorization is authoritative, we believe that they can be used as a source of scholarly reference, discussion, and debate.

The second method used was *cocitation*, which is a graph-theoretic approach frequently employed for identifying topics, clusters, and categories. The two main graph types studied were document and journal cocitation networks. While these graph types were useful in identifying key areas of research and important journals, the analysis also paints a larger picture. In the emerging era of "big data," M&S practitioners should consider data-centric methods (e.g., graph and network theory) for determining important questions about our discipline. What types of dynamic models are being used and can they be stratified over time or correlated with specific authors or disciplinary areas? What programming languages predominate in simulation? Answers to these questions at one time were mainly answered through survey articles; however, in the future with so much available data, we might consider new forms of data and graph analysis along the lines of the methods discussed in this article.

Future research directions could involve broadening the boundaries of our profiling data set. For example, we could utilize a more inclusive set of data that is representative of the whole domain and not restricted to a single publishing outlet with the purpose of capturing a more complete picture of the interrelationships between the key variables discussed in this paper. This would provide further insights into the M&S body of knowledge and would also make our results indicative of all M&S researchers and not only of the population of researchers who submit to the particular outlet, particularly since M&S communities are severely fragmented by application/technique area. Another area of future research is to use cocitation analysis to detect the emergence of new areas in M&S; one way to achieve this is through the detection of *citation bursts* which indicate articles that have received an extraordinary degree of attention from the scientific community. A further study could involve mapping the evolution of M&S research by performing cocitation analysis using well-defined time slices and noun phrases that are extracted from titles and abstracts, similar to the approach followed by Mustafee (2011).

M&S research during the last decade, as pictured in this study, brings out similarities but also differences to Ören's taxonomies and visions of the discipline. As Ören (2009) stated, researchers ought to continue advancing M&S methodology and solve problems in hundreds of traditional and new application areas. From our analysis, it is evident that M&S incorporates a plethora of techniques (Table 15.1), from fundamental ones (e.g., systems modeling) to contemporary techniques that have evolved with advancement in computing technology (e.g., parallel and distributed simulation, agent-based M&S). Moreover, these techniques have been applied in numerous application areas (Tables 15.2 and 15.3). Although these techniques have been predominantly skewed toward communications, engineering, and a few other sectors, other areas such as the environment, logistics, and supply chains make their appearance in M&S literature. However, the taxonomy that Ören provided for the organization of the M&S application areas (i.e., *Computers and Computation*, *Science Areas*, *Engineering Areas*, *Management Areas*, and *Education*) are more general than those identified in this study. Nevertheless, it may be possible to classify our application areas under the classification presented by Ören.

References

Bair LJ, Jackson JJ (2013) M&S professionals domains, skills, knowledge, and applications. In: Proceedings of the The Interservice/Industry training, simulation & education conference (I/ITSEC), vol 2013, no 1. Orlando, Florida. National Training Systems Association (NTSA)

Chen C (2004) Searching for intellectual turning points: progressive knowledge domain visualization. Proc Natl Acad Sci 101(Suppl 1):5303–5310

Chen C, Ibekwe-Sanjuan F, Hou J (2010) The structure and dynamics of co-citation clusters: a multiple-perspective co-citation analysis. J Am Soc Info Sci Tech 61(7):1386–1409

De Boer P-T (2006) Analysis of state-independent importance-sampling measures for the two-node tandem queue. ACM Trans Model Comp Simulat 16(3):225–250

Dupuis P, Sezer AD, Wang H (2007) Dynamic importance sampling for queueing networks. Ann Appl Probab 17(4):1306–1346
Fishwick PA (1992) An integrated approach to system modeling using a synthesis of artificial intelligence, software engineering and simulation methodologies. ACM Trans Model Comp Simulat 2(4):307–330
Fishwick PA (2014) Computing as model-based empirical science. In: Proceedings of the second ACM SIGSIM/PADS conference on principles of advanced discrete simulation (SIGSIM-PADS'14), Denver, Colorado. ACM, pp 205–212
Glasserman P, Kou SG (1995) Analysis of an importance sampling estimator for tandem queues. ACM Trans Model Comp Simulat 5(1):22–42
Kroese DP, Nicola VF (2002) Efficient simulation of a tandem Jackson network. ACM Trans Model Comp Simulat 12(2):119–141
Mustafee N (2011) Evolution of IS research based on literature published in two leading IS journals – EJIS and MISQ. In: Proceedings of the nineteenth European conference on information systems (ECIS 2011), Helsinki, Finland. Paper 228, Association for Information Systems
Mustafee N, Katsaliaki K, Fishwick P, Williams MD (2012) SCS – 60 years and counting! A time to reflect on the society's scholarly contribution to M&S from the turn of the millennium. Simulation 88(9):1047–1071
Mustafee N, Katsaliaki K, Fishwick P (2014) Exploring the modelling & simulation knowledgebase through journal co-citation analysis. Scientometrics 98(3):2145–2159
Oren T (2009) Modeling and simulation: a comprehensive and integrative view. In: Yilmaz L, Oren T (eds) Agent-directed simulation and systems engineering. Wiley-VCH Verlag GmbH & Co. KGaA, Weinheim
Ören T (2010) Simulation and reality: the big picture. Int J Model Simulat Sci Comput 1(01):1–25
Ören T (2011a) A basis for a modeling and simulation body of knowledge index: professionalism, stakeholders, big picture, and other BoKs. SCS M&S Mag 2(1):40–48
Ören T (2011b) A critical review of definitions and about 400 types of modeling and simulation. SCS M&S Mag 2(3):142–151
Ören T (2011c) The many facets of simulation through a collection of about 100 definitions. SCS M&S Mag 2(2):82–92
Ören T (2002) Future of modelling and simulation: some development areas. In: Proceedings of the 2002 summer computer simulation conference, San Diego, CA. Society for Computer Simulation, pp 3–8
Ören T (2005) Maturing phase of the modeling and simulation discipline. In: Proceedings of the 2005 ASC-Asian simulation conference, Beijing, P.R. China. pp 24–27
Ören T (2012) The richness of modeling and simulation and its body of knowledge. In: Proceedings of the 2012 international conference on simulation and modeling methodologies, technologies and applications, Rome, Italy
Ören T (2013) Modeling and simulation body of knowledge (M&S BoK) – index. http://www.site.uottawa.ca/~oren/MSBOK/MSBOK-index.pdf. Last accessed Sept 2014
Ören T, Waite B (2010) Modeling and simulation body of knowledge index: an invitation for the final phases of its preparation. SCS M&S Mag 1(4):1–5
Parekh S (1989) A quick simulation method for excessive backlogs in networks of queues. IEEE Trans Automat Control 34(1):54–66
Raghuram S, Tuertscher P, Garud R (2009) Mapping the field of virtual work: a co-citation analysis. Info Syst Res 21(4):983–999
SCS (2014) Simulation: mission & scope. Society for modeling & simulation international. http://www.scs.org/simulation?q=node/90. Last accessed Sept 2014
Tolk A (2010) Engineering management challenges for applying simulation as a green technology. In: Proceedings of the thirty first annual national conference of the American Society for Engineering Management (ASEM), Fayetteville, Arkansas. pp 137–147

ERRATUM

Concepts and Methodologies for Modeling and Simulation

Levent Yilmaz

L. Yilmaz (ed.), *Concepts and Methodologies for Modeling and Simulation,*
Simulation Foundations, Methods and Applications,
DOI 10.1007/978-3-319-15096-3

DOI 10.1007/978-3-319-15096-3_16

The original version of this book unfortunately contained a mistake. The name and affiliation of the author 'Gary R. Meyer' was incorrect. The author's correct name is 'Gary R. Mayer'. Furthermore, his correct affiliation is 'Department of Computer Science, Southern Illinois University, Edwardsville, IL, USA'.

The online version of the original book can be found at
http://dx.doi.org/10.1007/978-3-319-15096-3

Index

L. Yilmaz (ed.), *Concepts and Methodologies for Modeling and Simulation*, Simulation Foundations, Methods and Applications,
DOI 10.1007/978-3-319-15096-3

Printed in the United States
By Bookmasters